Springer-Verlag Berlin Heidelberg GmbH

P. E. S. Palmer

The Imaging of Tuberculosis

With Epidemiological,
Pathological, and Clinical Correlation

With contributions by S. J. Wambani and P. Reeve

Consultants: M. M. Reeder, D. H. Connor, and I. J. Dunn

With 104 Figures in 514 Separate Illustrations

 Springer

P.E.S. PALMER, MD, FRCP, FRCR
Em. Professor of Radiology
821 Miller Drive
Davis, CA 95616
USA

Originally published in P.E.S. Palmer, M.M. Reeder (2001) The Imaging of Tropical Diseases, VOL. 1,
Springer Berlin Heidelberg New York

ISBN 978-3-642-62610-4

Library of Congress Cataloging-in-Publication Data
Palmer, Philip E. S.
    The imaging of tuberculosis : with epidemiological, pathological, and clinical
correlation / P.E.S. Palmer, with contributions by S.J. Wambani and P. Reeve ;
consultants, M.M. Reeder, D. H. Connor, and I.J. Dunn.
    p.; cm.
  Includes bibliographical references and index.
  ISBN 978-3-642-62610-4   ISBN 978-3-642-56282-2 (eBook)
  DOI 10.1007/978-3-642-56282-2
  1. Tuberculosis--Imaging. 2. Tuberculosis--Epidemiology. I. Wambani, S. J. II. Reeve,
P. (Paul), 1957- III. Title
  [DNLM: 1. Diagnostic Imaging--methods. 2. Tuberculosis--diagnosis. 3.
Tuberculosis--epidemiology. 4. Tuberculosis--pathology. WF 220 P175i 2001]
  RC311.2 .P35 2001
  616.9'950754--dc21   2001040050

http//www.springer.de
© Springer-Verlag Berlin Heidelberg 2002
Originally published by Springer-Verlag Berlin Heidelberg New York in 2002
Softcover reprint of the hardcover 1st edition 2002
The use of general descriptive names, trademarks, etc.in this publication does not imply, even in the absence of
a specific statement, that such names are exempt from the relevant laws and regulations and therefore free for
general use.

Product liability. The publishers cannot guarantee the accuracy of any information about dosage and application
contained in this book. In every case the user must check such information ba consulting the relevant literature.

Cover-Design: Verlagsservice Teichmann, Mauer
Typesetting: K+V Fotosatz GmbH, Beerfelden
SPIN: 108 327 42   21/3130 – 5 4 3 2 1 0 – Printed on acid-free paper

Introduction

This book will serve as a reminder of the many different patterns of tuberculosis. Once again a major public health threat all over the world, tuberculosis can be a chronic and almost benign infection or an acute life-threatening tragedy. It may respond to treatment and leave no after-effects; or it may be resistant to any cure, with disastrous results for the individual and often the family or community as well.

Although originally written as a chapter in *The Imaging of Tropical Diseases* (P.E.S. Palmer and M.M. Reeder, Springer, Heidelberg, 2001), the clinical course and the images described are no longer found only in the tropics or in the developing world. The ways in which tuberculosis affects each individual are a reflection of the immune status of the patient and whether it has been compromised by AIDS, malnutrition, the continuous onslaught of too many parasites or other infections, chemotherapy, radiation or any other cause of immunosuppression. Added to this must be the sensitivity or resistance of the particular bacterium to the drugs which are available and how well the patient complies with the treatment regimen. Interpreting the patient's clinical status and the images is challenging: an acute fulminating tuberculous infection does not mean that the patient is HIV+ve, a favourable early response to treatment does not mean that the patient will be cured.

As a result of these differences in patients and organisms, tuberculosis presents a wide spectrum of images. Any part of the body may be infected, literally from the skin inwards. No tissue or organ is exempt. The infection is often multicentric and can mimic almost any other disease. Unfortunately, it is not possible to decide from one or two images whether the infection is active, healing or cured, nor can imaging immediately indicate drug resistance or recognise any of the many variants of the *Mycobacterium tuberculosis*.

There is a real need for close cooperation between clinicans and radiologists in the management of tuberculosis, a need which may be lost in the numerous notes and images which are the records of any patient with this infection. The main source of error for all who interpret the images is failing to remember tuberculosis as the possible cause of almost any abnormal patterns, however much the clinical illness may suggest some other origin. Most of the images in this book come from the tropics and the developing world, but similar images may be seen wherever the patient lives. Some might be better technically, because imaging equipment improves all the time. But their message is clear. No-one of any age is exempt from tuberculosis and this must not be forgotten.

Davis   P. E. S. PALMER

List of Contributors and Consultants

Daniel H. Connor, MD
11351 Morning Gate Drive
Rockville, MD 20852
USA

Ian J. Dunn, PHD, MD FRCP(C)
Clinical Assistant Professor
University of British Columbia
Burnaby General Hospital
3935 Kincaid Street
Burnaby, BC V5G 2X6
Canada

Maurice M. Reeder, MD
12646 Travilah Road
Potomac, MD 20854
USA

Paul Reeve, MB, MRCP (U.K.)
Specialist Physican
Health Waikato Ltd., Taumarunui Hospital
Private Bag 1002
Taumarunui
New Zealand
(previously in Malawi and Vanuatu)

S. J. Wambani, MB, M. Med. (R)
Department of Diagnostic Radiology
College of Health Sciences, Faculty of Medicine
Kenyatta National Hospital, University of Nairobi
P.O. Box 19676
Nairobi
Kenya

Contents

Synonyms 2

Definition 2

Geographic Distribution 2

Epidemiology and Pathology. 3

Laboratory Diagnosis 4

Clinical Characteristics. 5

Tuberculosis of the Respiratory Tract 5
    Tuberculosis of the Upper Respiratory Tract 5
    Primary (Nonimmune) Tuberculosis. 6
    Lobar Pneumonia 8
    Bronchopneumonia 13
    Hilar and Mediastinal Lymphadenopathy. 15
    The Destroyed Lung 19
    Pleural and Pericardial Effusions 24
    Miliary Tuberculosis 30
    Silicosis 33

    Immune Tuberculosis (Also known as Secondary, Hyperergic,
    Reactivation, or Adult Tuberculosis) 34
    Clinical Characteristics 35
    Imaging Diagnosis 35
    Tuberculoma 43
    Isoniazid Cysts. 45
    Chronic Pleural Disease 45
    Chest Wall. 45
    Congenital Tuberculosis. 45
    Immunization with BCG 45
    PPD Conversion. 47
    Clubbing. 47

The Other Mycobacterioses 48

Tuberculosis of the Alimentary Tract 49
    Tuberculosis of the Esophagus 49
    Tuberculosis of the Stomach 50
    Tuberculosis of the Duodenum and Small Intestine 51
    Tuberculosis of the Cecum
    (Hyperplastic Tuberculosis of the GI Tract) 58
    Tuberculosis of the Large Intestine 61
    Tuberculosis of the Rectum. 65
    Enteroliths 67

Tuberculosis of the Peritoneum and Abdominal Lymph Nodes   67
   Tuberculous Peritonitis   67
   Tuberculous Abdominal Lymph Nodes   69

Tuberculosis of the Liver, Spleen, and Pancreas   71

Tuberculosis of the Urinary Tract   75
   Kidneys and Ureters   75
   Bladder   81
   Adrenal Tuberculosis   81
   Genital Tuberculosis   81

Tuberculosis of Bone   84

Tuberculosis of the Spine   85
   Synonyms   85
   Clinical Characteristics   85
   Clinico-pathological-radiological Correlation:
   Spinal Tuberculosis   85

Imaging of Spinal Tuberculosis   86
   Vertebral Body   86
   Paravertebral Abscess   94
   Scanning, Paraplegia, and Myelography   101

Differential Diagnosis   104

Tuberculosis of Bones and Joints (Nonspinal)   104
   Tuberculous Arthritis   104
   Synovial Tuberculosis   107
   Osteoarticular Tuberculosis   107
   Tuberculosis of Bones (Nonspinal)   110
   Differential Diagnosis   120

Tuberculosis of the Central Nervous System   125
   Tuberculous meningitis   128
   Tuberculomas   128

Tuberculosis Involving Other Sites   134
   Tuberculous Lymphadenopathy   134
   Tuberculosis of the Breast   134
   Tuberculosis of the Parotid Gland   134
   Ocular Tuberculosis   135

Bibliography   137

Subject Index   143

I see a lily on the brow
With anguish moist and fever dew
And on thy cheek a fading rose
Fast withereth too.

La Belle Dame San Merci
John Keats, 1795–1821

There was a time in the eighteenth and nineteenth centuries when tuberculosis was a fashionable affliction. All the best poets, musicians, and writers, not to mention politicians and heroines of operas, suffered from consumption and died either beautifully or in interesting ways. Of course, the unknown poor simply died, often unpleasantly, of the white plague. There were many names for this common disease, which has been recognized and described for centuries: the earliest definitive record is of pulmonary and spinal tuberculosis in the mummy of a 5-year-old Egyptian child of about 3400 B.C. A later mummy of a young man, dated at about 1000 B.C., had a psoas abscess as well as spinal lesions. In the Americas, a naturally mummified middle-aged Peruvian woman who died about 1000 years ago had primary pulmonary tuberculosis with calcified hilar lymph nodes. Tuberculosis was identified in this case because a unique DNA segment was identified in a lung lesion, using the polymerase chain reaction.

The word "tubercle" was first used in the seventeenth century by a Dutchman, Franciscus Silvius, of Leyden, to describe the lung lesions. Later (1839), Johann Schönlein called the disease "tuberculosis." It was not until 1882 that Robert Koch identified and described *Mycobacterium tuberculosis*. With this discovery much of the romance and many of the myths vanished.

If the source had been found, the cure had not, and tuberculosis continued as a major cause of death and no respecter of social rank: it was familiar and fatal to everyone. Probably more than 25% of the graves in the cemeteries of Europe and America were filled by people of all ages who had died of tuberculosis. There is a large and interesting literature on the disease and the remarkable ways in which it was treated. Most were in some way or other uncomfortable for the patient, and none were very successful. Then in 1943, Selman Waksman, a microbiologist at Rutgers University, fortuitously discovered streptomycin while investigating a peculiar fungus which was killing chickens. This discovery, together with the development of two other drugs in the next few years, led to the apparent conquest of tuberculosis.

But the bacillus which had undoubtedly affected history by killing so many in their youth may well have the same power once again. In 1993 the World Health Organization (WHO) declared that tuberculosis was a global emergency and estimated that it would kill 30 million people in the next decade. In 1996 it was the leading cause of deaths due to micro-organisms: there are eight million new cases every year. One-third of the world's population (1.7 billion) have been infected at some time and 20 million currently have active tuberculosis. (On average, about 10% of those infected go on to develop clinically active disease.) More than half of all those who have been infected live in Asia and Africa. WHO estimates (1995) that only 10% of tuberculous patients also have AIDS, but in the next decade there may be seven million with the combined infection, rising to 14 million by the year 2010. Almost every patient with both diseases will have active tuberculosis, because the AIDS virus destroys the cell which normally controls the micobacteria.

The global rise in population, the wars and disputes which have caused refugees and migration, and the decline in the level of public health services are also to blame for the recent reemergence of tuberculosis. WHO has shown (in Tanzania) that it is possible to find and cure over 80% of infectious cases and that 6–8 months of proper treatment will achieve this goal. The sputum bacterial counts fall rapidly and the sputum becomes negative for bacilli within 2 months of adequate treatment, resulting in control of the epidemic spread. In 1997, WHO was a little more optimistic that tuberculosis might be leveling off but warned that failure to treat the infection promptly would result in drug resistance, which is already at a level of at least 7% and rising and often takes the form of multidrug resistance.

Against this background it is important to add that "tropical" tuberculosis is not a product of the AIDS epidemic. This chapter in the first edition of "The Radiology of Tropical Diseases" (1981) started by stating that "The causative organism of tuberculosis, Koch's bacillus, is morphologically and culturally the same in the tropics as in non-tropical countries. Yet the clinical disease "tuberculosis" appears in ways which may be unrecognizable to physicians trained in non-tropical countries. It is not a "chronic" disease in the majority of patients, but an acute and often fulminating infection whether it be in the chest, bones, joints or elsewhere. Moreover, it causes signs and symptoms and presents at sites which may be quite unexpected and contrary to the descriptions in many North American and European textbooks." To this it is now necessary to add, "unless those books refer to AIDS-related tuberculosis."

Both in the tropics and in HIV-positive patients, the difference lies in the immunological background, the state of nutrition, and the way of life; it is that mystical relationship between the bacillus and its host which changes the clinical course of the disease, aided by variations in the virulence of particular strains of bacilli. Tuberculosis in the tropics often mimics an infection in an immunosuppressed patient; it must be remembered as the possible etiology of almost any acute or unusual illness whether seen in the tropics or in a recent immigrant into nontropical countries, and this pattern of tuberculosis must not be assumed to be due to AIDS. There are many other common causes of immunosuppression. And in the same way, a person from a nontropical area moving into the tropics retains the nontropical pattern of disease; transplanting one's person does not immediately change one's nature. Only over several years may this phenomenon of "host-response" alter to the local pattern, unless a virus or other event intervenes.

Acute tuberculosis with unusual variations is not the monopoly of the tropics; the authors have seen cases following the severe "tropical" pattern in patients who do not have AIDS and who have spent their lives in North America or in Europe. Tuberculosis is a disease of infinite variation because it occurs in an infinite variety of people. The immunosuppression due to HIV infections has focused worldwide attention on the pattern of tuberculosis which has been the norm in the tropics, as it may well have been in the past wherever nutrition, hygiene, and living conditions were poor and parasites were plentiful.

Dr. Olive Shisana, Director General of the South African Department of Health and joint author of a 1996 WHO report, has summed up the situation: "Tuberculosis has become a disease more terrifying than AIDS or the Ebola virus." Worldwide, it is the principal cause of death amongst all adults. Referring to drug resistance, Dr. Donald Enarson, Director of the International Union Against Tuberculosis, based in Paris, has added, "This is the most frightening situation I have ever encountered. If we do not act now, we will have a situation that we cannot control."

Radiologists not only in the tropics but everywhere else, should consider tuberculosis in almost every differential diagnosis: its manifestations, like its sufferers, are legion.

Synonyms

TB. Koch's disease. Pulmonary tuberculosis: Phthisis. Consumption. Tabes pulmonalis. Spinal tuberculosis: Pott's disease. Abdominal tuberculosis: Tabes mesenterica. Tuberculous cervical lymphadenitis: Scrofula. Struma. General tuberculosis: Hectic fever. Asthenia. *Ger*: Tuberkulose. Schwindsucht. *Sp*: Tisis, Tuberculosis. *Fr*: Phtisie.

Definition

Tuberculosis is an infection with the *Mycobacterium tuberculosis* or *Mycobacterium bovis*. Both are gram-positive, acid and alcohol fast, aerobic, non-spore forming rods, classified with the actinomycetes. *M. tuberculosis* is a facultative intracellular parasite, which, many believe, is capable also of extracellular growth and can remain dormant for years or even decades.

Geographic Distribution

Tuberculosis is a worldwide infection from which some three million people die annually, more than from any other single infectious disease. It is probable that, because of AIDS, drug resistance, and population stresses, the epidemic will increase in the future unless brought under control. The countries with the largest number of patients with tuberculosis are Bangladesh, Brazil, China, India, Indonesia, Nigeria, Pakistan, the Philippines, and Vietnam. The rate of disease is highest in sub-Saharan Africa. In some parts of the world, 80% of the population react positively to a tuberculin skin test by the age of 25 years and many react much earlier (unless they are HIV-positive). In developing countries tuberculosis accounts for 26% of avoidable deaths.

The reported incidence, however, may not be accurate: it is often a reflection of the efficiency and methods used to detect the disease rather than a true indication of its actual prevalence. Public health facilities vary greatly, but, for example, it was estimated East Africa (1971) that half a million persons were newly infected each year and that 75 000 would develop clinical tuberculosis, of whom only 25 000 would be diagnosed and 16 000 treated. Such figures have improved in some countries, but not in all. A person, once infected, is likely to harbor the tubercle bacillus for the rest of his or her life and it may remain dormant unless there is a change in general health or immunity, as occurs with AIDS, diabetes, leukemia, or malnutrition. Much less commonly, there may be reinfection.

Epidemiology and Pathology

Tuberculosis affects everyone from infants to the very aged. It is spread by droplet infection or by contaminated foods, particularly milk from herds infected with *M. bovis*. In many countries bovine tuberculosis has been almost completely eliminated following the pasteurization of milk and the control of infected cattle. Unfortunately, there is a reservoir in wild species and it is known to have infected buffalo and rhinoceros in East and southern Africa, for example. Reinfection of domestic cattle is thus always possible.

The first infection is most commonly in the lungs, but can be in any part of the body. Both *M. tuberculosis* and *M. bovis* can directly infect the skin, presenting as a serpiginous ulcer, although this is uncommon. Lupus vulgaris is a secondary (immune) infection. Transplacental infection is possible but very rare: there are about 150 well-documented cases in the literature, but it may become more frequent in HIV-positive mothers.

Even if AIDS is not taken into account, tuberculosis is most prevalent in crowded cities, yet there is also a high incidence in rural areas. In the tropics it is a household infection and WHO has reported an incidence five to ten times higher in household contacts than among the general population. Living and sleeping in proximity to an infected person in a crowded house or hut is more risky than using public transport or working in ill-ventilated buildings.

The susceptibility of a person to tuberculosis is influenced by genetic and ethnic factors, malnutrition, the quantum of bacteria entering the body, and other illness, such as diabetes, pneumoconiosis, infection with multiple parasites, and anemia, as well as AIDS. There is an individual and herd variation in immunity. The Moslem community of southern India, the North American Indians and Eskimos, many Africans, and Malays have a high susceptibility and morbidity. Extensive campaigns by WHO utilizing mass radiographic surveys, BCG vaccination, and education were very successful, but now the disease is once again a major health problem throughout the tropics and is getting more common because of AIDS and drug resistance, as well as population movements.

Modern chemotherapy has made it possible for patients with active pulmonary tuberculosis to stay at home and have treatment without endangering other inhabitants of the house. As mentioned above, a bacillus-positive sputum will become negative after about 2 months of therapy. The real danger of spreading infection occurs before the disease is diagnosed. Ambulant treatment has encouraged many patients to accept the necessary regimen; previously, isolation or sanitarium treatment met with considerable resistance. Now the difficulty is to ensure that treatment is continuous and adequate.

There are also variations in the sensitivity of the organism to different drugs; for example, in southern India, where tuberculosis is common, the bacilli seem to be isoniazid-sensitive, whereas in the United Kingdom the response is not so good. p-Aminosalicylic acid is better tolerated by patients in the tropics. Thiacetezone is not tolerated by HIV-positive patients and may even be fatal. The atypical (nontuberculous) mycobacteria are also a cause of treatment failure, but the major therapeutic problem is the development of drug-resistant organisms, the frequency of which now varies from 7% to 20%. This is usually the result of inadequate and incomplete treatment: many patients do not feel the need to continue therapy for 6 or 8 months and abandon their pills as soon as they feel better. Poverty, the nonavailability of drugs in the regular market at normal cost, interference in the treatment by unqualified practitioners, and the side-effects of the drugs (e.g., tinnitus, deafness, neuropathies, hepatitis, gastroenteritis) all offer reasons for the failure of continuous treatment. Black market drugs at inadequate dosage add to the problem. Yet WHO reckons that 80% of patients can be cured with proper supervision and full treatment, and that even in the poorest countries a case reduction of 50% is possible.

A patient's resistance to tuberculosis is a very complex process, much of which is not yet fully understood. In very general terms, no two mycobacterial strains are genetically identical (as is the case with people), but if a person is infected with a virulent strain of tuberculosis, the outcome depends on the state of competence of the patient's cell-mediated immunity (CMI). With a competent CMI the body produces derivatives of the monocyte-macrophage cells that have the ability to kill and sequester mycobacteria. These derivatives are the activated macrophage, the epithelioid cells, and the Langhans giant cell. All these cells produce hydrogen peroxide and oxygen in forms which are toxic, such as hypochlorous acid and toxic enzymes: all of these kill mycobacteria. However, the effects are unfortunately not limited to the bacteria and these toxic products kill much more host tissue than mycobacteria, which is why the disease becomes chronic, resulting in persistent destruction as seen in fibrocaseous tuberculosis. This is known as hyperergic tuberculosis, a response which is characterized by the granulomatous reaction. It is the only effective defense and the only hope that the patient has of survival. A positive tuberculin skin test is an indication that CMI has developed, and there will be a sharp reduction in the number of mycobacteria and the development of a layer of palisaded epithelioid cells around the necrotic foci. Eventually there will be an outer layer of granulation tissue surrounding the necrotic foci and scars will form that contain any remaining mycobacteria.

When CMI is suppressed, becoming too weak or nonexistent, the mycobacteria can proliferate unchecked and destroy host tissues directly. This is anergic tuberculosis, which, unchanged, is progressive and soon fatal. It is characterized by broad areas of contiguous noncaseous necrosis, large numbers of bacilli, and very little or no granulomatous reaction. Caseous necrosis is flecked with mineral deposits, hence calcification, and retains for a short time a ghost outline of previously vital tissue. The necrosis of anergic tuberculosis does not have calcification or recognizable tissue. The suppression of CMI may occur for unexpected reasons, such as grief, trauma, exposure to cold, cytotoxic agents, x-radiation in large doses, and corticosteroids, as well as the more recognized causes such as malnutrition, alcoholism, diabetes, malignancy, and, of course, the AIDS virus. The problem with this oversimplified explanation is that it is difficult to confirm in the laboratory and there is no general explanation why patients who have apparently successfully overcome the infection should suffer a breakdown of this protective immunity, allowing the disease to surge again. One of the suggested explanations is the difference in the genetic background of the host, but this is also very controversial. For all these reasons, the terms "hypersensitivity," "immunity," and "acute" are used less often. Much remains to be clarified, especially about the cell envelope and the associated and secreted protein. It is not even clear how the effective immune response in 90% of people succeeds in killing tubercle bacilli. As so often happens in medicine, the natural history of the disease is fully understood, the explanation of why it happens is not. Research continues, and is particularly oriented toward the development of a vaccine against tuberculosis.

Laboratory Diagnosis

The only real proof of tuberculosis is the isolation on smear or culture of *M. tuberculosis* from the patient's sputum, other body fluids, or tissues. The histological mark is the epithelioid granuloma with mononuclear and Langhans giant cells, together with caseating necrosis. Clinically the tuberculin test is almost universally accepted as a fairly reliable way to detect a current or previous tuberculous infection, but there are sources of error: When the tests are properly applied and material is fresh and active, it is an effective screening method unless the individual is immunosuppressed. Another source of error, in which the test may be positive, is previous BCG vaccination. Sensitivity to tuberculin may also be related to unrecognized infection with similar micro-organisms which are not tuberculous. This "false" sensitivity is particularly common in the tropics and subtropics and has a variable geographic distribution both in Africa and in India; it is more common in low-altitude populations than in those who live on the mountains. The responsible organism has not yet been identified. Thus, when there is a weak reaction to low doses of tuberculin and a strong reaction to a high dose, it may indicate a nonspecific reaction rather than tuberculosis. During severe and overwhelming generalized infections (including HIV) the skin tests may remain negative, as seen in very sick, undernourished children. A similar misleading result can occasionally occur in bone tuberculosis. It is not common, but is well documented. Thus, a negative skin test does not totally exclude tuberculosis, nor in the tropics does a positive test necessarily prove its presence. The new γ-interferon blood test, which is thought to be 98% specific and is without observer error, may solve many clinical problems.

Other laboratory tests do not differ in the tropics from those accepted elsewhere in the world, but unfortunately the host of parasitic infestations which are endemic in the tropics may confuse routine investigations. An elevated erythrocyte sedimentation rate (ESR), a raised eosinophil count, lymphocytosis, and anemia may all be due to the patient's "underlying normal" ill-health. Sputum or gastric aspirates remain the most reliable method of diagnosis, particularly in children. Under the age of 1 year, it is the lungs which are most commonly infected and positive recovery of the bacilli is possible in about 75% of cases. In older children, even when very ill, the recovery rate of bacilli may be only 40%. Examination of effusions in the pleura, peritoneum, or joints, or of the pus from a spinal or other abscess, may demonstrate the organism on smear or culture, but failure to recognize the bacillus does not exclude active tuberculosis. Unfortunately, culture is a slow process, expensive, and not always readily available. Apart from the classical histological findings, the polymerase chain reaction, a method amplifying specific DNA sequences, is successful in detecting mycobacterial DNA in clinical specimens, including fluids.

The cerebrospinal fluid (CSF) will only contain tubercle bacilli in about 10% of initial samples. More information can be obtained from the ELISA assay and the polymerase chain reaction. There is a latex particle agglutination assay which is inexpensive and suitable for use in developing countries. The spinal fluid will usually be under pressure and clear, with an increased protein content and reduced glucose level: there is often an increased CSF white blood cell count, with 90%–95% monocytes.

Examining urine for the bacillus was once suggested as a way of confirming generalized tuberculosis, but this is positive in less than 7% of samples (except when there is active genitourinary tuberculosis), and is too inaccurate to be used. General hematological examination is usually normal in tuberculosis, but can be confusing when there is acute tuberculous pneumonia because the white blood cell count may be increased. When there is bone marrow involvement, there may be leukopenia. A patient who has multiple parasites is also likely to have an abnormal hematological background.

In many patients the disease is a severe, acute infection and no time can be wasted in waiting for laboratory results which may be of dubious value. Even when the skin reaction is negative, if the clinical suspicion of tuberculosis is high then therapy is often started and the patient reviewed, particularly as to the radiological findings after about 2 months of treatment. Hypercalcemia is an infrequently recognized complication of tuberculosis, particularly in small children and those with disseminated infection; it probably occurs because vitamin B_3 metabolites have a role in immunoregulation, and the level of these metabolites is regulated in granulomata by T-small cells.

Clinical Characteristics

Clinical tuberculosis is so well described in so many textbooks that there is little reason to repeat the findings. To the usual picture of malaise, tiredness, loss of weight, night sweats, and mild pyrexia must be added two important variants: the acute form of the disease and the background of ill health which is so common in many of the tropical patients.

Ill health, anemia, malnutrition, pyrexia, and dysentery are so much a normal part of the low standard of living and hygiene of many countries that these conditions cannot be used as an indication of possible tuberculosis. For example, in Egypt and elsewhere in Africa, many adults in rural areas normally have several varieties of parasites and this is equally true for much of India, Asia, and South America. This background of ill health makes the patient more susceptible to tuberculosis and the delicate balance of borderline ill health may have been upset by a recent tuberculous infection which has brought the patient to the physician.

In other patients, acute tuberculosis, whether in lungs or in joints or elsewhere in the body, may closely mimic an acute bacterial infection. The classical presentation of a low-grade illness is not the only way in which tuberculosis behaves. The history of the illness may be short, as when a child who has a temperature of 39.5°C (103°F) seems to have acute bacterial lobar pneumonia but actually has primary tuberculosis. Initial examination of the sputum (or gastric lavage) in such cases may be confusing because of secondary infection: the diagnosis of tuberculosis may not be made because the possibility is often forgotten. Similarly, acute osteomyelitis, septic arthritis, mastoiditis, and peritonitis may all be due to tuberculosis and not an "ordinary" pyogenic infection. There must be a high index of suspicion and clinical sixth sense, and tuberculosis should be included in almost every differential diagnosis.

Tuberculosis of the Respiratory Tract

Even before AIDS swept throughout the world, pulmonary tuberculosis has been one of the major causes of illness and death: it is not likely to relinquish this leading role for many years, even in the millions who are HIV-negative.

Tuberculosis of the Upper Respiratory Tract

The possibility of tuberculosis of the upper respiratory tract should always be considered in a patient from whom the sputum has been found to contain tubercle bacilli and yet the routine chest radiograph is normal. Clinically there will be excess sputum, hoarseness, and, usually, general symptoms of tuberculosis. Endoscopy, computed tomography (CT), and, in some cases, a contrast tracheogram will demonstrate the tuberculous ulceration and granulation tissue (when CT is not available, standard tomography can be helpful) (Figs. 1, 2). The trachea may be narrowed, the larynx swollen, and the cartilage eroded, all of which can be demonstrated on CT. Active tuberculosis is always much more extensive on CT scanning than it appears on endoscopy. Quite often the whole length of the trachea is thickened and it may extend into the major bronchi. The thickening is irregular, and where there is an intraluminal granulomatous mass, the degree of narrowing can be quite significant. The extent of the infection is most clearly seen on three-dimensional imaging, but actual images often show the relationship of the neighboring structures.

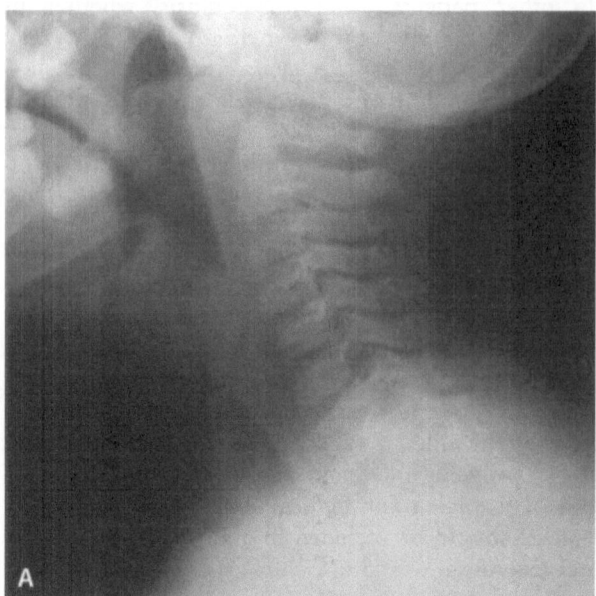

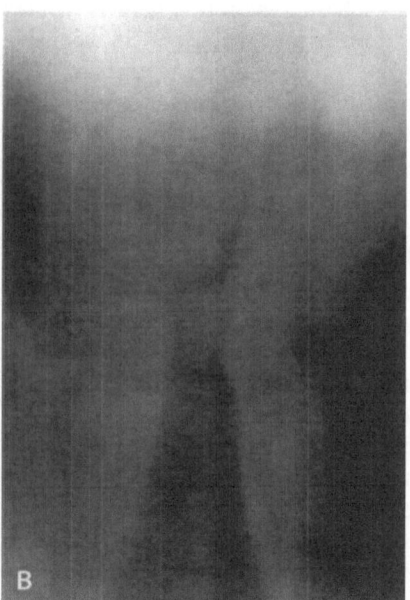

Fig. 1 A, B. A Laryngeal and subglottic tuberculosis in a 3-year-old girl from one of the Pacific Islands. There is narrowing and anterior displacement of the airway. The diagnosis was confirmed by bronchoscopy: both lungs showed miliary tuberculosis and there was tuberculous thickening of the ileocecal region also. Upper airway infection is usually part of generalized, hematogenous spread and this case was no exception. (Courtesy of Dr. Cheryl Sisler, Hawaii) B Linear tomograms of laryngeal tuberculosis which presented in an adolescent as upper airway obstruction. There was a granulomatous mass lesion in the larynx. (From Cremin and Jamieson 1995)

The fibrotic stage of tuberculosis usually results in smooth narrowing of the lumen and the thickening of the wall is less marked. When the disease spreads down to the main bronchi, the left main bronchus is often more involved than the right.

Although pulmonary tuberculosis is common, laryngeal and tracheal infection are not, even as a complication in patients with heavily infected sputum. As an isolated infection, without pulmonary tuberculosis, laryngeal or tracheal tuberculosis is rare. While the granulomas can be imaged, there is no satisfactory radiological method to confirm that the etiology is tuberculosis.

Fig. 2 A–F. Tuberculosis of the trachea and major bronchi is ▶ best shown with multiplanar and three-dimensional reconstruction. A An irregular mass narrowing the trachea, with marked thickening of the whole length of the tracheal wall. B Three-dimensional CT scan of the same patient confirms the irregular and diffuse narrowing of the whole length of the trachea and the proximal main bronchi. C Thin-section CT, at the level of the thoracic inlet, clearly showing the marked thickening of the wall of the trachea and the narrow lumen. In the fibrotic stage, the distortion may be even more marked. D Three-dimensional CT scan of a 29-year-old woman shows marked irregular narrowing along the length of the left main bronchus. E A thin-section CT scan of the same patient, just below the carina, confirms the narrowing of the lumen and shows the marked thickening of the wall of the bronchus. F The coronal scan shows the full extent of the bronchial fibrosis. (Courtesy of Dr. Kyung Soo Lee, Seoul)

Primary (Nonimmune) Tuberculosis

The prevalence of acute primary infection at any age, even without AIDS, is one of the major variations of acute tuberculosis. In one series in the tropics, more than 50% of the patients with the primary pattern of tuberculosis were over 18 years of age. Clinically patients may present with an acute illness which may be so severe that its true cause is not easily recognized. Other patients may seem to have a simple upper respiratory infection, perhaps one which does not improve clinically, and primary lung tuberculosis is an unexpected finding on a chest radiograph. There are five ways in which primary tuberculosis presents radiographically:

- Lobar pneumonia
- Bronchopneumonia
- Hilar and mediastinal adenopathy
- Pleural or pericardial effusion
- Miliary tuberculosis

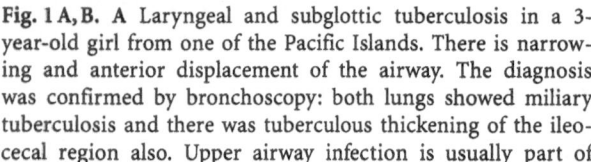

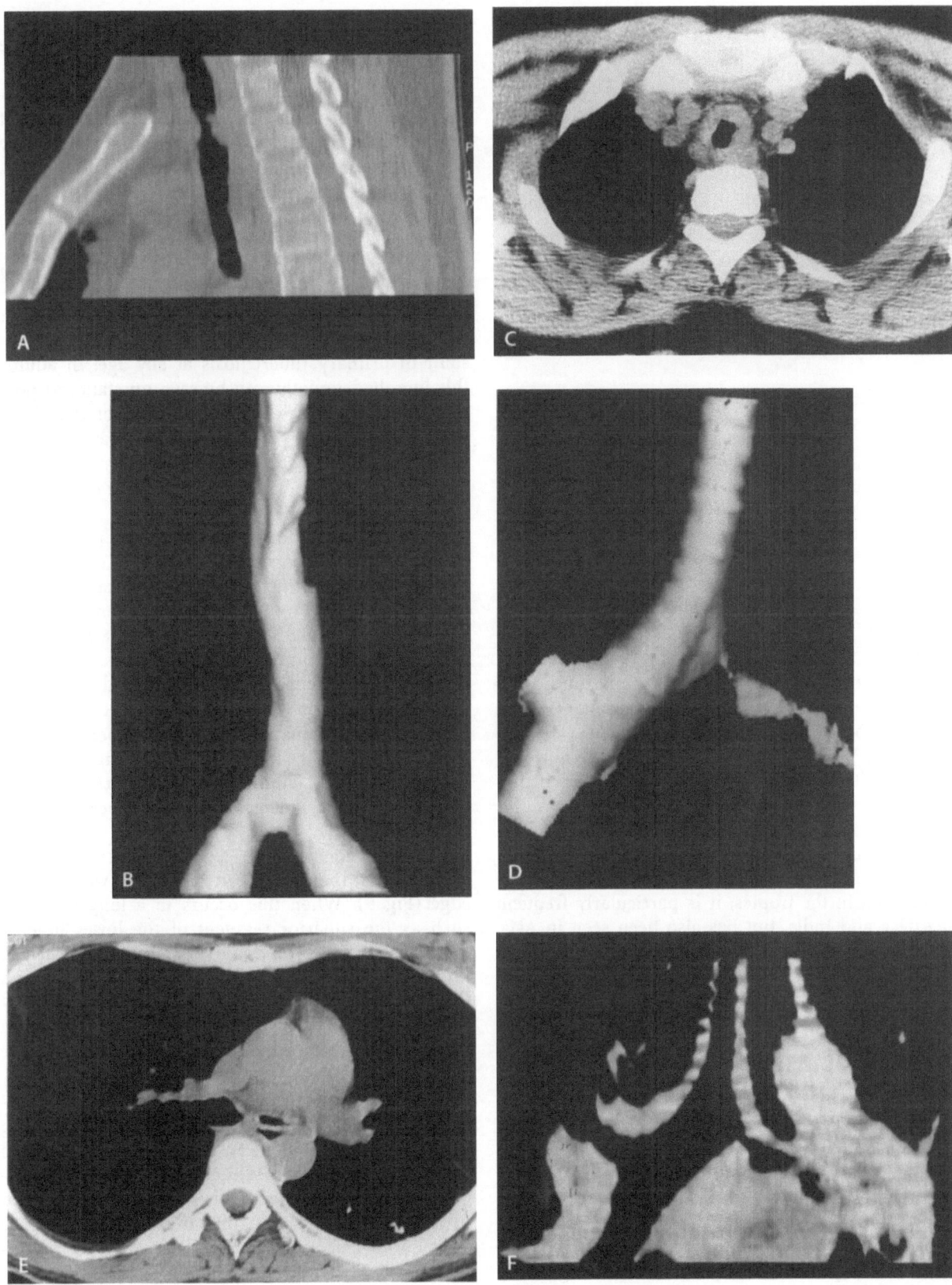

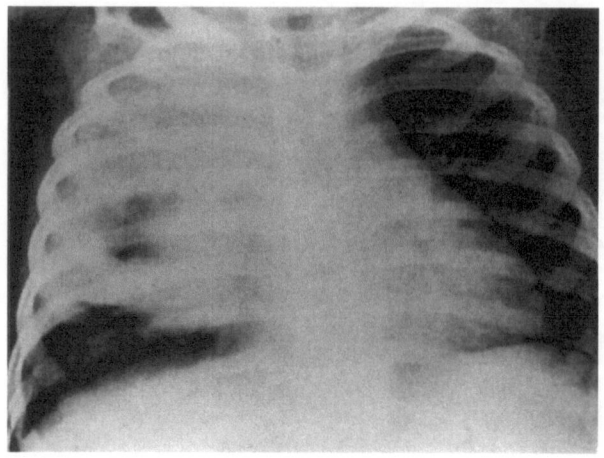

Fig. 3. Progressive primary tuberculosis. Right upper lobe tuberculous pneumonia has progressed, with central necrosis, and become a large tuberculous lung abscess. There is a small overlying pleural reaction. Shortly after this film was taken, the child developed an acute pneumothorax that required surgical decompression. (Courtesy of *Semin Roentgenol*, 1979)

Any of these patterns may appear on the first chest radiograph, separately or in any combination. In patients with poor immunity for any reason, the primary infection may never heal but continue on to severe cavitating, spreading pneumonia (Fig. 3). This is progressive primary tuberculosis (see also p. 13). The majority of primary infections, however, heal with little residual scarring; only a few will imperceptibly change to the secondary, or immune, pattern of tuberculosis with resulting fibrosis, distortion, and calcification. There is one particular end result, a "destroyed lung" or lobe. This outcome is common in the tropics; it is particularly frequent in Africa and India, but has also been seen in Asia, Australia, Hawaii, and South America. It occurs in children as well as in adults (see p. 19).

Lobar Pneumonia

The classical "primary" lobar pneumonia of nonimmune tuberculosis may affect any part of the lung, usually a whole lobe but sometimes only a segment (Fig. 4). It is probably a little more common in the upper lobes, but no part of the lung escapes and being influenced by the site of the lobar infection is not a safe way to exclude tuberculosis. Clinically the patient may present at any age, with a high temperature and the physical findings of lobar pneumonia. Quite contrary to this acute presentation, in other patients the lobar consolidation may be seen on a chest radiograph when there is little or no clinical ill health and the pneumonia is quite unsus-

pected. There is no atelectasis in the early stages, nor is there a pleural effusion associated with the lobar pneumonia (see p. 24).

There may be *M. tuberculosis* in the sputum or gastric lavage, but failure to find the bacillus does not invalidate the diagnosis. A tuberculin skin test may be negative or change to positive during the illness, usually in 2–3 weeks after the onset.

The lobar consolidation is definite but seldom dense, yet it is more opaque than the hazy and ill-defined consolidation which is usually seen in a viral infection. The distinctive feature of primary pulmonary tuberculosis is enlargement of the hilar and mediastinal lymph nodes, which is nearly constant in primary tuberculosis at any age; in adults, this lymphadenopathy can be seen on standard posteroanterior (PA), high kV, grid films of the chest. In children it may be necessary to obtain a lateral view as well as the frontal film to reliably identify the enlarged nodes: as always, full inspiratory radiographs are essential. Lymphadenopathy is a way to distinguish tuberculosis from pyogenic pneumonia, in which hilar adenopathy recognizable on standard radiographs is distinctly unusual (although it may be visible with CT). "If the nodes are enlarged and can be seen on standard radiograph in a case of lobar or segmental pneumonia, then suspect tuberculosis." This is a reliable dictum. It was the Viennese pathologist, Anton Ghon, who in 1912 first described the combination of a focal tuberculous lesion in the lung with regional lymphadenopathy. Radiologists have continued to recognize the Ghon complex as important evidence of a tuberculous infection.

Tuberculous consolidation resolves from the periphery towards the hilum, often with a well-defined edge (Fig. 5). When this occurs in a lung segment, such as the superior segment of the lower lobe, it may radiologically resemble a tumor, particularly if this is the appearance on the first chest radiograph (Fig. 6). In children this will be an unlikely diagnosis; malignant disease of the lung (apart from metastases) is uncommon in many parts of the world at present.

Tuberculosis in one-third of babies may progress radiologically for as much as 3 months, even during intensive antibiotic therapy. In addition to extension of the lobar consolidation, there may also be increasing lymphadenopathy, thought to be the result of a high-sensitivity reaction during the first 2–10 weeks after infection. In most patients, the resolution of the radiographic changes of tuberculous pneumonia is slow: it often takes as long as 9 months even when adequate therapy has been given and when the clinical improvement has been dramatic (as is the case in many patients).

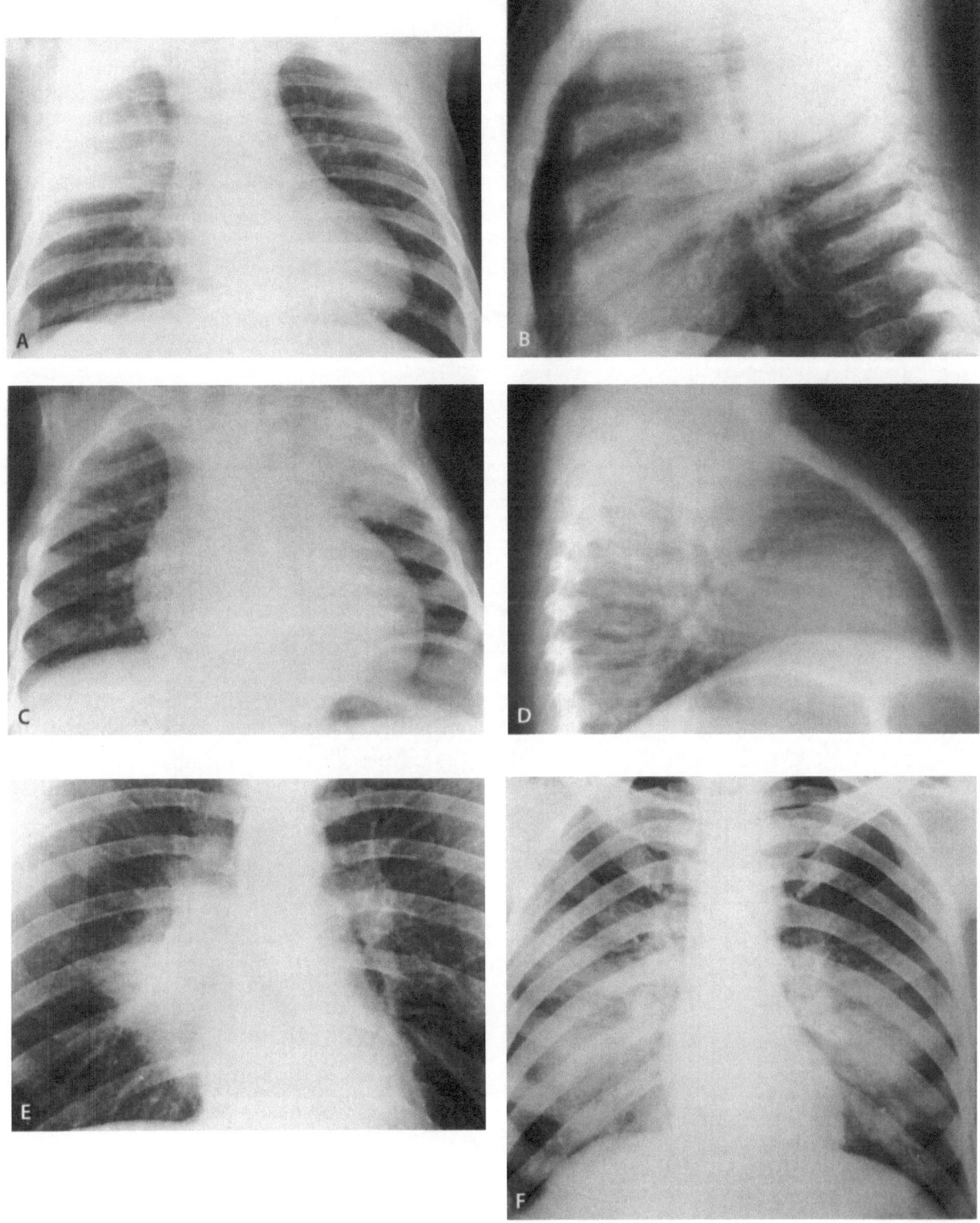

Fig. 4A–F. Tuberculous primary lobar pneumonia can occur in any lobe or segment of lung. There is almost always hilar lymphadenopathy. A, B Primary pneumonia in the posterior segment of the right upper lobe, with right hilar lymphadenopathy. C, D Left upper lobe tuberculous pneumonia: the enlarged lymph nodes are only clearly seen in the lateral projection. There is also tuberculous infection in the lower thoracic spine. E Right lower lobar pneumonia. There is probably lymphadenopathy, partially obscured by the lung consolidation. F Bilateral primary pneumonia in the right middle lobe, the lingula, and the anterior segment of the right upper lobe. This child had had a mild but persistent cough and a low fever for 2 weeks. (A, B, E, F courtesy of *Semin Roentgenol*, 1979)

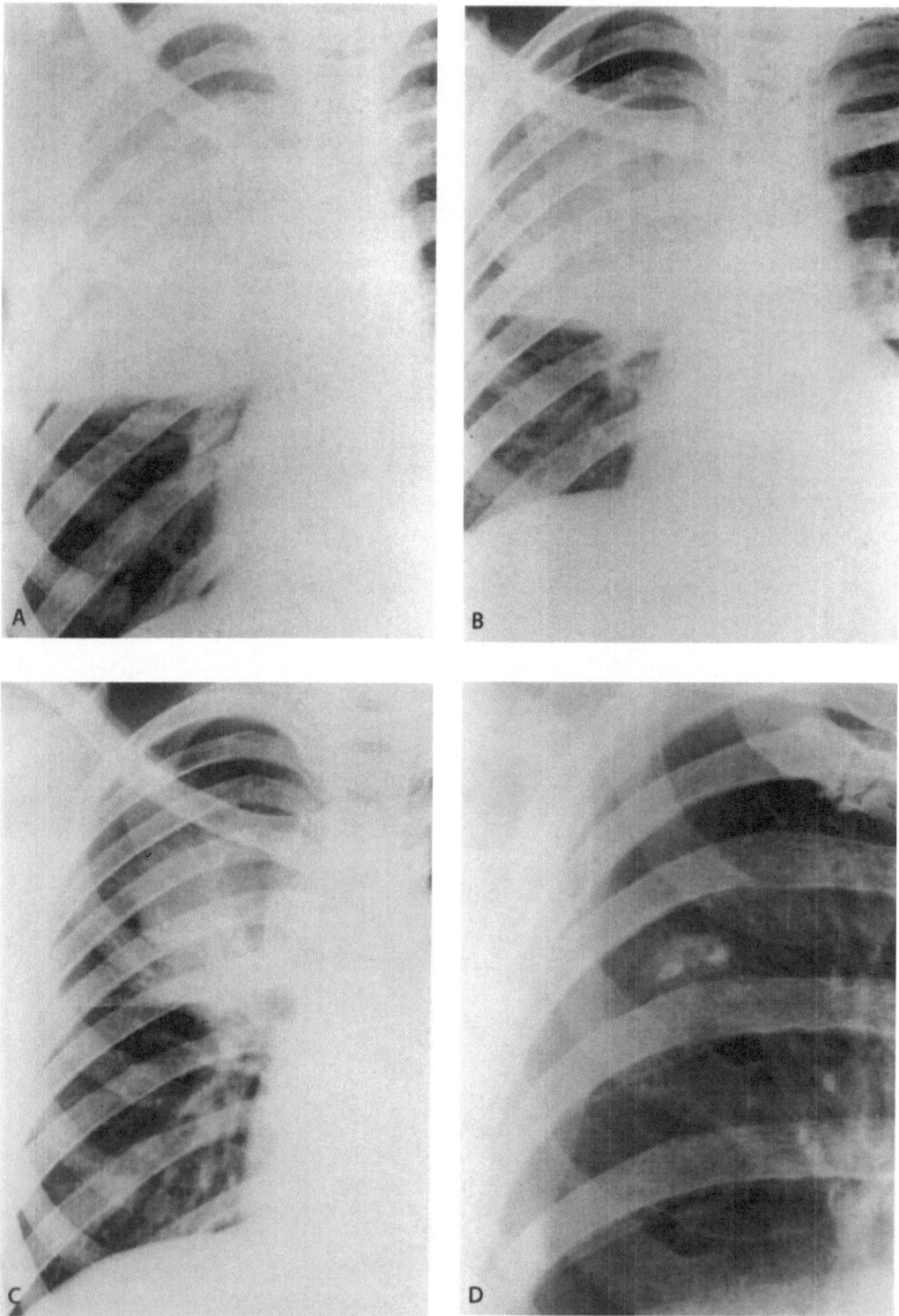

Fig. 5 A–D. The resolution of primary tuberculous pneumonia usually occurs from the periphery inwards towards the hilum. **A** Right upper lobe pneumonia, moderate pyrexia and cough, and a negative PPD. The patient was given penicillin, after which there was some clinical improvement but persisting pyrexia (December). **B** By January the PPD had converted and the lobar pneumonia had begun to resolve from the periphery towards the hilum. There were tubercle bacilli in the sputum and antituberculous therapy was started. **C** By March consolidation had decreased further towards the hilum. **D** One year later, after continuous treatment, there is calcification in the small residual granuloma. The patient had remained clinically fit. This pattern of healing can be seen in a patient of any age. (Courtesy of *Semin Roentgenol*, 1979)

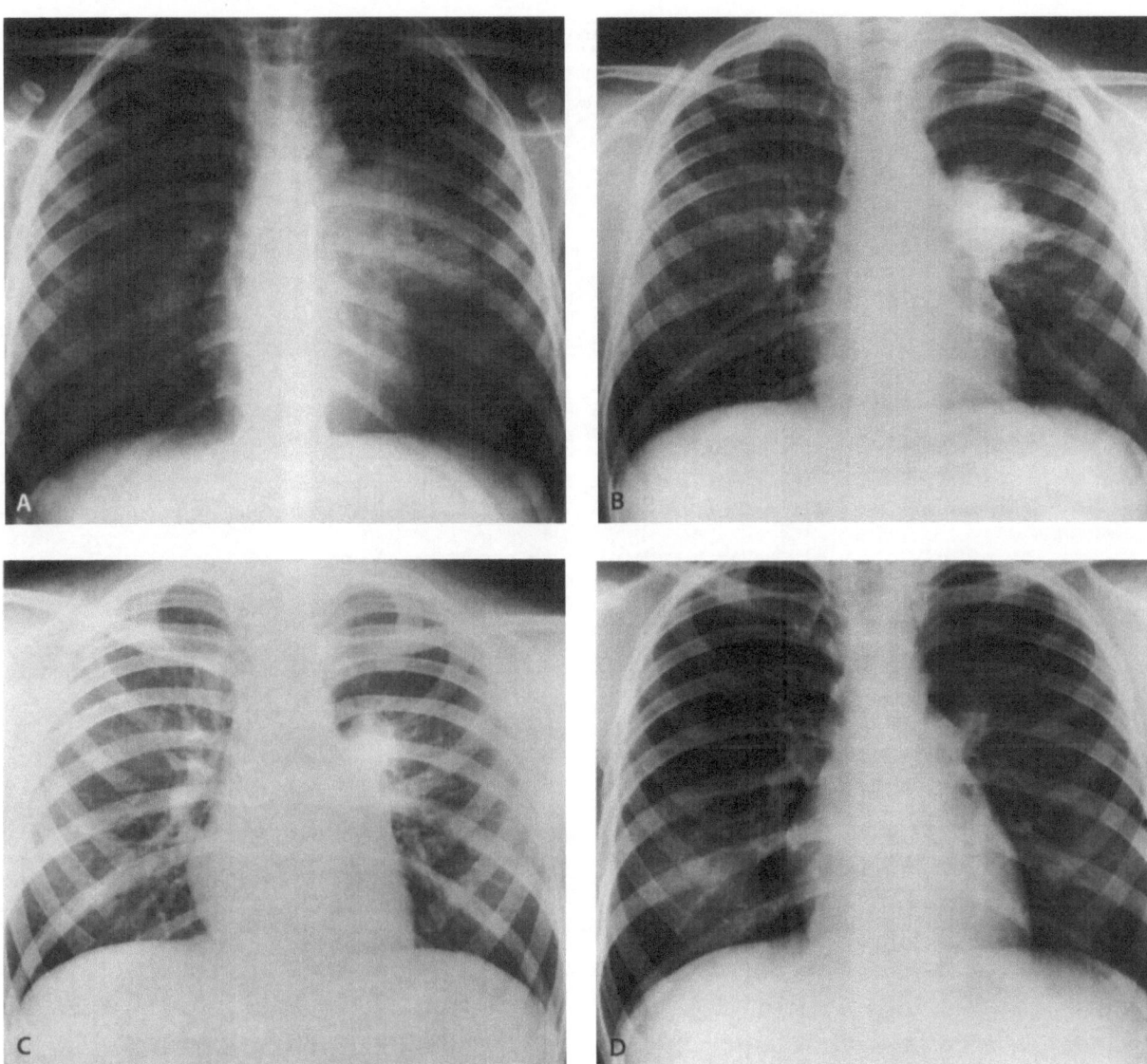

Fig. 6 A–D. The resolution of primary (nonimmune) tuberculosis in a young male adult. **A** July: Persistent loss of weight and cough due to a tuberculous infection in the superior segment of the left lower lobe, with perihilar lymphadenopathy. PPD was positive and there were *M. tuberculosis* in the patient's sputum. Treatment was started with three drugs. **B** Four months later consolidation is less pronounced but the lymph nodes are still enlarged and appear much more dense.

C After a further 3 months' treatment the consolidation and nodes are again much smaller: shrinkage has continued from the periphery inwards. If the patient had been radiographed at this stage, or at the stage depicted in B, the "mass" could have been mistaken for a neoplasm. **D** After another 14 months of treatment there has been much improvement, but the lymph nodes are still slightly enlarged and the lung is a little more dense than normal

The considerable discrepancy between the x-ray appearance and the well-being of the patient may be useful confirmation of the diagnosis when there has been no other more positive proof of tuberculosis (see Figs. 5, 6). Any lobar pneumonia which disappears within 2 or 3 weeks, leaving a normal chest radiograph, is probably not tuberculous.

Radiologically visible calcification occurs quite early in tuberculosis, both in the peripheral focus in the lung and in the hilum. As the calcified lung focus shrinks, it may seem to be more dense; the end result is the calcified, healed primary complex (Fig. 7). In other patients, the peripheral granuloma will disappear altogether, leaving only residual calcification in the lymph nodes. Calcification seen radiologically is evidence that healing is occurring; unfortunately, it does not always mean that the infection has been conquered. It is possible for a partially calcified lymph node to rupture into a bronchus and cause inhalation bronchopneumonia. Because the lymph nodes, although starting to calcify, may still be large and soft, they may rupture

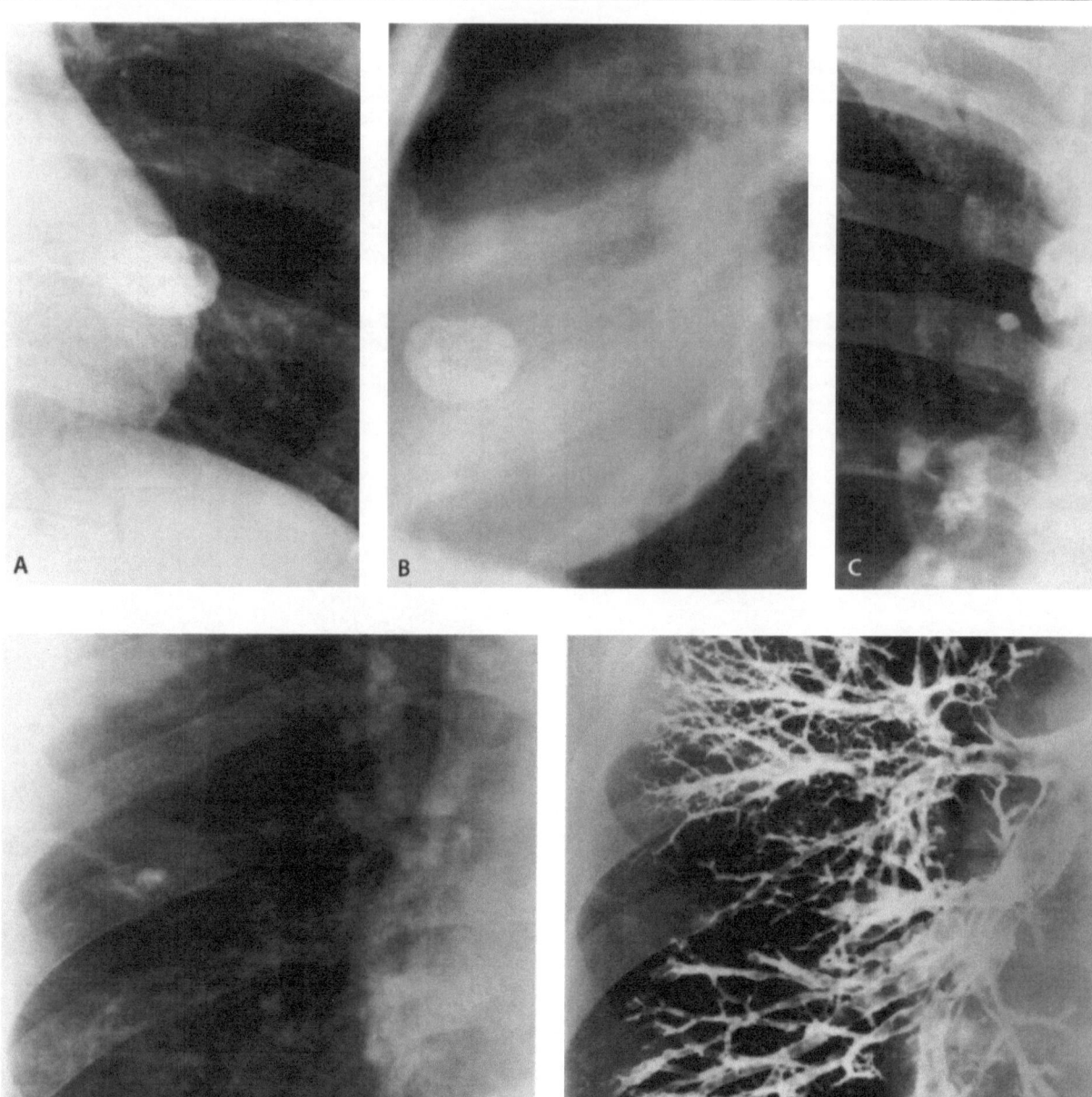

Fig. 7 A–E. In the majority of patients of any age, the end result of a primary tuberculous infection is a calcified granuloma: this is often a chance finding and seldom of clinical significance. **A, B** A well-defined, moderate sized calcified granuloma in the lingula of a physician, known to have been present for years. The PPD was positive and the patient was healthy. **C** Calcified hilar and right paratracheal lymph nodes. There is a very small calcified granuloma in the apical segment of the right lower lobe, seen alongside the paratracheal calcification. **D** Linear fibrosis can be seen around a small calcified peripheral granuloma. **E** Bronchography (performed for another reason) confirms the distortion of the bronchial pattern, localized emphysema, and some fibrosis, which is in a lower subsegmental branch of the right upper lobe. Such distortion is seldom of clinical significance, although occasionally it is the site of repeated infection. (Courtesy of *Semin Roentgenol*, 1979)

into a bronchus (Fig. 8C–F) or cause external pressure resulting in peripheral atelectasis (see "The Destroyed Lung" p. 19).

The radiological differential diagnosis of primary lobar tuberculosis includes pyogenic or other lobar pneumonia, especially in those patients in whom the enlarged lymph nodes cannot be seen. Where there is consolidation in the lower lobe particularly, amebiasis will have to be excluded; both paragonimiasis and melioidosis can mimic tuberculosis, although usually at the later, cavitary stage of the disease.

Finally, it must be emphasized that because most tuberculous infections are airborne, a pulmonary lesion may subsequently develop wherever the tubercle bacilli happen to lodge in the lung. Thus, distribution is by chance. The majority of such contacts with bacilli leave no evidence or, at the most, microscopic foci which are not visible radiologically. Careful histopathology of many "normal" lungs has shown that most of these invisible lesions are in the lower lobes and seldom develop further. It is the minority of foci that cause the majority of radiological and clinical findings. Nevertheless, it is very important to recognize that no part of the lung is exempt from any pattern of tuberculosis and in the nonimmune patient there is usually a poor correlation between the radiological pattern and the clinical state of the patient.

Bronchopneumonia

Tuberculous bronchopneumonia presents radiologically as multiple, often bilateral, patchy, "cottonwool" densities. These occur when tubercle bacilli are forced or inhaled into multiple terminal bronchial segments during caughing, either when an open pulmonary lesion (cavity) communicates with a bronchus, or when an infected lymph node has ruptured into a bronchus (Fig. 8). In nontropical countries, this radiological pattern of bronchopneumonia usually suggests staphylococcal infection and the acute clinical presentation of tuberculosis may be very similar. The tuberculous patient, usually a young child or baby but sometimes an adult or elderly person, can be severely ill clinically. The sputum invariably contains many tubercle bacilli, but the tuberculin skin test is often negative, especially in the early stages or when the patient is HIV-positive.

Radiologically the infection is usually bilateral and widespread, but not always symmetrical. There may be multiple thin-walled cavities (lung abscesses), together with the fluffy ill-defined densities. The appearances change daily, and the thin-walled cavities may vary in size, be empty or contain fluid, expand, and develop a surrounding pulmonary reaction. This sequence depends not only on the severity of the infection and the resistance of the patient, but also on the accumulation of secretion in the cavity and how frequently and how well it is coughed out. In progressive primary tuberculous infections there will be marked bilateral adenopathy in almost every patient, the nodes being large and ill defined. (When a similar bronchogenic spread occurs during the secondary, or immune stage of tuberculosis, there is no adenopathy (see Fig. 26F, p. 37).) In this acute bronchopneumonic, cavitating pattern of tuberculosis, any of the peripheral tuberculous lung abscesses may rupture into the pleura or pericardium, resulting in a tuberculous empyema or pericarditis (Fig. 20, p. 28).

Histological examination at this stage shows that the lungs contain multiple thin-walled abscesses which are filled with caseous pus. This can be demonstrated by CT, particularly with high resolution and thin sections. It must be differentiated from bronchiectasis by the finding of relatively normal bronchi and the absence of atelectasis.

Calcification in this pattern of infection is uncommon, in either the lung foci or in the nodes. In some patients infection is so acute that it may cause death: if the patient recovers, the end result can be a remarkably normal chest radiograph or one showing only a little scarring.

The differential diagnosis includes staphylococcal pneumonia. Adenopathy is extremely rare in similar cavitating pyogenic infections, most of which are the result of septicemia with a visible source of infection, such as osteomyelitis or an infected skin ulcer. In both tuberculous and staphylococcal infections there can be hepatosplenomegaly.

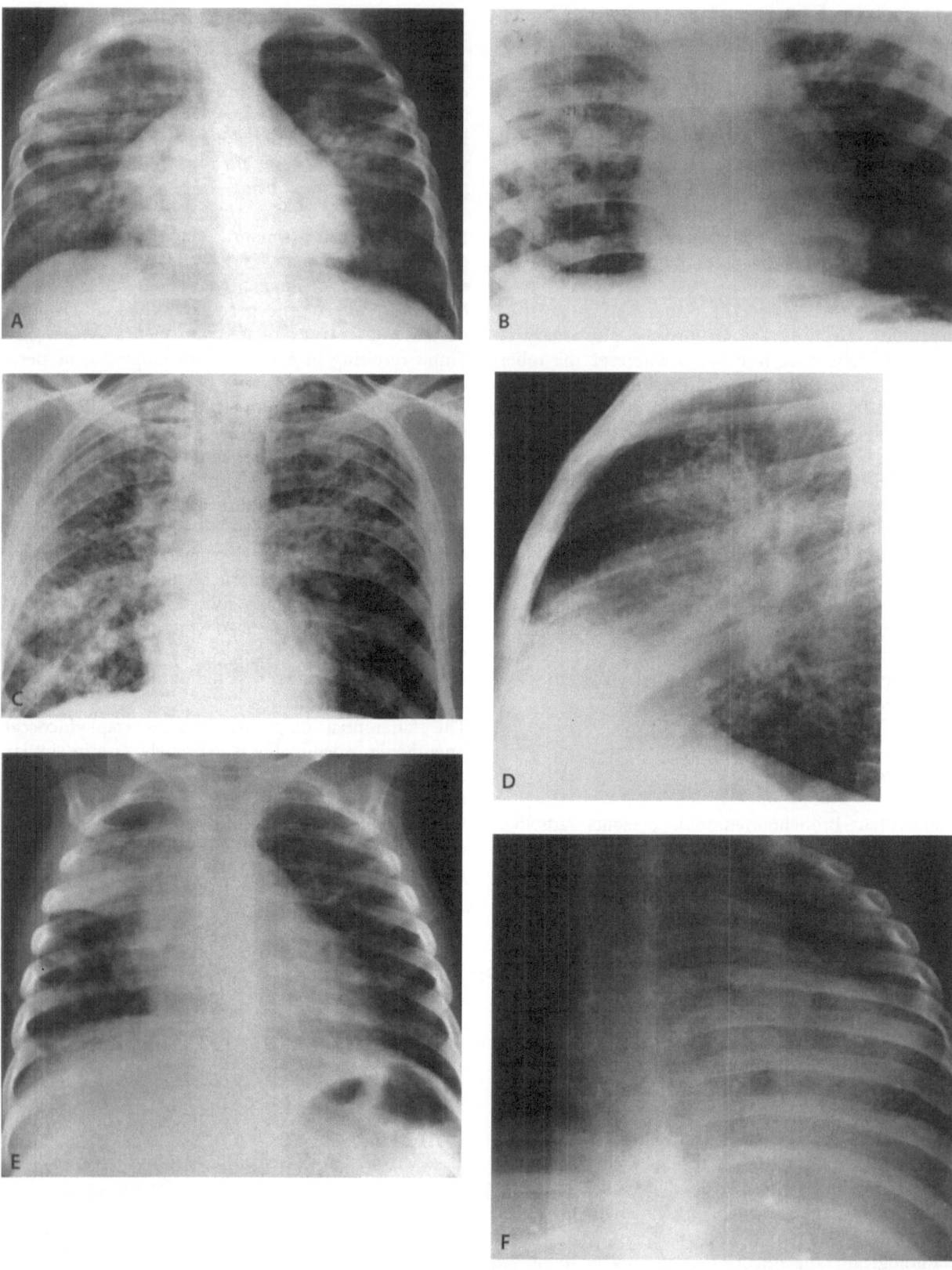

◀ **Fig. 8 A–F.** Complications of primary tuberculosis: the progressive primary infection. **A** An African infant from Kenya with bilateral bronchopneumonia and lymphadenopathy due to acute primary tuberculosis. **B** Bilateral "fluffy" bronchopneumonia in a young Indian adult. In the right lung some of the consolidation is cavitating. Clinically this was an acute illness resembling staphylococcal pneumonia, but the sputum contained many *M. tuberculosis*. **C, D** This adult has a 24-hour history of cough, sputum, headache, and feeling unwell. There is bilateral bronchopneumonia and hilar lymphadenopathy. The lateral view shows well-marked calcification in the paratracheal lymph nodes. There were many *M. tuberculosis* in the sputum. The response to antituberculous therapy was clinically dramatic, but much slower radiologically. Presumably this is a primary infection, unknown to the patient and healing satisfactorily as shown by the nodal calcification. One of the lymph nodes had ruptured into a bronchus and caused the acute bilateral inhalation bronchopneumonia and his acute illness. **E** This child was being treatment for primary tuberculosis and improving clinically. After 3 months treatment, she suffered an acute upper respiratory infection and became acutely ill, with a high temperature, severe cough, and obvious ill health. There is resolving right upper lobe pneumonia and enlarged right hilar and paratracheal lymph nodes complicated later (**F**) by bronchogenic spread throughout both lungs. The coarse and irregular pattern is unlike the fine and softer appearance of miliary tuberculosis. (C–E courtesy of *Semin Roentgenol*, 1979)

Hilar and Mediastinal Lymphadenopathy

The lungs may be normal but enlargement of the mediastinal lymph nodes may be the only presenting abnormality in tuberculosis, discovered on a radiograph of the chest taken because the patient is complaining of a cough, other pulmonary symptoms, tiredness, or weight loss (Fig. 9–12). Thoracic lymphadenopathy in tuberculosis occurs only in the primary (anergic nonimmune phase), and is not seen in the secondary (hyperergic immune) pattern (see p. 34).

The majority of patients with tuberculous hilar lymphadenopathy will have a positive tuberculin skin test; if it is negative it should be repeated after a short interval to exclude sarcoidosis. This is uncommon (although it does occur) in many parts of the tropics despite its frequency among Afro-Americans in the United States.

At any stage of the infection the nodes may enlarge sufficiently to rupture into a bronchus, causing an acute inhalation tuberculous bronchopneumonia. This may lead to severe clinical symptoms or, alternatively, may produce little clinical response. It should be suspected radiologically when fresh segments of patchy, consolidation appear unexpectedly during any primary tuberculous infection (see Fig. 8).

Enlarged mediastinal nodes can also rupture into the pericardium, causing acute tamponade or an acute tuberculous pericarditis. In some patients this causes sudden deterioration, but in others the clinical course is surprisingly benign (see p. 24).

Hilar and mediastinal lymphadenopathy may resolve completely, but this may take many months. There may be residual calcification, although the incidence of calcified nodes is variable in different parts of the world, possibly related to the state of nutrition. Statistically, in the tropics hilar calcification is more common in children than in adults, for which there are two possible explanations. Either calcification has occurred and subsequently been absorbed, or the background immunity and the state of nutrition of children and young adults is improving compared with their elder generation. AIDS may reverse this statistic, as it has many others.

The differential diagnosis of enlarged mediastinal and hilar lymph nodes is largely dependent on the local pattern of disease; in the tropics, tuberculosis usually will be the most likely etiology, especially when the more peripheral peribronchial and perihilar nodes are enlarged. Sarcoidosis occurs in the tropics, but not commonly and with considerable variation geographically; in sarcoidosis there is usually enlargement of the mediastinal and paratracheal nodes in particular, as well as bilateral symmetrical hilar node enlargement. Lymphoma, especially in children, may present with enlarged nodes in the chest and elsewhere, and is indistinguishable from tuberculosis. Similarly, Kaposi sarcoma in HIV-ve children in the tropics may present as an acute lymphadenopathy with very few of the characteristic pigmented skin nodules. (There is often lacrimal involvement.) But tuberculosis is also a common cause of peripheral lymphadenopathy: for example, the majority of enlarged cervical nodes in Uganda in patients under the age of 50 years are tuberculous (see p. 134). In adults, malignant disease is often the initial diagnosis when mediastinal nodes are enlarged, but it must be remembered that even old people may have "primary" tuberculosis. When the patient is HIV-positive, intrathoracic lymphadenopathy is as likely to be tuberculous as malignant; for example, in West Africa over 80% of patients with combined TB and AIDS have mediastinal lymphadenopathy. At any age, biopsy and histopathology may be necessary to establish the correct diagnosis.

(Text continues on p. 19)

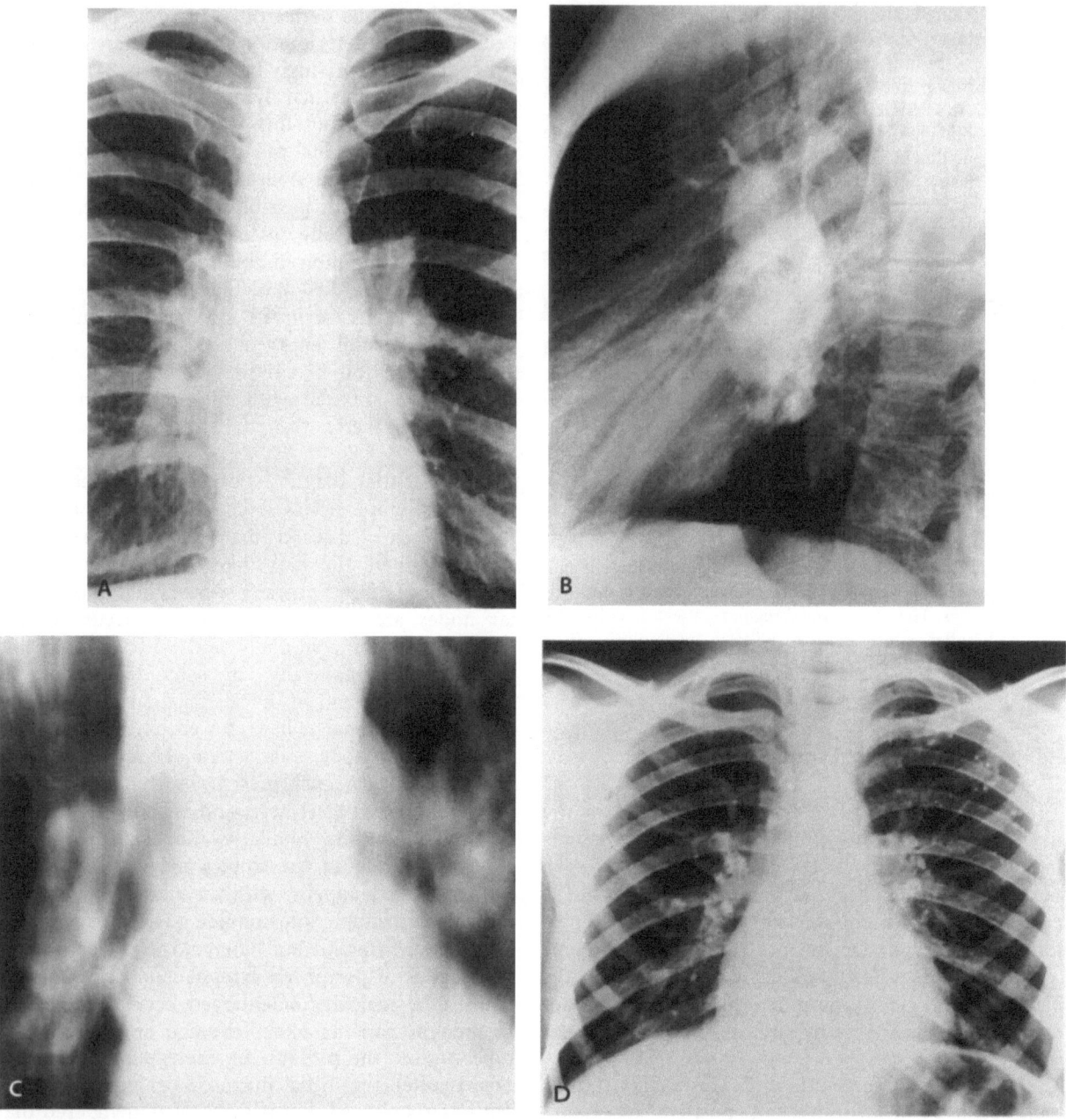

Fig. 9 A–D. Tuberculous lymphadenopathy. This is only found in the primary (anergic, nonimmune) phase. The pattern and density of the lymph nodes is very variable. **A, B.** PA and lateral radiographs of the a 32-year-old woman who had been "unwell" for some weeks, without specific complaints or cough. Her PPD was positive and she had a mild evening pyrexia. There is marked bilateral peribronchial and perihilar lymphadenopathy. **C** Standard tomography of a different patient shows large lymph nodes around the main branch of the major bronchi on both sides. **D** Multiple calcified granulomas in both hilar and right paratracheal lymph nodes. There is a calcified complex in the left upper lobe. (D courtesy of *Semin Roentgenol*, 1979)

Fig. 10 A–F. Tuberculous lymph nodes may enlarge and cause pressure on different parts of the bronchial tree. **A** Marked enlargement of the peritracheal nodes: no lung focus is visible. Only biopsy or therapeutic trial could differentiate between tuberculosis and lymphoma. **B** Standard tomography shows compression of the right main and the right upper lobe bronchi. There is also compression of the intermediate and right middle lobe bronchi, but a little less marked than that of the upper lobe. There is a large thick-walled cavity beyond the almost blocked upper lobe bronchus: the tuberculin test was positive, but the sputum did not contain *M. tuberculosis*. The adenopathy was tuberculous but the cavity was pyogenic, the result of the blocked upper lobe bronchus. **C, D** Partial atelectasis accompanying infection of the right middle lobe, with markedly enlarged hilar and mediastinal lymph nodes. **E, F** Enlarged mediastinal lymph nodes (best seen in **F**) which have caused obstructive emphysema, with over-expansion of the left lung and displacement of the mediastinum to the right. This was a young patient from the Pacific Isles. (E, F courtesy of Dr. Cheryl Sisler, Hawaii)

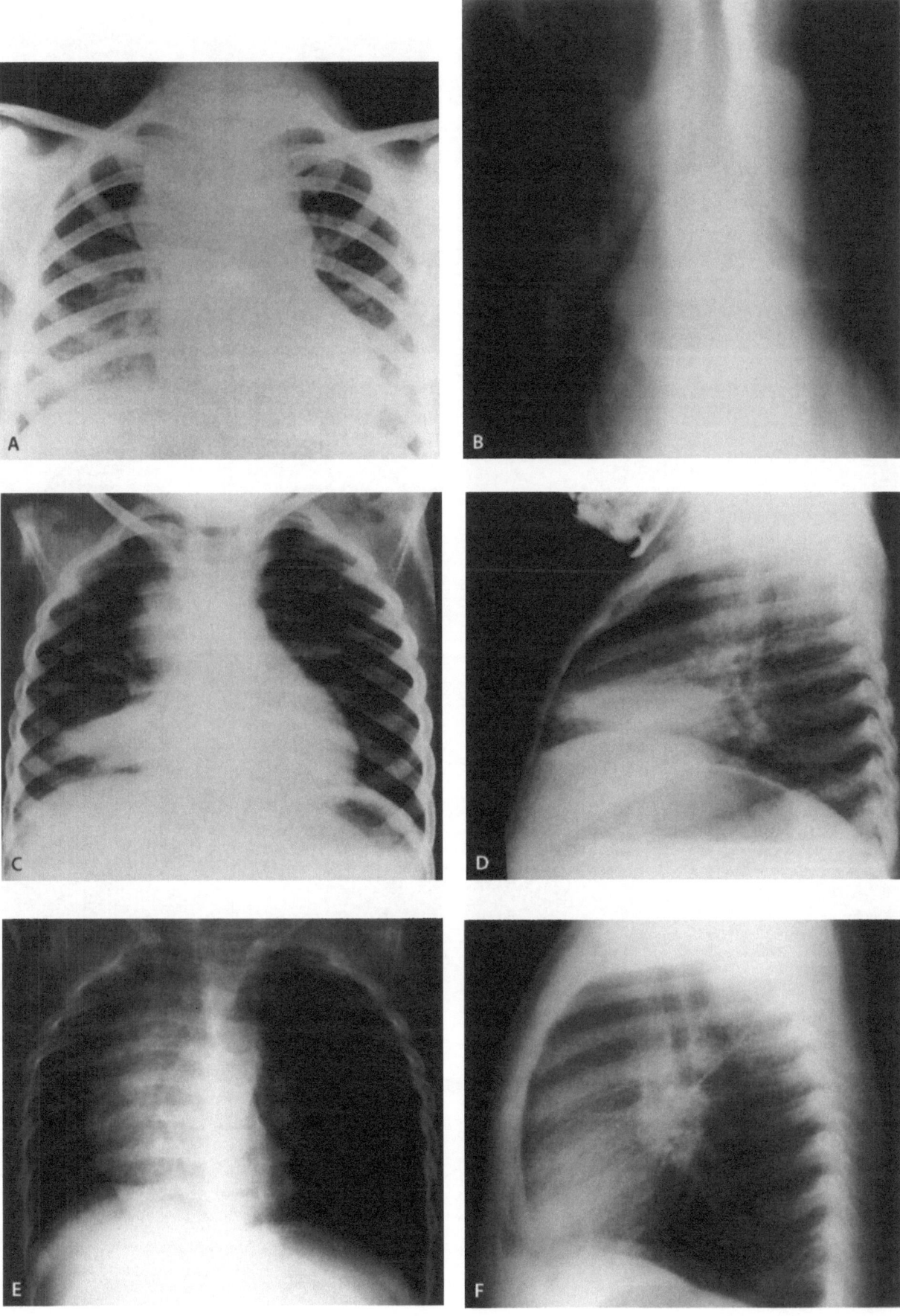

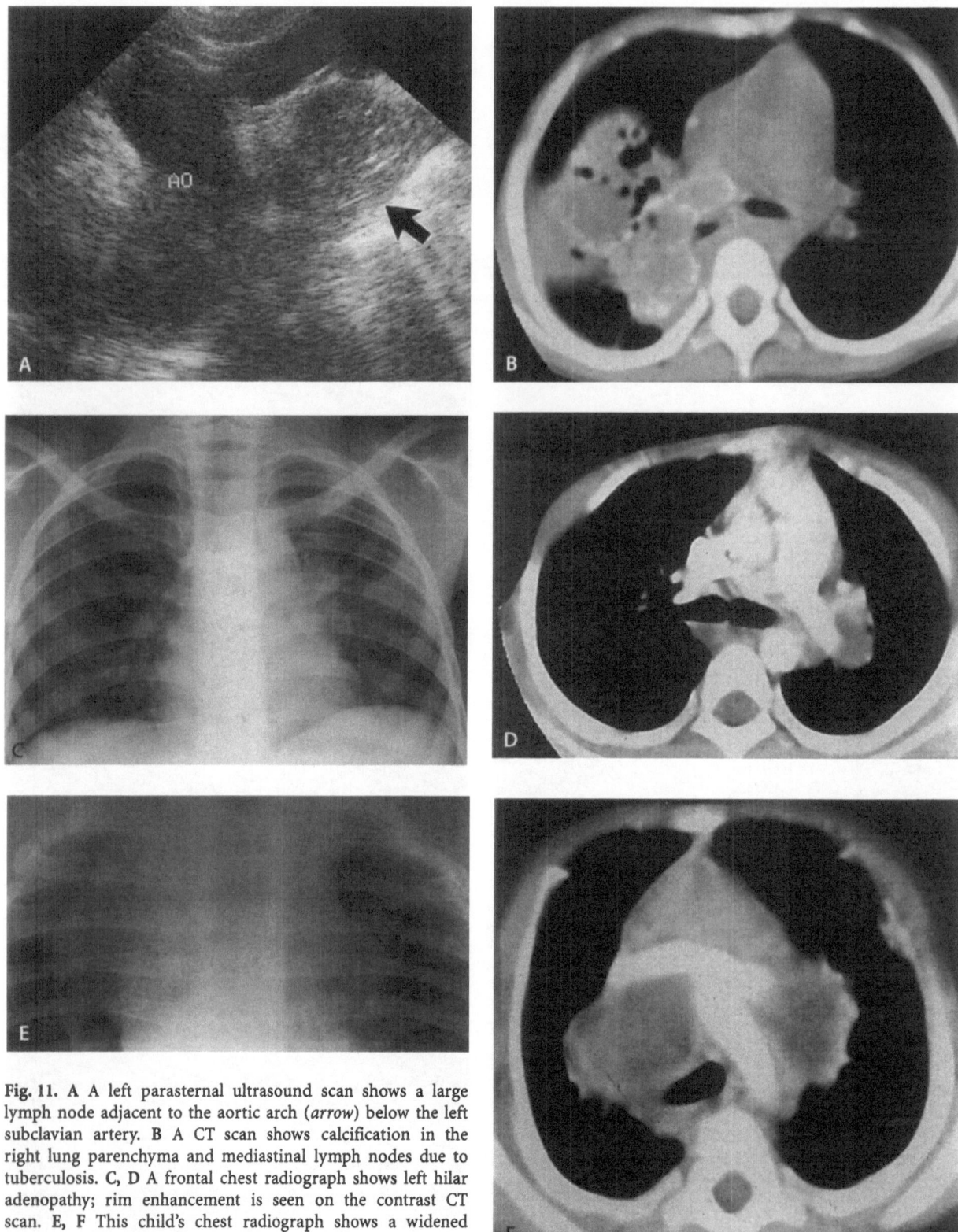

Fig. 11. A A left parasternal ultrasound scan shows a large lymph node adjacent to the aortic arch (*arrow*) below the left subclavian artery. **B** A CT scan shows calcification in the right lung parenchyma and mediastinal lymph nodes due to tuberculosis. **C, D** A frontal chest radiograph shows left hilar adenopathy; rim enhancement is seen on the contrast CT scan. **E, F** This child's chest radiograph shows a widened superior mediastinum which might be mistaken for the thymus; the contrast-enhanced CT scan shows large tuberculous lymph nodes, with rim enhancement. (From Cremin and Jamieson 1995)

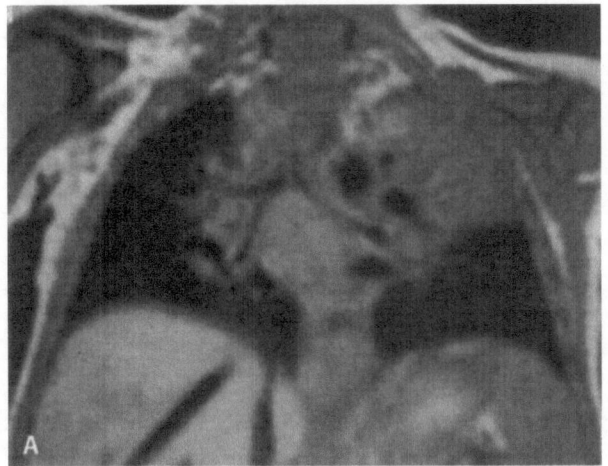

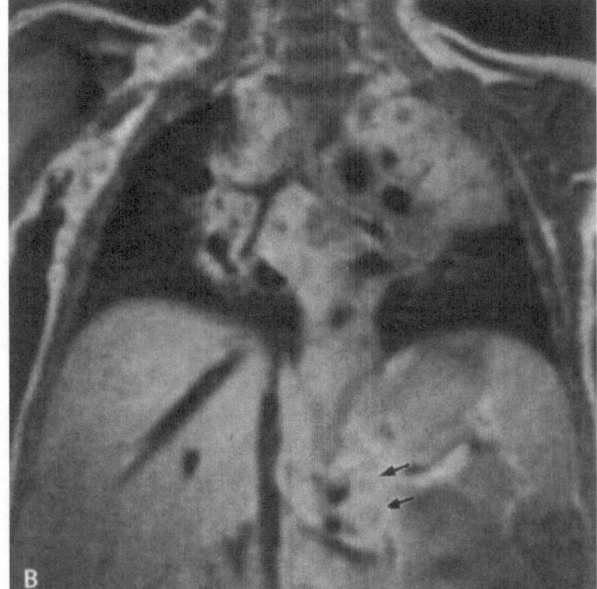

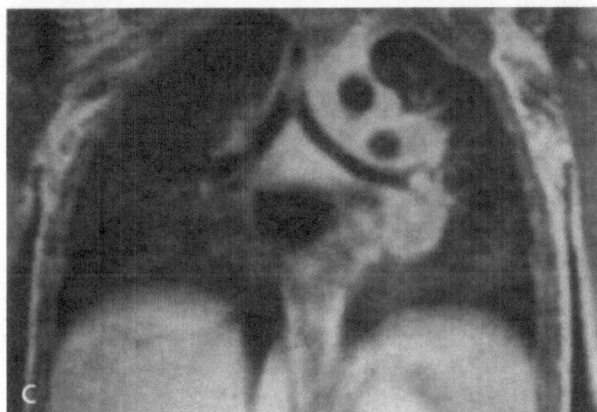

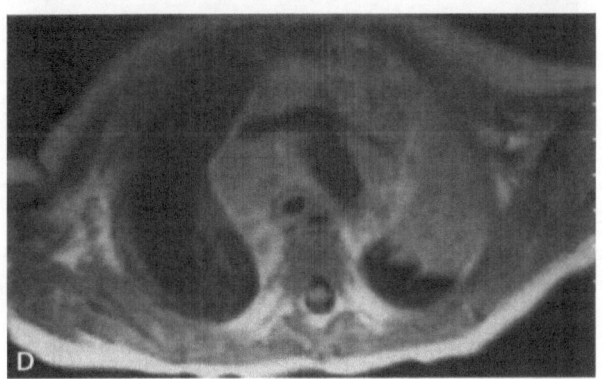

Fig. 12 A–D. MRI can also demonstrate both the pulmonary and nodal changes of tuberculosis. **A** The coronal T 1-weighted and **B** coronal i.v. gadolinium-enhanced T 1-weighted scans showing consolidation in the left upper lobe, which is enhanced, and nodal masses in the right superior mediastinum, right hilum, and below the carina, with rim-enhancement. There are also contrast-enhanced para-aortic nodes (*arrows*). **C** A coronal i.v. gadolinium-enhanced T 1-weighted scan showing rim-enhancing nodal masses in the left hilum. **D** An axial T 1-weighted scan showing superior mediastinal adenopathy and consolidation in the left upper lobe. (From Cremin and Jamieson 1995)

The Destroyed Lung

The "destroyed lung" describes the very characteristic radiographic appearance of a common end result of primary tuberculosis, or less often, of a nontuberculous mycobacterial infection, e.g., *M. avium-intracellulare* (Fig. 13).

The pathophysiology of the destroyed lung is easy to understand (Fig. 14). There are lymph nodes at the division of each major and segmental bronchus which are enlarged by a primary tuberculous infection. With such a close anatomical relationship, any swelling results in pressure on the bronchus, causing narrowing often to the extend of blockage. This may resolve as the nodes subside or the bronchial wall may be so involved and damaged that the carti-lage is destroyed. If there is pressure or fibrosis sufficient to cause blockage of the bronchus, there may be collapse of the segment or even the lobe of the lung peripherally.

Enlarged tuberculous lymph nodes are the usual cause of the destroyed lobe or lung, which is a very common finding in Africa or India, but is also seen wherever tuberculosis is common (e.g., in the Australian aborigines). Statistically the left upper lobe is more frequently affected; in one extensive survey of admission chest x-rays in a large African hospital, left upper lobar pneumonia in young adults was more likely to be tuberculous. The right upper lobe is a little less often involved, and none of the other lobes escape. (The "right middle lobe syndrome" described in England results from similar lymph

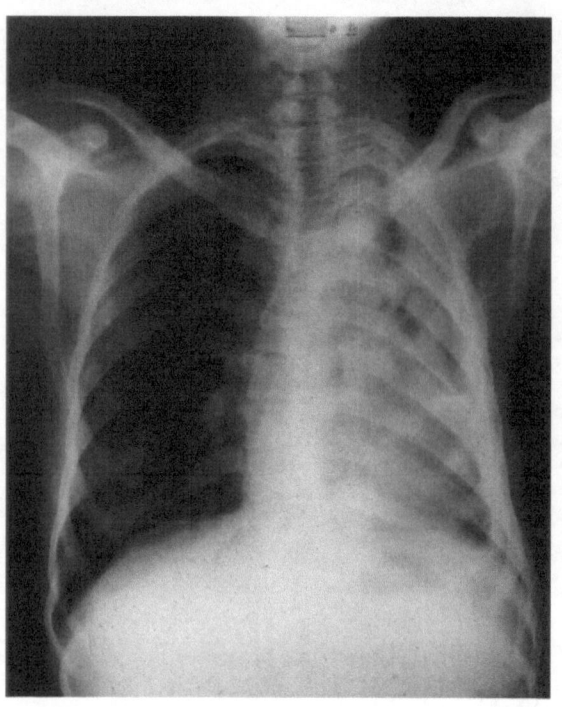

◀ **Fig. 13.** The characteristic radiograph of a "destroyed lung" in a patient from Malawi. There is flattening of the left chest wall, with displacement of the mediastinum to the left. There are cavities in the fibrotic left lung. Some pleural thickening is present on the left. The right lung is over-expanded, but otherwise normal. (The nodular shadow on the right lower lung field is the nipple.)

Fig. 14 A–F. The development of a "destroyed lung." **A** Primary left upper lobe tuberculous pneumonia. **B** Standard tomography shows the enlarged mediastinal and perihilar lymph nodes, compressing not only the apical posterior bronchus but also the other bronchi to the left lung. **C** Because of the compression and resulting obstruction, the left lung is shrinking and the mediastinum is moving to the left. The left chest wall is being flattened and there is some pleural reaction. **D** A bronchogram about a year later shows widespread bronchiectasis with overlying pleural thickening. There are adhesions along the diaphragm and widening of the lower half of the trachea. The right lung is over-expanded but is otherwise normal. **E** A similar destroyed lung, but on the right side. The mediastinum is shifted to the right and the right chest ist contracted. **F** The bronchogram shows the underlying shrunken right lung, with bronchiectasis, and the overlying pleural thickening

node fibrosis.) The outcome of the fibrosis and bronchial constriction is often secondary pyogenic infection of the atelectatic segments. Bronchiectasis then develops, often being both tuberculous and pyogenic in origin (Fig. 15). Because of the underlying lung collapse and fibrosis, the chest wall flattens, particularly over upper lobe disease, and the mediastinum is pulled by the fibrotic lung towards the same side. The lung damage is slowly progressive, with recurrent infection and further fibrosis. There may be cavity formation, and as the disease heals it is very difficult to decide clinically or radiologically whether a fresh cavity is the result of recurrent tuberculous infection or secondary infection or whether there are fibrotic bullae only. Although no longer commonly performed where CT scanning is available, bronchography in these patients is dramatic and demonstrates the extent and pattern of lung damage. When the underlying sequence of lymphadenopathy, atelectasis, infection, bronchiectasis, fibrosis, contraction, and scarring is understood, it is so characteristic that it is possible to look at the plain chest radiograph and describe the bronchogram (or CT scan; Fig. 16) without actually performing it (which is much appreciated by the patient).

During the process of healing, local pleural effusions frequently occur over the destroyed area of lung, especially over the apex. Such fluid is usually trapped by adhesion over the upper lobe, so that the costophrenic angles frequently remain clear. Fluid does not have any prognostic significance but

may resemble a mass: CT or ultrasonography may help to distinguish between pleural fluid (mobile) and pleural thickening.

Subsequently, the major clinical problem in patients who have a "destroyed lung" is recurrent infection. This may be tuberculous or pyogenic, or even both, and can be contaminated by a fungus or other saprophyte. Examination of the sputum is very misleading; it almost always contains a multiplicity of organisms and it is difficult to identify tuberculous bacilli. Management of these patients presents a problem; some control of the pyogenic infection may be obtained by antibiotics, but such treatment is not always successful, the outcome depending on the extent of the fibrosis and on the drainage from the affected area. Surgery is sometimes attempted but can be extremely difficult because the pleura is thickened and adherent, and the underlying lung fibrotic and friable. The end result may be a disastrous tuberculous empyema. Localized thoracoplasty has helped to enhance the drainage in some patients, but masterly inactivity is another option.

In summary, a destroyed lung is easy to recognize and describe radiologically but there are no satisfactory criteria to distinguish the etiology of subsequent infections. Clinical recognition is equally straightforward, but treatment remains a recurring problem in some patients.

(Text continues on p. 24)

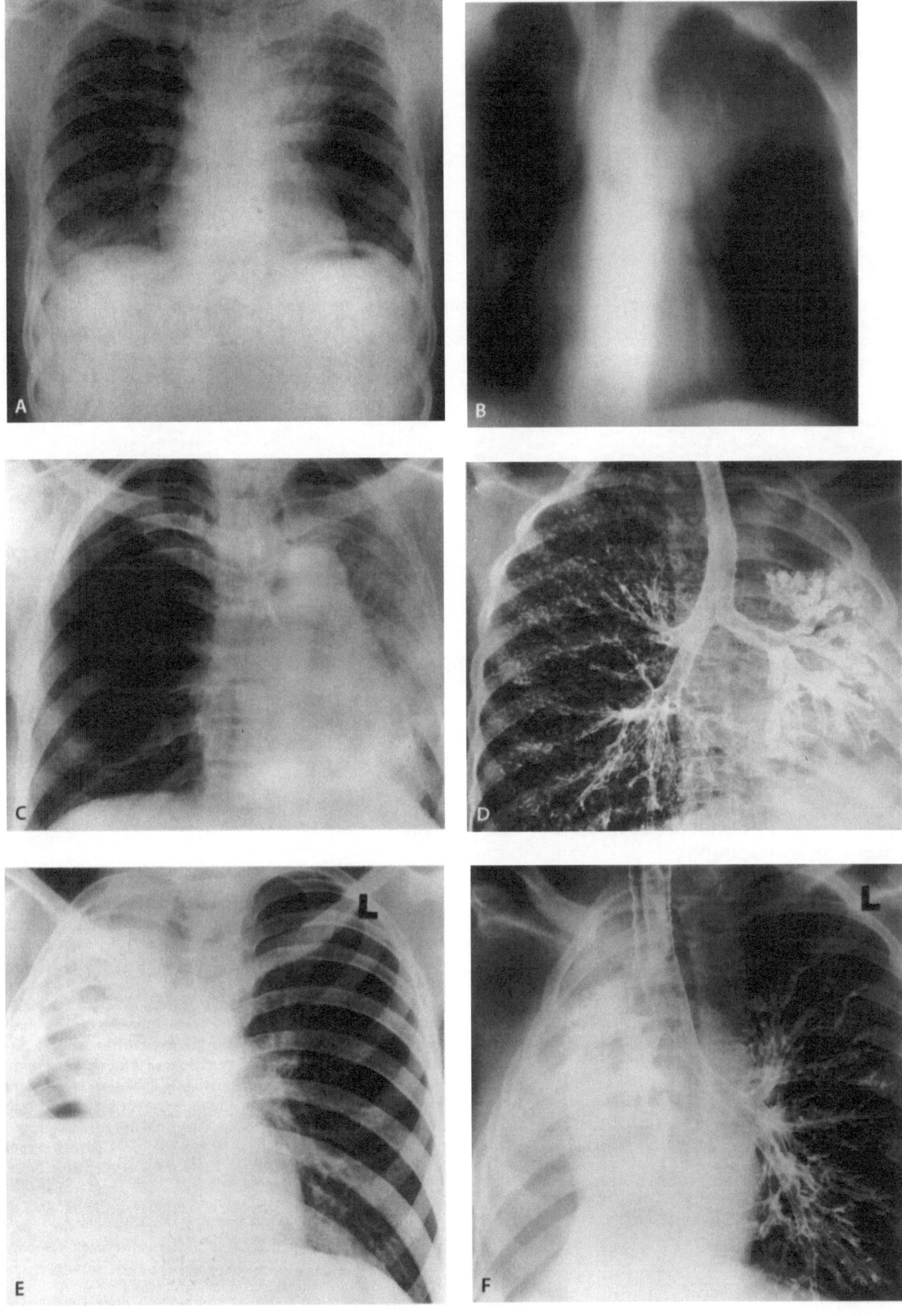

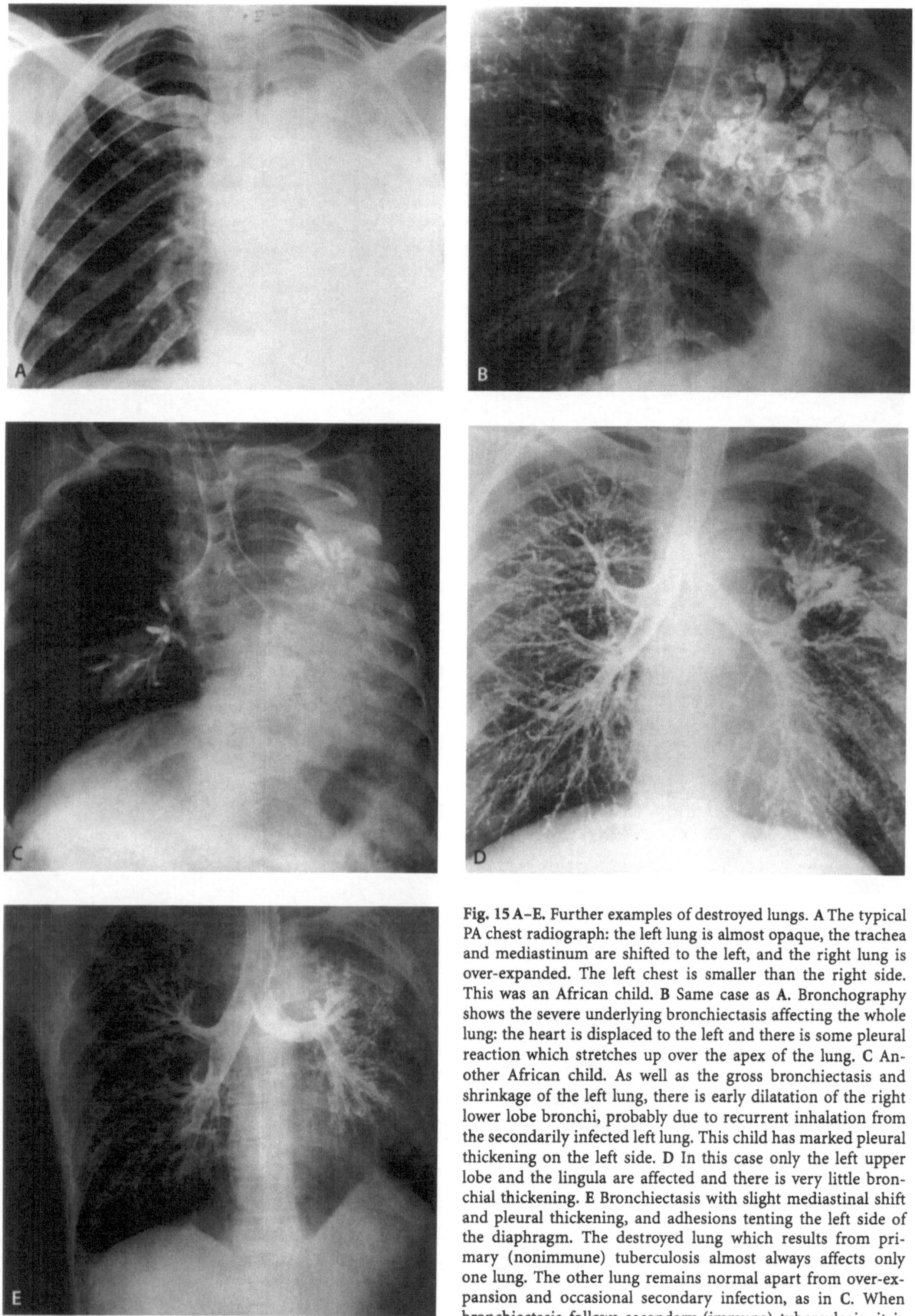

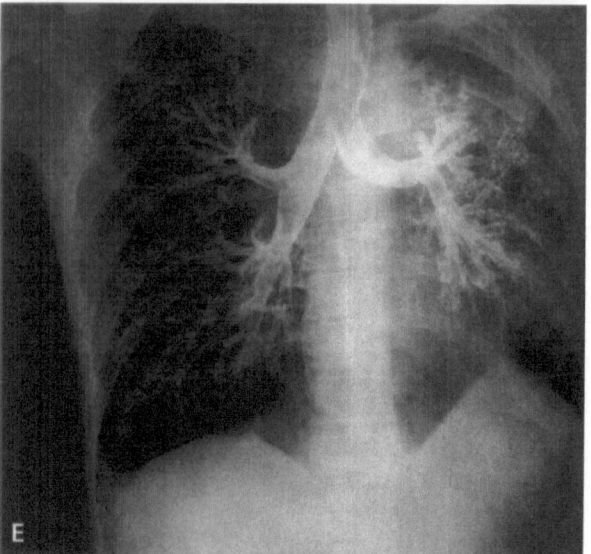

Fig. 15 A–E. Further examples of destroyed lungs. **A** The typical PA chest radiograph: the left lung is almost opaque, the trachea and mediastinum are shifted to the left, and the right lung is over-expanded. The left chest is smaller than the right side. This was an African child. **B** Same case as **A.** Bronchography shows the severe underlying bronchiectasis affecting the whole lung: the heart is displaced to the left and there is some pleural reaction which stretches up over the apex of the lung. **C** Another African child. As well as the gross bronchiectasis and shrinkage of the left lung, there is early dilatation of the right lower lobe bronchi, probably due to recurrent inhalation from the secondarily infected left lung. This child has marked pleural thickening on the left side. **D** In this case only the left upper lobe and the lingula are affected and there is very little bronchial thickening. **E** Bronchiectasis with slight mediastinal shift and pleural thickening, and adhesions tenting the left side of the diaphragm. The destroyed lung which results from primary (nonimmune) tuberculosis almost always affects only one lung. The other lung remains normal apart from over-expansion and occasional secondary infection, as in **C.** When bronchiectasis follows secondary (immune) tuberculosis, it is nearly always bilateral

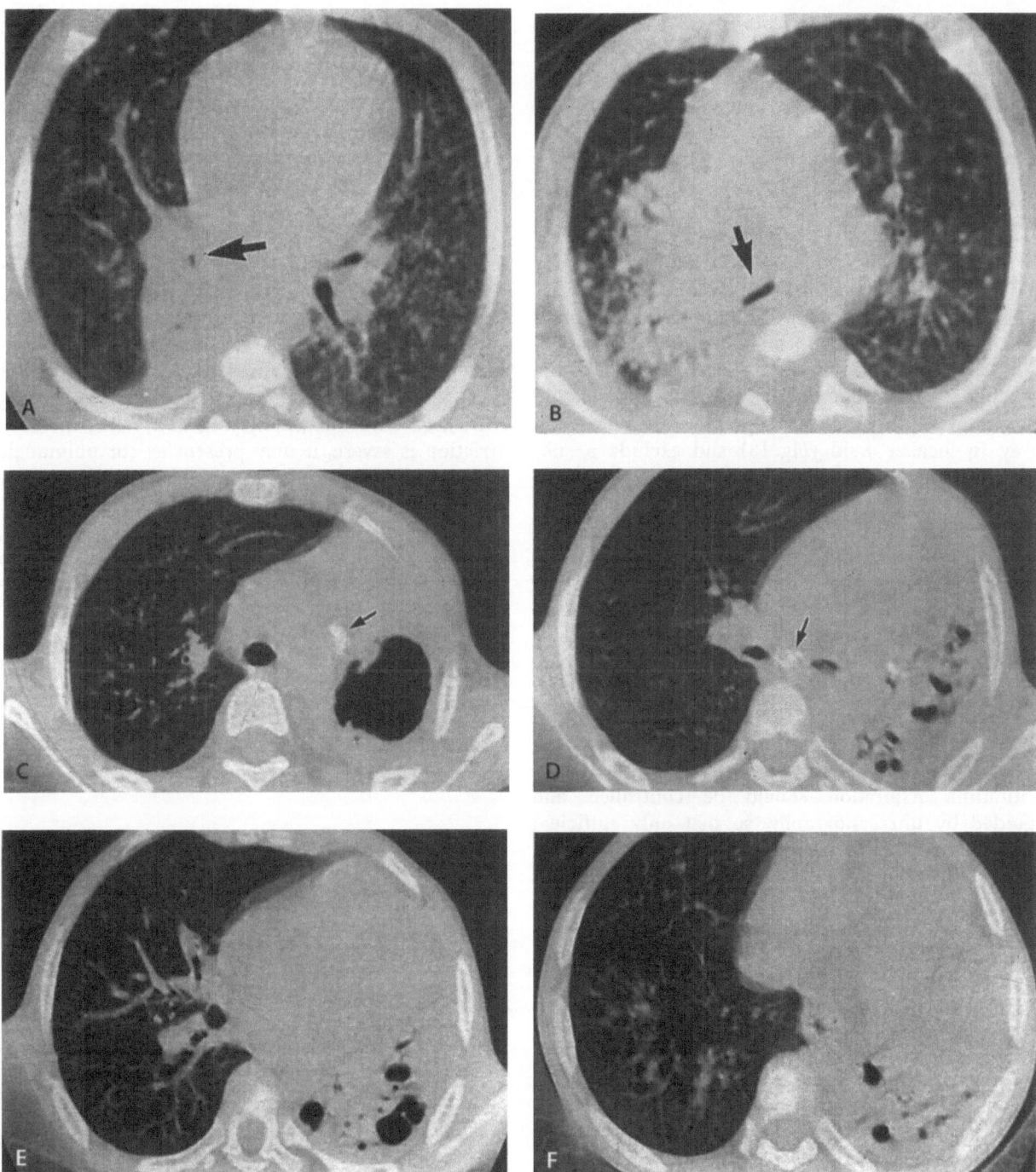

Fig. 16 A–F. CT of the destroyed lung. **A** Marked compression of the right lower lobe bronchus (*arrow*) by enlarged nodes. There is segmental consolidation and compensatory emphysema in the remainder of the right lung. **B** Severe compression of the trachea (*arrow*) due to mediastinal lymphadenopathy: there is right upper lobe consolidation. **C–F** Scans at different levels in the same patient showing destruction of the left lung, with volume loss, cavitation, and consolidation. There is early calcification in some of the left hilar and subcarinal nodes (**C, D** *arrows*). There is no gross pathology in the right lung. The patient eventually had a left pneumonectomy. (From Cremin and Jamieson 1995)

Pleural and Pericardial Effusions

Tuberculous pleural effusions at any age are usually a manifestation of primary, anergic (nonimmune) tuberculosis. The fluid may be serous, protein-aceous, bloody, or occasionally purulent (Fig. 17). Pleural biopsy and DNA fingerprinting are reliable methods of establishing the diagnosis. Mediastinal lymphadenopathy often can be demonstrated by CT or tomography, even when there is no radiologically visible pulmonary parenchymal lesion. The majority of tuberculous effusions will be unilateral, but bilateral effusions are not uncommon in tuberculosis in the tropics. (Whenever there is a bilateral effusion, a tuberculous infection of the thoracic spine should also be excluded.) Ultrasonography is an excellent way to localize fluid (Fig. 18) and exclude a subphrenic cause, such as an hepatic amebic abscess or ascites.

Tuberculous effusions may loculate anywhere, including over the upper zones of the lung or in the pleural fissures. The radiological appearance may then suggest local consolidation or a tumor, and decubitus radiographs and lordotic views may be necessary for differentiation. Ultrasonography will demonstrate the septate adhesions and pleural thickening.

Removal of a large amount of pleural fluid in a malnourished patient may cause sudden clinical deterioration. Aspiration should be controlled and guided by ultrasonography, so that only sufficient fluid is removed to permit a functioning lung volume; if the fluid is purulent, total and continuous drainage will be required. If this is not adequate, pleural thickening and lung encasement will follow. Most tuberculous pleural effusions respond well to appropriate drug therapy.

Pleural calcification may be the sequel to a tuberculous pleuritis (Fig. 19). The calcification may have a feathery or lacy pattern, or be linear around an encysted effusion. Calcification is not evidence of cure, and decalcification may occur, particularly if the infection becomes active again. Calcification in the diaphragm is almost always due to asbestosis, and not to tuberculosis. It must be distinguished from calcification of the diaphragmatic pleura; but this also is uncommon following tuberculosis and is much more likely to be due to asbestosis. Of course, asbestosis and tuberculosis may coexist, and in some parts of the world frequently do.

Tuberculous pericarditis (Fig. 20) may be a primary anergic tuberculous infection or may follow rupture of adjacent mediastinal lymph nodes into the pericardium. The posterior aspect of the pericardium is in close apposition to the mediastinal lymph nodes beneath the carina. Rupture of these nodes may cause an acute clinical crisis, with tamponade, or may be a subclinical event. Subsequent healing can be complete, leading to an apparently normal pericardium, or may result in severe thickening and calcification. Because both layers of the pericardium are involved, adhesions are frequent and may be extensive; the epicardium may be involved. When there is restrictive (constrictive) pericarditis there may be a need for surgical decompression; this procedure can be hazardous because the cardiac muscle and small epicardial coronary arteries can be damaged. Careful ultrasonography may delineate the extent of the fibrosis before surgery.

Constrictive pericarditis may exist without any radiologically visible calcification, and may be better recognized with CT, magnetic resonance imaging (MRI), or ultrasonography (Fig. 21). When the constriction is severe, it may present as cor pulmonale with a radiographically normal heart size. In many patients extensive pericardial calcification has little physiological effect on the heart, while in others, minimal calcification seen on a plain radiograph may be the only sign of extensive pericardial fibrosis.

A solitary pericardial tuberculoma has been reported, forming a small mass along the left heart border. It is not easy by any method to distinguish a pericardial tumor from pericardial thickening.

Fig. 17 A–F. Tuberculous primary (nonimmune) pleural effusion. **A** A young African man with a right-sided pleural effusion and right hilar adenopathy. With antituberculous treatment, the fluid resolved more rapidly than the lymph nodes, which eventually calcified. **B** A teenage girl with a left-sided pleural effusion which extends up into the interlobar fissure. There is probably left hilar adenopathy, partially obscured by the fluid. **C** Localized pleural effusion on the right side which has been present for years. No pleural effusion is "safe". This type of effusion may become reactivated quite unexpectedly. **D** An encapsulated effusion along the right chest wall. The central density over the right chest is due to pleural thickening and fluid lying posteriorly. There was no change over many years and the patient remained quite healthy. **E** Bilateral pleural effusions in a young African girl. The lateral views clearly showed mediastinal lymphadenopathy; tuberculosis was confirmed by pleural biopsy. **F** A large pleural effusion almost filling the left side of the chest and partially collapsing the underlying lung. Although this resembles a destroyed lung, the mediastinum is shifted away from the density and the left side of the chest has not collapsed significantly. The main left bronchi can be seen clearly and are not obstructed. (**A, B, D** courtesy of *Semin Roentgenol*, 1979)

(Text continues on p. 30)

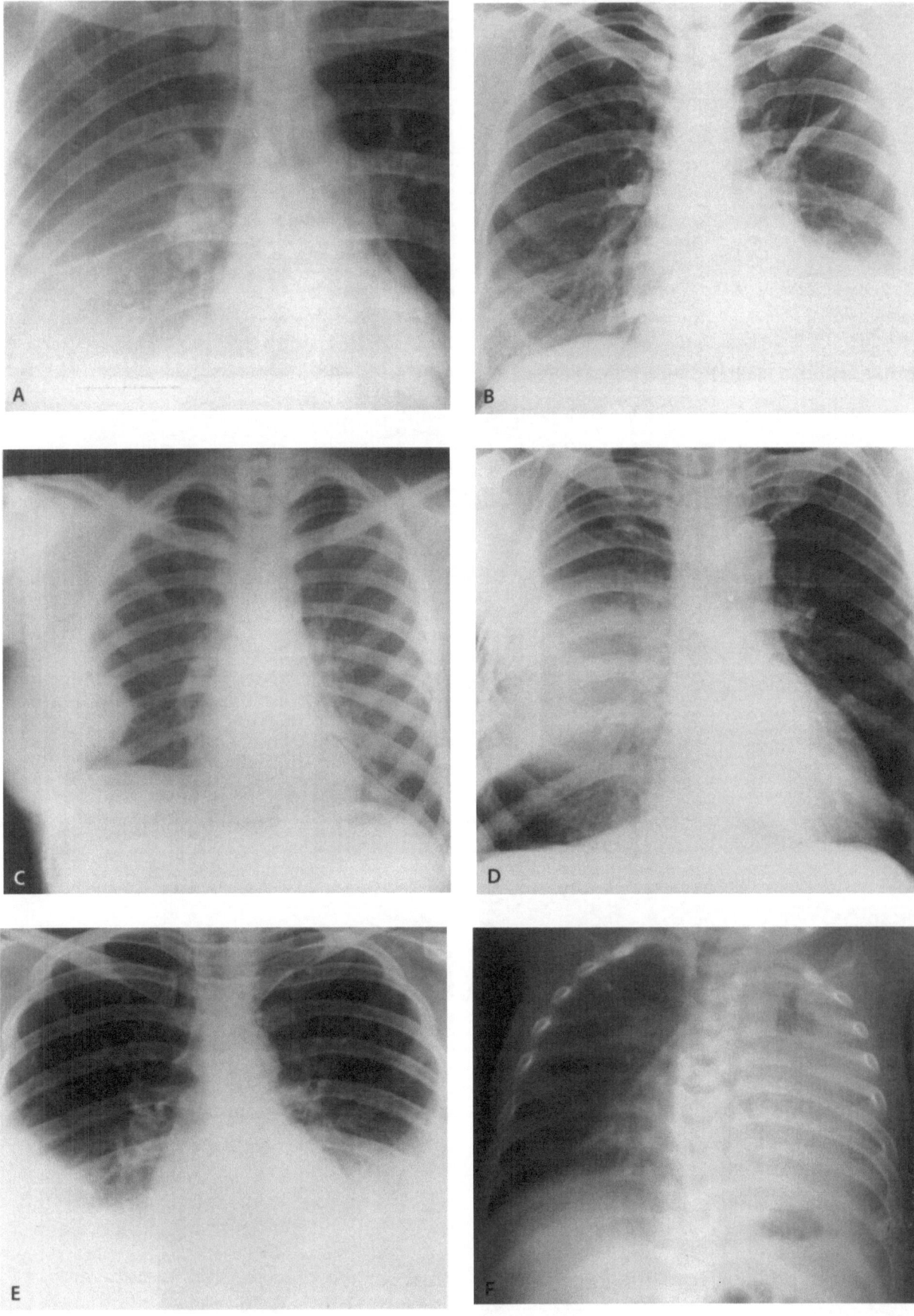

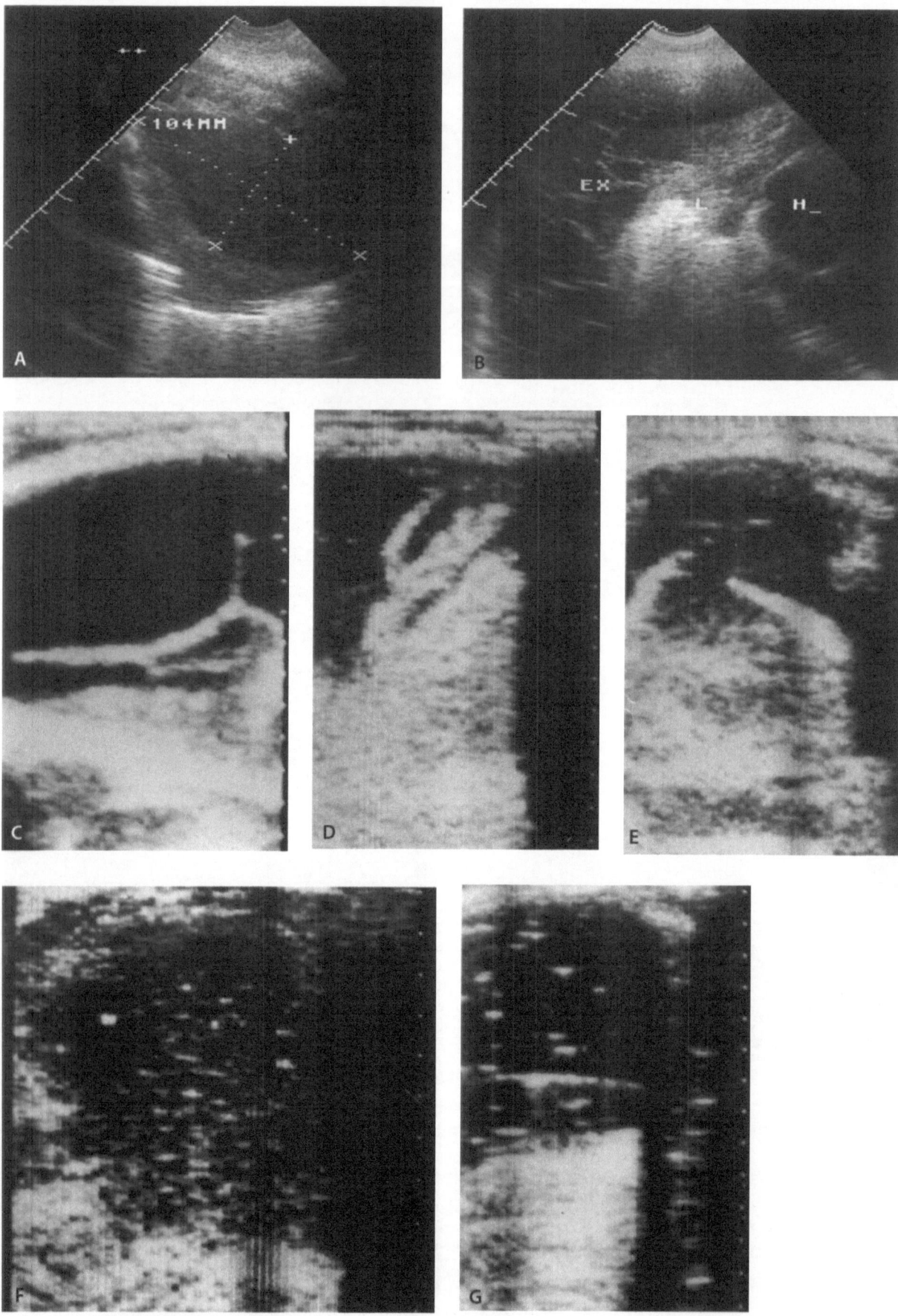

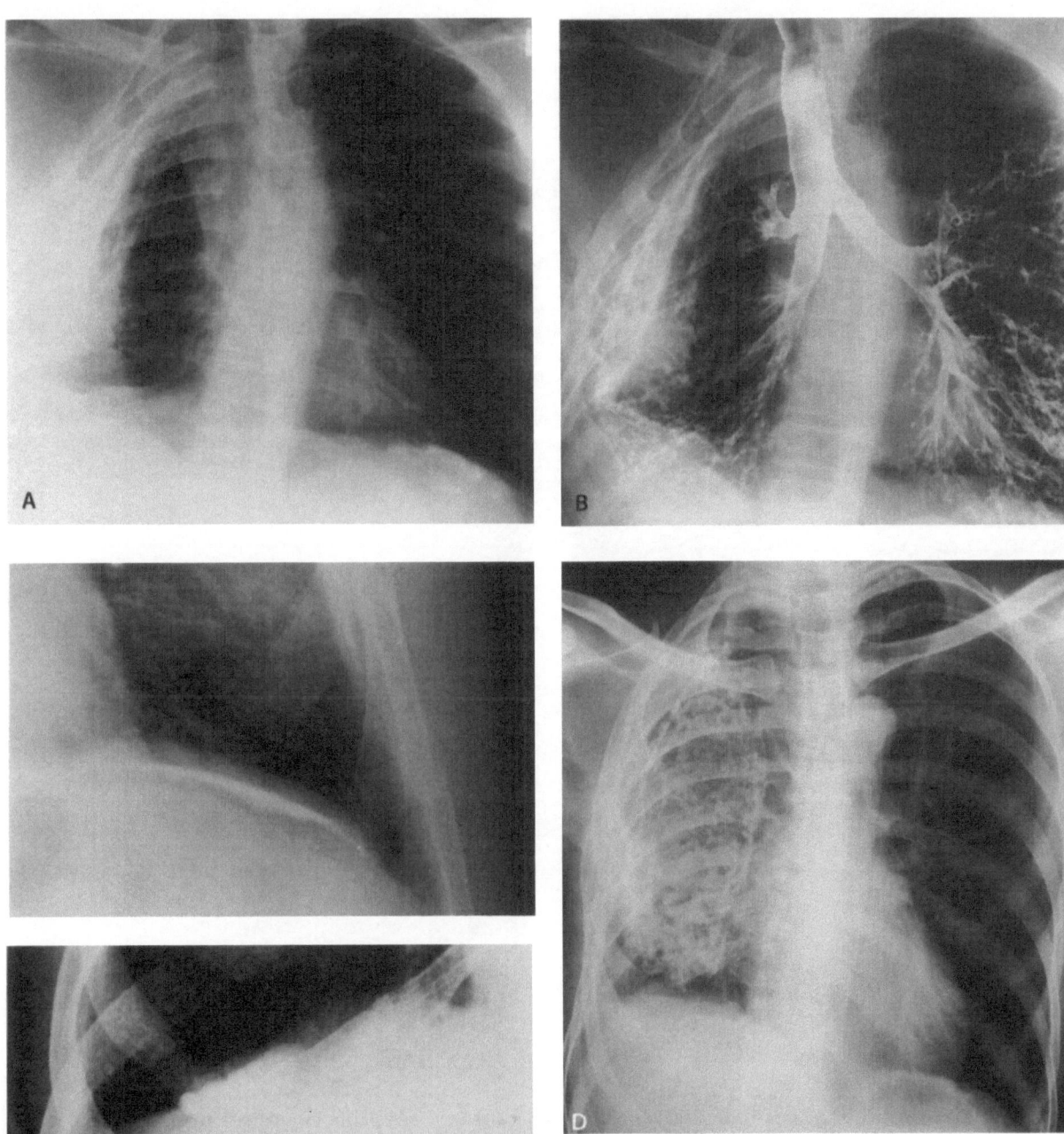

◀ **Fig. 18 A–G.** Ultrasonography in tuberculous pleural effusions.
A Fluid in the left chest; this could be tuberculous or pyogenic.
Aspiration and biopsy would be necessary to discover the etiology. **B** A multiseptate right-sided pleural exudate due to tuberculosis. **A** and **B** are patients from Zimbabwe. (Courtesy of Dr.
Sam Mindel) **C–E** Three different patients with septate tuberculous pleural effusions: the serous bands are adherent to both the
visceral and parietal pleura. **F, G** Many tuberculous pleural effusions will have multiple internal linear echoes, which move freely
when the patients changes position. This is a characteristic of
tuberculosis. These patients are from Uganda. [Courtesy of
Dr. L. Belli, Turin, and *Radiol Med* (*Torino*), 1992]

Fig. 19 A–D. Pleural calcification is often the end result of tuberculous pleurisy and has many patterns. **A, B** This African
patient had a pleural effusion many years previously which resulted in fibrotic shrinkage of the right chest and mediastinal
shift to the right. The bronchogram showed the underlying
contracted lung but there is no bronchial obstruction or saccu-
lar bronchiectasis as is found in a destroyed lung. The pleural
fibrosis has caused the right side of the chest to contract. The
left lung is emphysematous but otherwise normal. **C** Typical
diaphragmatic calcification due to asbestosis, seen in a patient
who has never had tuberculosis. This can be bilateral or unilateral, and is seldom symmetrical. Both costo-phrenic angles are
clear; had the calcification been the result of a tuberculous effusion there would have been pleural thickening. **D** This Indian
had a right pleural effusion which healed, leaving pleural thickening laterally and calcification posteriorly. Although the right
chest is a little contracted, there is no mediastinal shift

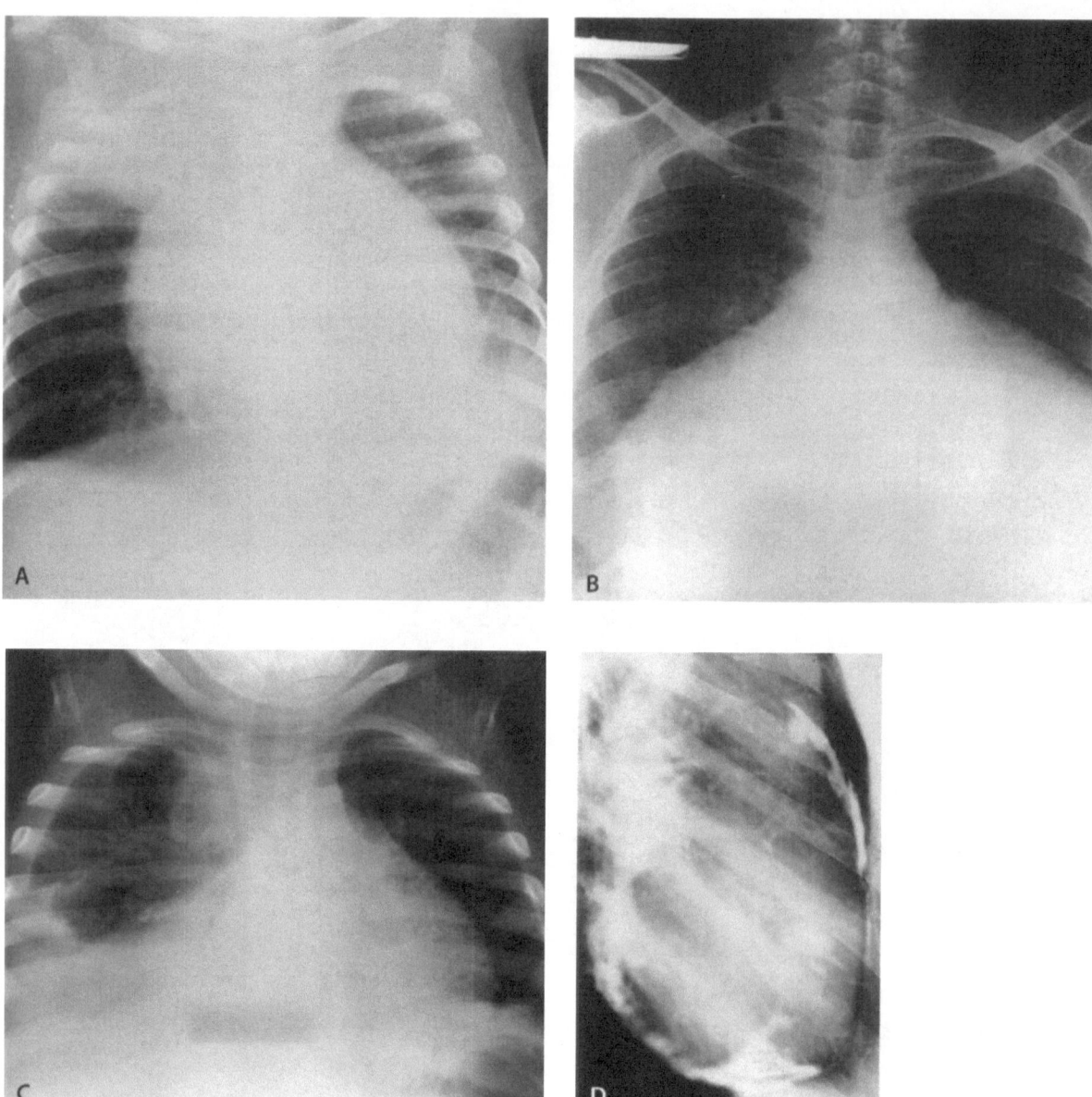

Fig. 20 A–D. Tuberculous pericarditis may be a primary tuberculous infection of follow rupture of an adjacent lymph node. **A** This African child had a right upper lobe primary tuberculous pneumonia and marked lymphadenopathy; she then developed a pericardial effusion and a left pleural effusion. **B** A huge cardiac outline which results from a large pericardial effusion. There is no evidence of cardiac failure or pleural fluid at this stage. **C** An African child from Zimbabwe with a large right pleural effusion and right lower lobe pneumonia. The right paramediastinal shadow is caused partly by pleural fluid and partly by enlarged lymph nodes. Other enlarged lymph nodes can be seen through the heart shadow around the left hilum. She had a pericardial effusion which needed to be removed, but which did not contain *M. tuberculosis*. However, the bacilli were present in her sputa. **D** Extensive calcification of the pericardium seen in a lateral view of the heart. The amount of pericardial calcification does not indicate the degree of pericardial constriction. This patient did not suffer from constrictive pericarditis. (**A, C** courtesy of *Semin Roentgenol*, 1979)

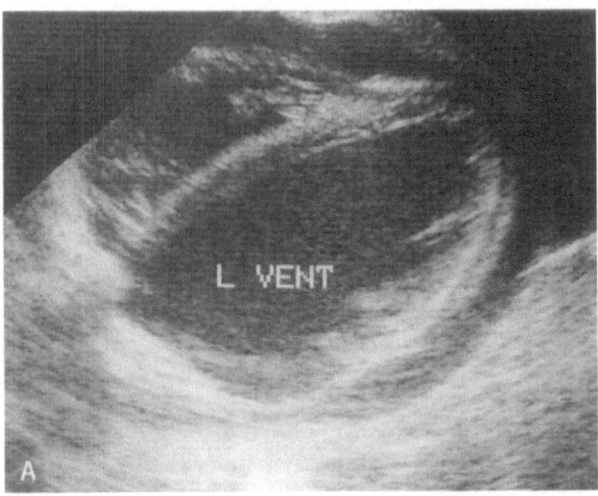

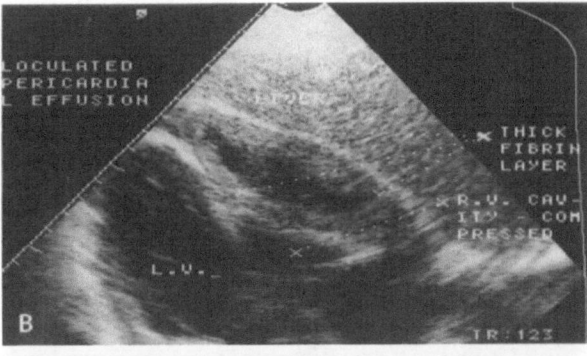

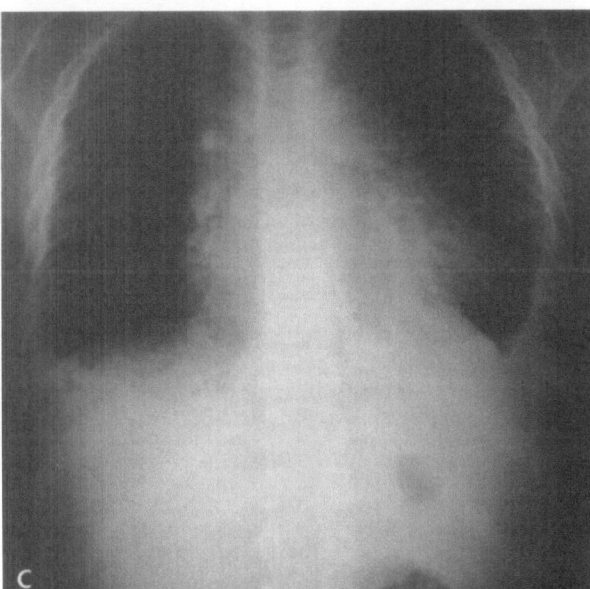

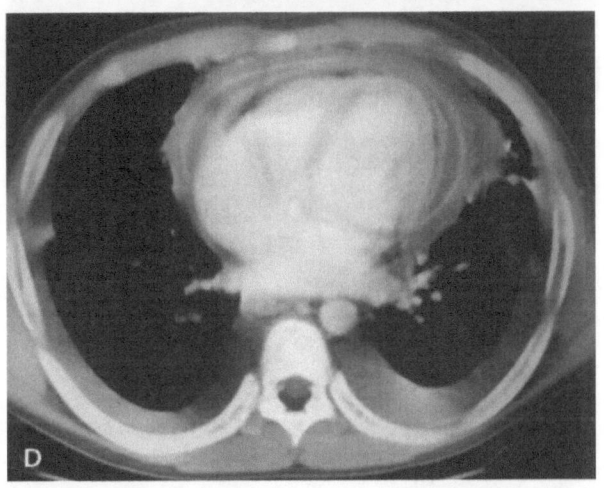

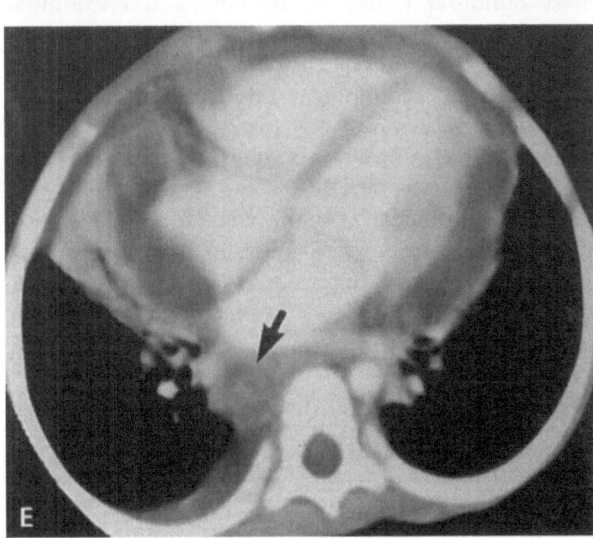

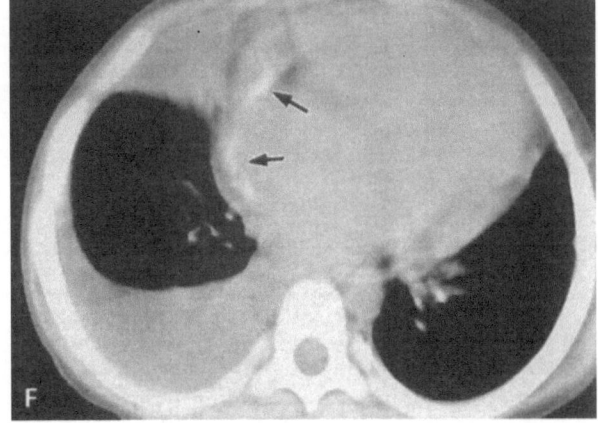

Fig. 21. A Ultrasonography is an excellent way to identify pericardial fluid, as in this child from southern Africa. **B** A loculated pericardial effusion with fibrous septa and marked thickening of the pericardium. There is some compression of the right ventricle (R.V.). This was an African from Zimbabwe with known tuberculosis. **C** A 13-year-old Polynesian boy with a pericardial effusion and bilateral pleural effusions. There is mediastinal lymphadenopathy. **D** The CT scan of the same patient, confirming the pericardial fluid, bilateral pleural effusions, and moderate lymphadenopathy. **E** A large tuberculous pericardial effusion in a child from South Africa, shown by contrast-enhanced CT: there is a large tuberculous lymph node (*arrow*) and consolidation in the right middle lobe. **F** Pericardial calcification seen on a CT scan (*arrows*). There is a right pleural effusion and right middle lobe consolidation. **G–I** see p. 30

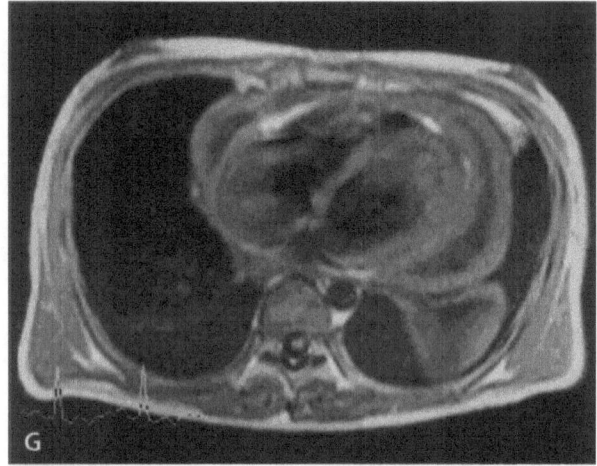
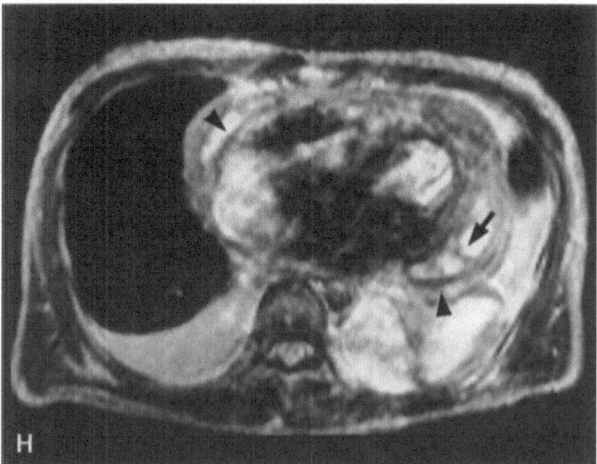
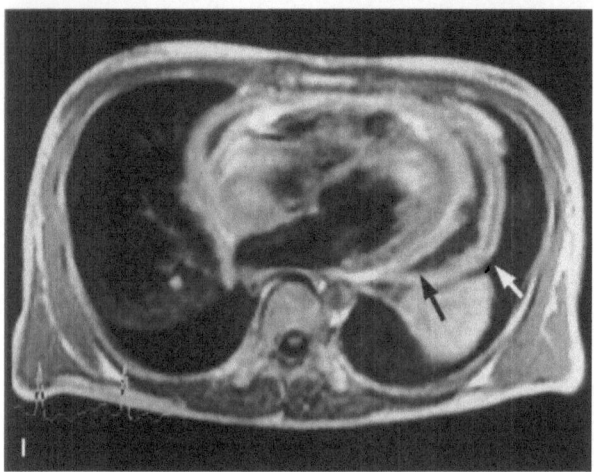

Fig. 21 (*continued*). G–I. MRI of tuberculous pericarditis shows marked percardial thickening. **G** T1-weighted image showing the pericardium with the same signal intensity as the myocardium. **H** T2-weighted image demonstrating low signal intensity on the inner surface of the thickened pericardium (*arrowhead*). The pericardial effusion also shows a low signal (*arrow*). – The lower scan is gadolinium enhanced, showing enhancement of thickened parietal and visceral pericardium (*arrow*). (**A, E, F** from Cremin and Jamieson 1995; **B** by courtesy of Dr. Sam Mindel; **C, D** courtesy of Dr. Cheryl Sister, Hawaii; **G–I** courtesy of H. Hayashi and *Br J Radiol*, 1998)

Miliary Tuberculosis

Miliary tuberculosis is an acute disseminated infection, the result of hematogenous spread; the primary focus may not be recognizable and may be extrapulmonary (Fig. 22). In other patients, miliary spread occurs during active pulmonary tuberculosis when there is rupture of a caseating lymph node or a cavity into a blood vessel. Even with good treatment there is a mortality of 13%–50%. Only about 30% of patients with miliary tuberculosis will have the bacillus in their sputum; up to 60% may have negative skin tests. Miliary tuberculosis is always evidence of a life-threatening infection. While it is usually easy to recognize the miliary pattern on a chest radiograph of a patient of any age, there is no correlation between the number or size of miliary nodules and the clinical health of the patient. Often those who are very ill show fewer nodules. The patient's general health and state of nutrition are very important factors.

Clinical correlation is essential to differentiate miliary tuberculosis from other causes of miliary nodulation, many of which may be found during the investigation of any debilitating illness. The initial chest radiograph may be normal, and reexamination of the chest should be requested "if symptoms persist," because miliary tuberculosis may only become apparent radiologically up to 10 days or more after the clinical illness has started; it may be recognized earlier, however, with high-resolution CT. There is also considerable variation in the "miliary" pattern radiologically, again with little correlation with the patient's clinical status. In some patients, at first there is only interlobular septal thickening, which can be very difficult to recognize, particularly on the chest x-ray of a small child: high-resolution CT is more reliable. The traditional miliary (millet seed) nodules are tiny, discrete, and all about the same size, less than 2 mm. These nodules may coalesce into patchy and more irregular opacities, and high-resolution CT will show even more variation (Fig. 23). The distribution of the nodules is not symmetrical and there are differences in size and density, which may represent different episodes. High-resolution CT may also show cavitation which

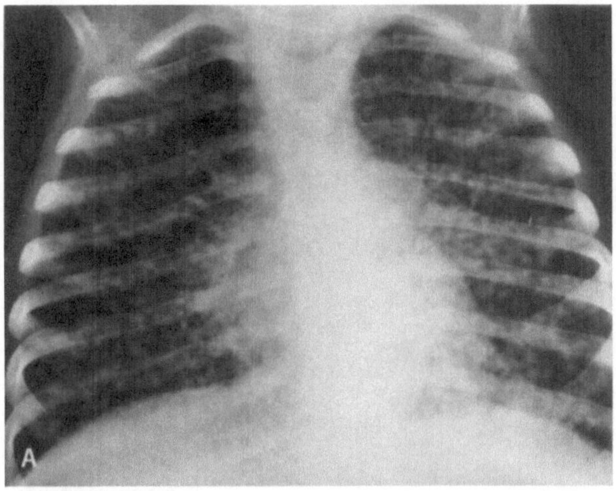

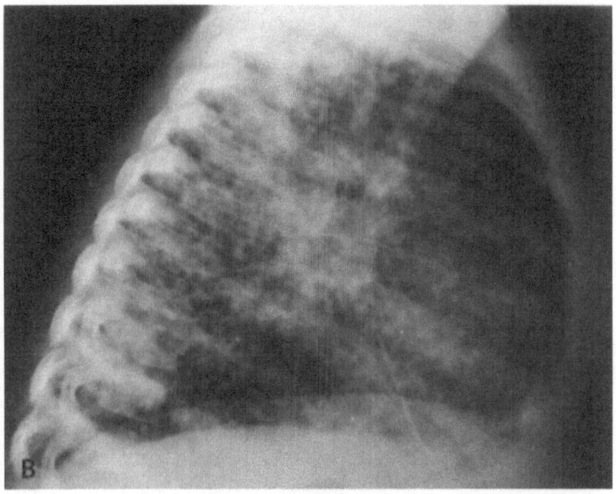

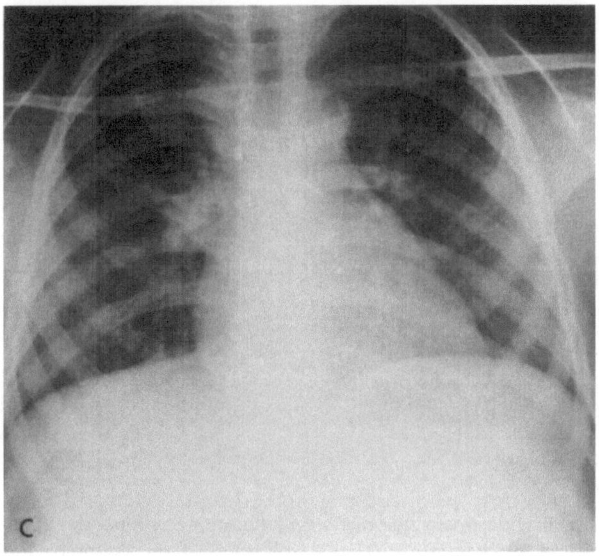

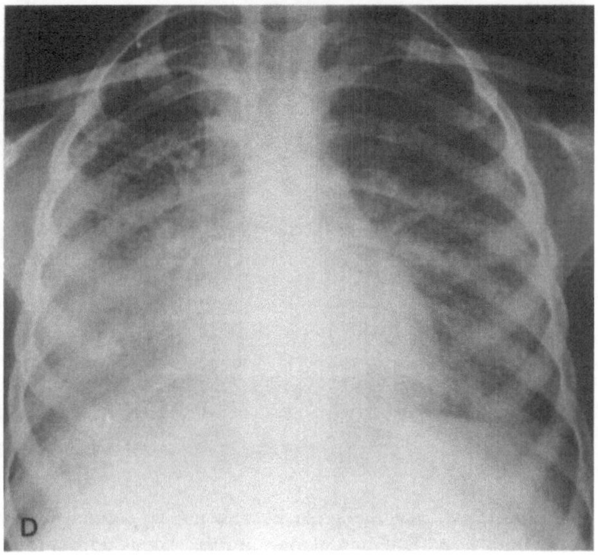

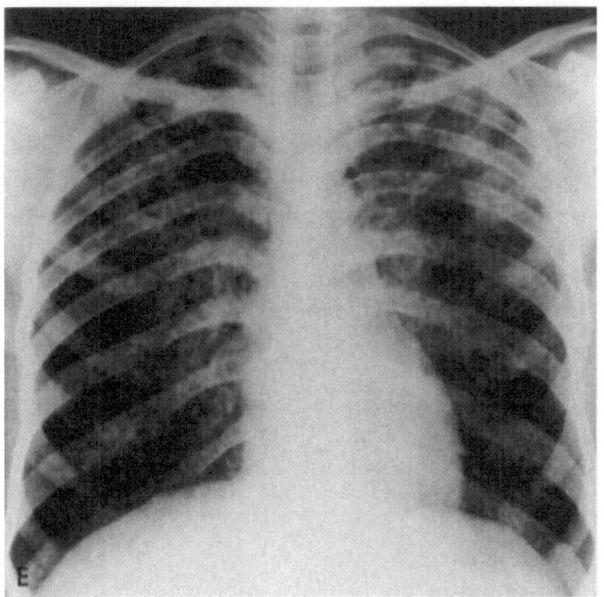

Fig. 22 A–E. Miliary tuberculosis. **A, B** Acute miliary tuberculosis in an African child. The lateral view shows enlarged mediastinal nodes. **C, D** This child was undergoing treatment for primary tuberculosis (**C**). There were enlarged lymph nodes in both hila and in the right paratracheal region, all showing early calcification. At no time was there a visible pulmonary lesion and the child had no symptoms. A few weeks later (**D**) for no apparent clinical reason, her general condition deteriorated and there was hematogenous spread to both lungs. There was also some patchy nodulation in the right lower lung suggesting that a lymph node had ruptured into a vessel and a bronchus. The calcification and the lymph nodes provided evidence only of local healing, and not of cure. **E** Miliary tuberculosis with cavitation in both upper lobes and enlarged hilar lymph nodes on both sides. This indicates a primary tuberculous infection in a 21-year-old African, whose only clinical complaint was a sore throat and ear ache. He was found to have tuberculous right tonsil and mastoiditis. As so often happens in developing countries, at this stage the patient disappeared. (**C, D** courtesy of *Semin Roentgenol*, 1979)

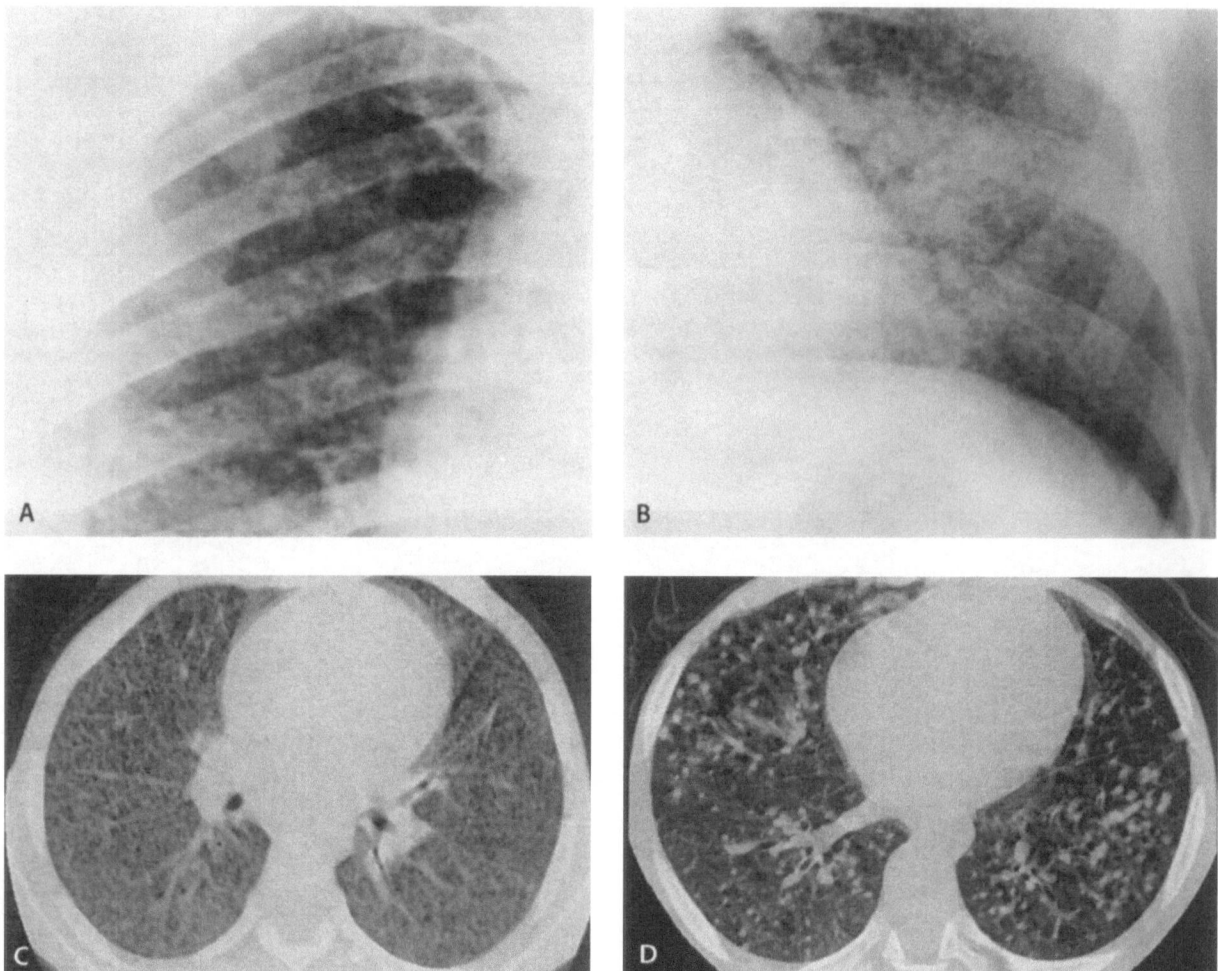

Fig. 23 A–D. Miliary tuberculosis. The miliary nodules vary considerably. **A** Fine nodulation in the right upper lobe, evenly distributed but with some nodules beginning to coalesce. **B** The left lower lobe of another patient with more numerous nodules varying a little in size. In this patient the nodules did not coalesce. **C** High-resolution CT showing small widespread nodules with interlobular septal and interstitial thickening. **D** High-resolution CT in another child from South Africa, with multiple nodules of two sizes, suggesting two episodes of hematogenous dissemination. Early diagnosis of hematogenous spread is important to reduce the high mortality. (**C, D** from Cremin and Jamieson 1995)

has not been suspected on the chest radiographs. In some patients there may be hilar lymphadenopathy, but in others the miliary disease is so acute that the lymph nodes are normal.

Hepatosplenomegaly often occurs in association with miliary infection, which on ultrasonography is seen as a granular echoic appearance in the liver or spleen. In some patients there may be multiple small hypoechoic nodules, some of which have central echogenic necrosis. On CT the small nodules are of low density and usually do not enhance with contrast, except occasionally around the rim of the nodule. These granulomas may become confluent, appearing as hypoechogenic or echogenic masses up to 2 cm in size. Both hepatic and splenic nodules may calcify (see p. 71).

Miliary tuberculosis may occur at any age, including in the elderly, even when the patients are HIV-negative. But patients who have altered immunity (e.g., from AIDS, malnutrition, recurrent infection, diabetes, leukemia, lymphoma or other malignancy) are at risk when exposed to tuberculosis and are particularly liable to develop miliary spread. The differential diagnosis includes the miliary pattern caused by the passage of larvae of various parasites through the lungs, during which there will in most cases be a peripheral blood eosinophilia; unfortunately, this can be misleading, because miliary tuberculosis may be superimposed on the parasitic background of the patients, and is not excluded by eosinophilia. Fortunately, in the majority of parasitic infections the radiological appearances change rapidly, whereas in miliary tuberculosis the nodules may become more apparent but the overall pattern does not fluctuate within a short period. Miliary tuberculosis may take 2 or 3 months to fade, even with adequate therapy. Only rarely does it result in miliary calcification.

There are many miners (for coal, tin, gold, etc.) throughout the world in whom silicosis and silicotuberculosis must be considered in the differential diagnosis of miliary patterns in the lungs. Silicotuberculosis also occurs in women (see below).

Silicosis

Any pneumoconiosis, but silicosis in particular, carries an increased risk of tuberculosis. Routine chest radiography is required by law in many countries before anyone may work underground and is repeated every year. The increased risk of tuberculosis is such that many countries provide a pension, not only for silicosis but whenever a miner develops tuberculosis while at work or within 1 year of leaving the mines or high-risk occupation. The radiological differential diagnosis between silicosis and miliary tuberculosis can be extremely difficult, and depends on the distribution of the miliary nodules in the lung. In silicosis the nodules are almost invariably first recognized in the intraclavicular, upper mid-zone of the frontal chest film; over a period of months they become visible in the apices and the lower lung sequentially, and the original nodules become more prominent. Miliary tuberculosis has no such differential distribution and may occur everywhere (both lungs) at the same time. If both diseases occur together (silicotuberculosis) the radiologist is quite often unable to recognize the underlying pathological process and must resort to careful study of the preceding films and a sound clinical knowledge of the patient. High-resolution CT may help in the differential diagnosis. Lymphadenopathy can suggest tuberculosis, but also occurs in silicosis, developing a significant and almost unique "eggshell" pattern (sarcoidosis is the only other likely cause of this calcification). The development of a large, irregular, fibrotic "conglomerate mass" (massive fibrosis) in an upper zone is indicative of silicosis, but some authorities believe that there must also be infection before this occurs.

Silicosis can occur in unexpected places and persons, and not only in those who work in high-risk occupations, e.g., miners, quarriers, stone workers, and even sculptors in stone. In the Transkei in the Republic of South Africa, approximately half a million women developed silicosis because, from childhood onwards, the girls and women ground corn between two stones which had a high silica content (Fig. 24). The stones slowly disintegrated and as

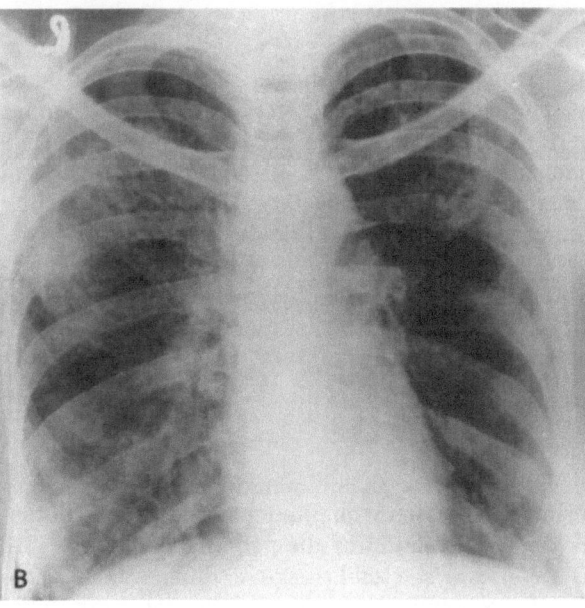

Fig. 24 A, B. Nor every miliary pattern on a chest radiograph is due to tuberculosis. This African woman had been grinding corn between stones since she was 9 years old. She was under treatment for miliary tuberculosis until it was realized that the distribution of the miliary nodules within her chest was not compatible with this diagnosis, and there had been no response to antituberculous therapy. The sputum did not contain *M. tuberculosis* nor did her clinical condition suggest active miliary tuberculosis. She had silicosis and was lucky, because many of the women who ground corn in this way developed silicotuberculosis. The possibility of some alternative diagnosis must always be considered when the patient's clinical condition does not match the radiographic appearance. (Courtesy of *S Afr Med J*, 1967)

the women leaned over to add pressure to powder the corn finely, they inhaled large quantities of the silica-containing dust. Their children, strapped to their backs or running around the huts, were exposed to silica in the same way. The dust in these huts became full of silica of the correct particle size and characteristics to cause silicosis; as a result many of the young girls and women not only developed silicosis but there was an increased incidence of tuberculosis (more accurately, silicotuberculosis). This had a strange collateral effect; their husbands, many of whom worked underground in the gold mines of the Transvaal but under good health surveillance, went home and then returned to the mines from the rural areas with a higher incidence of tuberculosis than their fellow miners whose wives made flour by more modern methods! The whole population became at risk because of this domestic occupation. Moreover, even those who produced the grindstones were likely to suffer from silicosis as, for example, in a village north of Kano in Nigeria, West Africa, where sandstone is quarried in pits from which grindstones are hewn and then widely used throughout Nigeria for grinding guinea corn. A total of 126 workers were examined and 49 had silicosis, 17 of them in an advanced stage of massive fibrosis. Among these, seven also had pulmonary tuberculosis. No doubt similar hazards exist elsewhere in the tropical world for equally unexpected reasons (and elderly American Indian women have been shown to have faced the same hazards from grinding corn).

Summary

Primary, anergic or nonimmune tuberculosis occurs in any combination of the patterns described. This is significant in establishing the differential diagnosis; for example, it is unusual to see both lobar pneumonia and bronchopneumonia on the same chest radiograph in a pyogenic infection. Similarly, patchy bronchopneumonia with a small pleural effusion, or lobar pneumonia with pericarditis, is more likely to be tuberculous than pyogenic. Lobar pneumonia which does not clear in a few weeks is probably not pyogenic. Any chest infection at any age which does not respond appropriately to antibiotic therapy may be tuberculous, unless, particularly in adults, it is secondary to bronchiectasis or underlying neoplasm. Multiple cavities in young children may be tuberculous or staphylococcal. Enlarged hilar and mediastinal lymph nodes with any form of pneumonia, abscesses, or pleural or pericardial fluid should always suggest tuberculosis. Pleural effusions in young teenage patients, especially if bilateral and with no cardiac or renal disease, are probably tuber-

culous. Miliary lung disease, especially when accompanied by lymphadenopathy and effusions, also suggests tuberculosis.

It is very important to recognize that the lung changes in primary (nonimmune) tuberculosis go through a progressive spectrum, and that the first radiograph may be taken at any stage in the sequence. In some there may be lobar pneumonia and/or hilar lymphadenopathy, while in others there may be a healing peripheral focus or bronchogenic spread or pleural effusions; furthermore, there may be a combination of any of these events. Primary tuberculosis is not a static process, and the progress shown radiographically may provide the diagnosis when it is not clear on the initial film. Unlike a pyogenic infection, tuberculosis seldom simply fades away. The rate at which these changes develop and heal depends on what is called, in broad terms, the "resistance" of the individual patient, together with the success of any treatment. It is this whole repertoire of change which may allow the correct diagnosis, hopefully before it is too late for the patient.

Immune Tuberculosis
(Also known as Secondary, Hyperergic, Reactivation, or Adult Tuberculosis)

The term "adult" tuberculosis is misleading because this pattern of disease can occur in any patient at almost any age beyond infancy. The majority of cases of "adult" pattern tuberculosis develop as a separate illness, or as a complication of other ill health, e.g., AIDS, diabetes, or malnutrition. There are a few patients whose immunity alters during the course of a primary tuberculous infection and an "adult" pattern develops without an interval. Whatever the sequence, the radiologist must not, therefore, hesitate to diagnose "adult" tuberculosis in a child, particularly one who has received BCG vaccination.

Secondary or immune tuberculosis is usually bilateral. In one survey only 4% of patients had unilateral disease. Cavitation is the rule rather than the exception, particularly during the active stage of the disease. When there is bilateral infection in the upper lobes, it is almost certainly due to tuberculosis; if it is bilateral and in the lower lobes, it may well be pyogenic bronchiectasis, following pertussis (whooping cough) or chronic sinus infection. If the infection involves one upper lobe and the opposite lower lobe, it is likely to be tuberculous. In Asia, all these appearances may also be due to paragonimiasis or to melioidosis.

Clinical Characteristics

The clinical symptoms of patients in the tropics are similar to those of patients in nontropical areas, but may be more severe. Hemoptysis, cough, fever, night sweats, and general debility are common. Many patients have clinical evidence of lung fibrosis, which causes asymmetrical flattening of the chest wall and shift of the trachea (mediastinum): there are often associated physical signs of cavitation. Clinically the differential diagnosis must include pyogenic pneumonia, chronic lung abscess, bronchiectasis, fungal infections, paragonimiasis or melioidosis (where appropriate), amebic infection of the lung (usually but not invariably associated with a large liver), and malignant disease.

Imaging Diagnosis

In the majority of patients the radiological diagnosis of hyperergic secondary (immune) tuberculosis is not difficult. Tuberculosis is a necrotizing infection in which cavitation and fibrosis can occur together or separately; both are usually bilateral in the upper lobes, especially tending to involve the apical and posterior segments. The superior segment of a lower lobe is another common site, but no lung segments escapes. Any multilobar infection with fibrosis and cavitation is likely to be reinfection tuberculosis, or an atypical mycobacterial infection. If the fibrosis is sufficient to shift the mediastinum, tuberculosis is even more likely. The mediastinal and hilar lymph nodes are not usually enlarged and may or may not be calcified. Sometimes it is possible to identify a primary parenchymal granuloma (such granulomas are usually calcified), but failure to do so does not exclude tuberculosis. There will often be pleural thickening, particularly in the costophrenic angle or over the lung apices. The rate of progress is very variable, but is usually slow and insidious, except when tuberculosis complicates some other severe illness, such as AIDS.

The initial lesion of reactivation tuberculosis is usually an alveolar haze with a few small pulmonary nodules, from which there may be some interstitial lines, often radiating to the hilum (Fig. 25). Further ill-defined opacities develop, with commencing distortion of the interstitial pattern due to fibrosis. At this stage the appearances often change if the patient is radiographed each month, with further distortion of the normal vascular and interstitial patterns. Contraction can progress until the interlobar fissure and hilum are displaced, usually upwards until one or both hila are much higher than normal.

Even without treatment, the infection may be arrested at any stage. This leaves residual scarring but in many cases the healing process is complicated by endobronchial involvement and there is stenosis or kinking of one or more peripheral bronchi. This results in secondary pyogenic infection and bronchiectasis, superimposed on the contracted fibrotic lung. In some patients this destructive process progresses to such an extent that it resembles the destroyed lung of primary infections (see p. 19). Unfortunately, at any stage the infection may be reactivated, sometimes after many years. Reactivation has been recorded from 35 to 51 years after the initial diagnosis. It is therefore not possible to describe any patient as being "cured." It may be better to report "probably old tuberculosis" or "apparently well healed." Unfortunately, clinicians are not much better at judging the future course of the disease.

Contrary to what is often taught, cavitation (Fig. 26) has diagnostic significance but no prognostic value, because necrosis and liquefaction are the normal course of tuberculosis. Cavities appear in 37%–75% of all cases of active secondary tuberculosis. They may be thick walled or thin walled and may be surrounded by a parenchymal reaction or appear to cause no local disturbance. They are more common in the upper lobes and the superior segments of the lower lobes than elsewhere, but may occur in any part of either lung. Cavities fluctuate considerably during treatment; radiographically, they may increase or decrease in size, may contain fluid, and may show a fluid level or apparently be empty, depending on the reabsorption, expectoration, or bronchial spread of the contents.

When there is obvious cavitation on standard radiographs, the sputum will be positive for bacilli in 80% of patients, but not always at the first examination; a negative sputum does not therefore exclude tuberculosis or indicate cure. Cavities may not disappear: 7%–10% of all cavities become bacteriologically sterile under treatment but remain patent. If, following surgery, cavities are examined histologically, part of the cavity wall may show healing while in another part there is active infection. Histopathologically, tuberculous cavities are abscesses and the wall of the cavity is lined with necrotic tissue. The only slight help which may be obtained in identifying active infection is when the inner wall of the cavity becomes radiographically ill-defined and the outer wall becomes less distinct. Unfortunately, a secondary infection, particularly fungal, can have exactly the same effect. Each and every radiological characteristic of the cavities can alter month by month, while the clinical condition of the patients may improve or worsen without any relationship to any radiological change in the cavities.

(Text continues on p. 39)

Unfortunately, space does not permit the illustration of all the vagaries of pulmonary tuberculosis: the most important differences are the rapidity with which tuberculosis may change or progress, and how slowly it may heal even during adequate treat- ment. Equally important is the often acute and un- expected pattern of the infection, which may make it difficult to accept that it may be due to tuberculo- sis, whether in the primary (nonimmune) or in the secondary, reactivation (immune) phase.

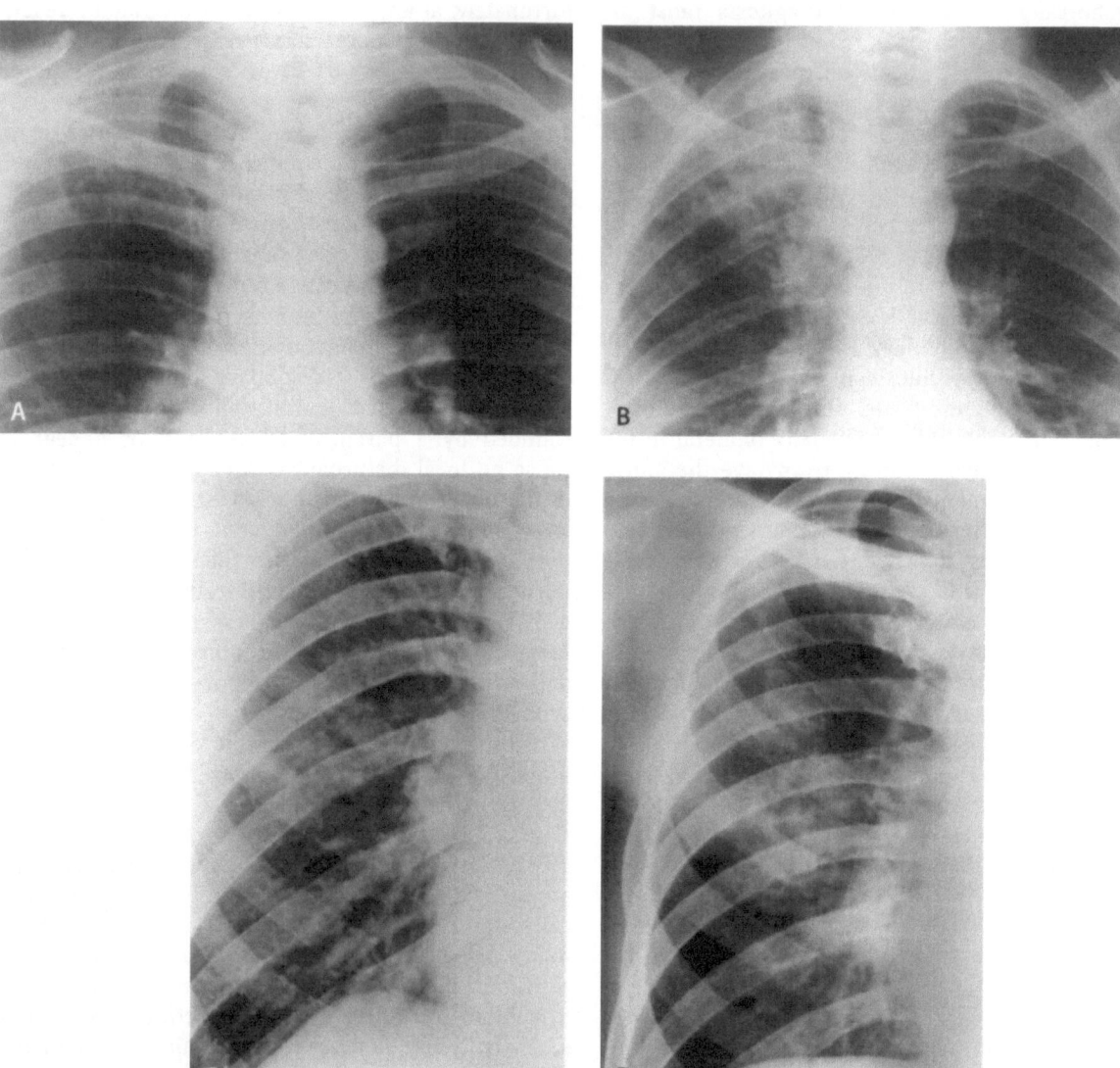

Fig. 25 A–D. The earliest evidence of immune (reactivation) tuberculosis is a soft alveolar haze, usually in the apex or upper lobe. This African man worked in a hospital and had routine chest radiograph which was normal 3 weeks before he became ill, with a cough, sputum, and minimal hemopty- sis. His next radiograph (**A**) showed the clouding in the right apex with a few small tubercles. There were *M. tuberculosis* in the sputum. Although treatment was started, 12 weeks later (**B**) the infection had spread and was beginning to cavi- tate. The patient was found to be diabetic. **C** Another patient, who had lost weight and was unwell. There is a soft alveolar haze with small nodules in the periphery of the lung, prob- ably in the anterior segment of the right upper lobe. The spu- tum contained *M. tuberculosis* and treatment was started: the patient improved clinically but after 7 months (**D**) the radio- graphic appearances seemed to have deteriorated

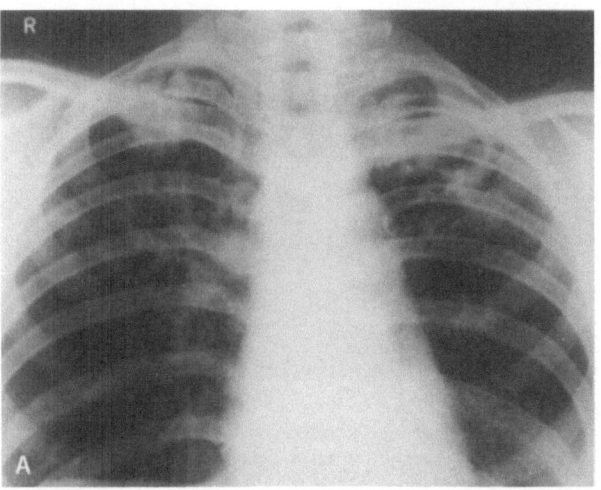

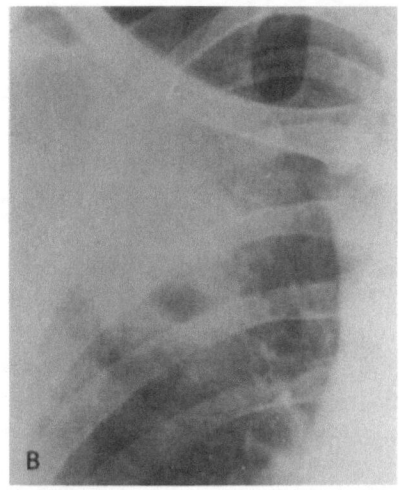

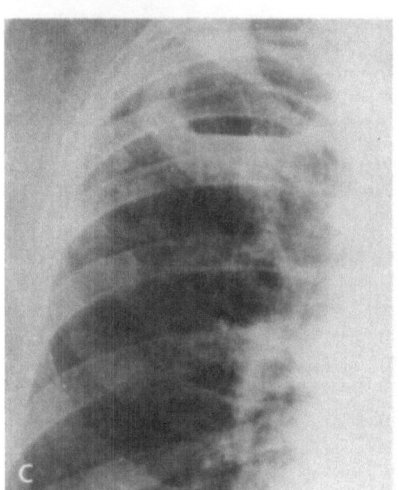

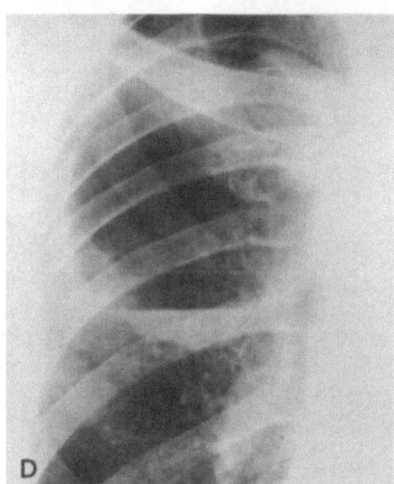

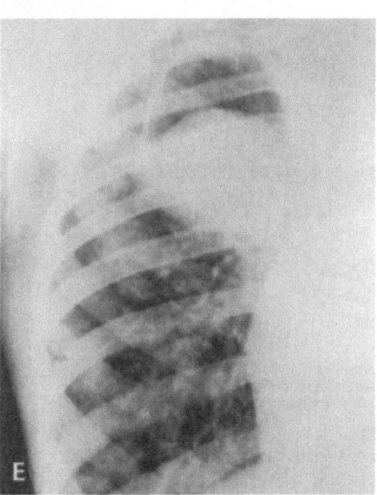

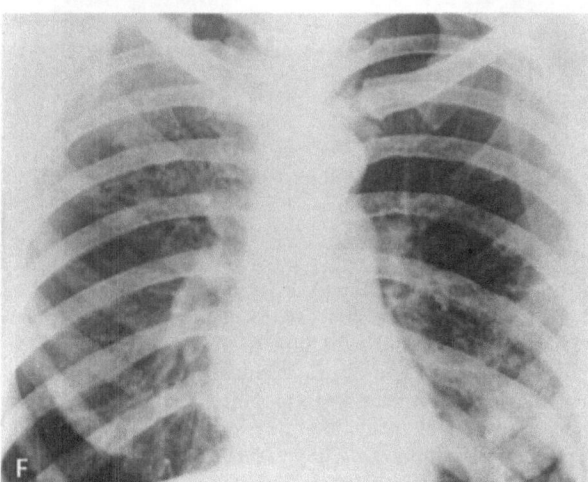

Fig. 26 A–F. Many patients present with bilateral upper lobe infection, often as in A with asymmetrical cavitation on both sides. **B** Cavities are not a reliable way to judge the progress of healing: they are evidence of necrosis, which is part of the natural history of the infection. This cavity in the right upper lobe has developed within an area of consolidation. **C** This cavity has a thick wall and a fluid level, which means that it communicates with a bronchus. There is nodulation in the lung below it. **D** In another patient there is a much larger thin-walled cavity with less fluid, probably because some of the contents of the cavity have been coughed up. **E** The solid "mass" lying in this thin-walled cavity could be either a mycetoma or a large blood clot, or both. Hemorrhage into a cavity with subsequent infection occurs quite often. There are fibrotic changes throughout the right lung because this cavity appeared unexpectedly when the infection seemed to be quiescent. **F** In this patient there were several small cavities in the right upper lobe which must have communicated with the bronchi because there is now inhalation bronchogenic spread into the left lower lobe. This pattern, with infection of an upper and a lower lobe, is very common

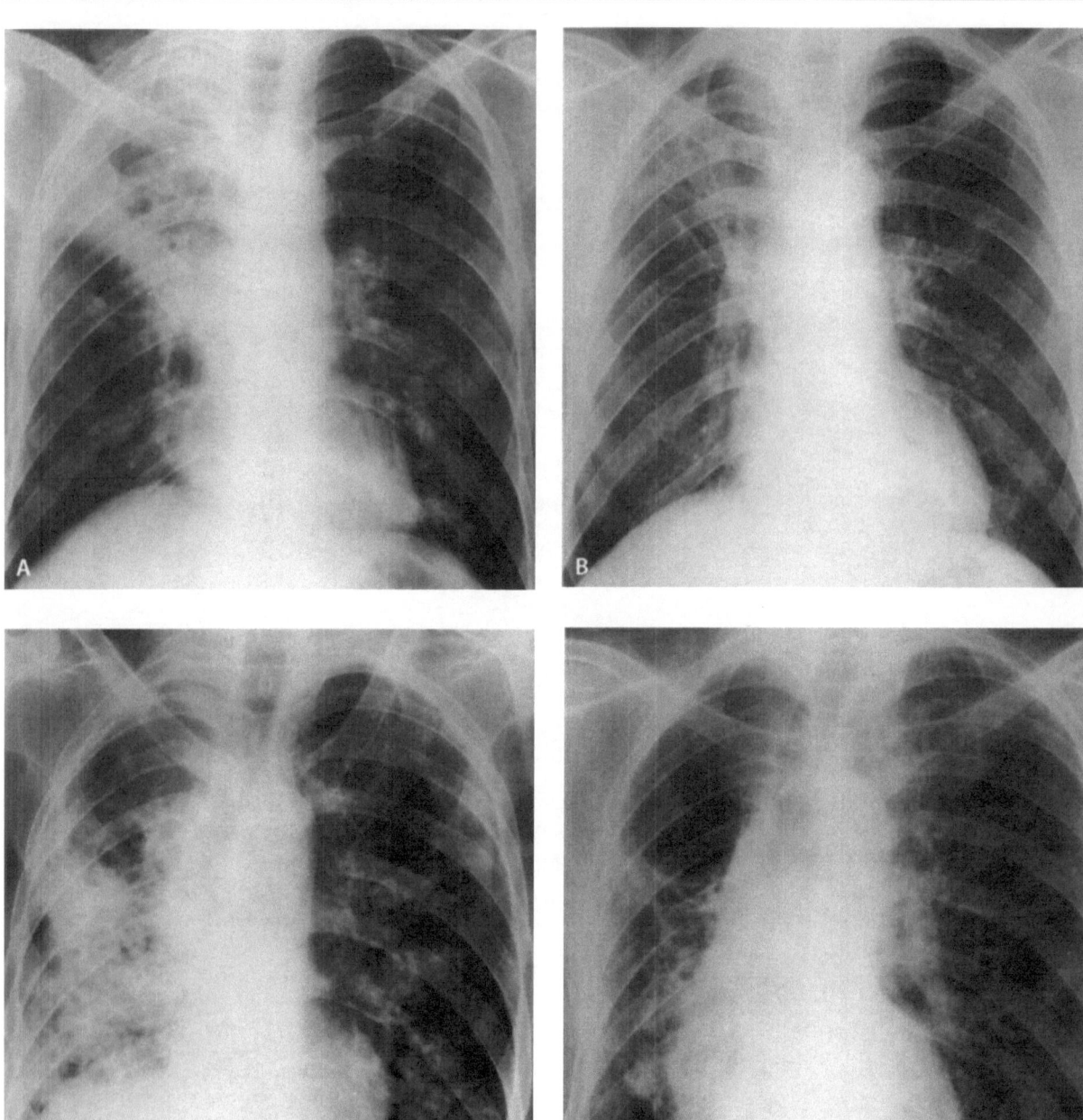

Fig. 27 A–D. The success, and failure, of treatment for tuberculosis. This African patient had started treatment in November for active tuberculosis. **A** Two months later the sputum was still positive for tuberculosis. Cavitation and consolidation are still present in the right upper lobe, with some overlying apical fluid. Some shrinkage of the upper part of the right lung has occurred, with slight mediastinal shift to the right. Infection is present in the left lung, particularly in the lower lobe but also in the apex. **B** By June, after 7 months' treatment there has been remarkable improvement. The right upper lobe has contracted and become fibrotic. The cavity has disappeared and the fluid over the apex of the lung has been absorbed. The left lung has also improved, leaving some minimal fibrosis only. The sputum was now negative. **C** Unfortunately, 1 year later there has been a severe relapse; a tuberculous focus has probably ruptured into a bronchus, with resulting widespread bronchopneumonia in both lungs. There is now consolidation in the right lung and a huge cavity in the right upper lobe. There is active disease and cavitation in the left lung. The mediastinum is still shifted a little to the right because of the underlying fibrosis and the right side of the chest is smaller than the left. Treatment was continued with slow improvement until (**D**) (4 years after the original radiographs. A large emphysematous bulla is present in the right upper lobe and the right lower lobe is also emphysematous. The middle lobe is severely contracted. This lung is also fibrotic and shows some small bullae. The sputum had been negative for about a year, but the fibrotic contraction and formation of bullae had continued until the right lung was virtually destroyed. Though fibrotic, the left lung is the only functioning remainder. The right lung is bronchiectatic and will often be secondarily infected. From this stage onwards it will be very difficult to decide whether any clinical illness is evidence of further reactivation of the tuberculosis

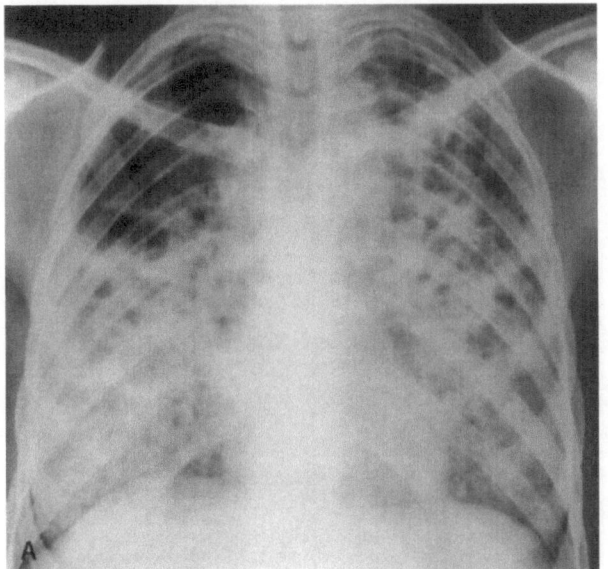

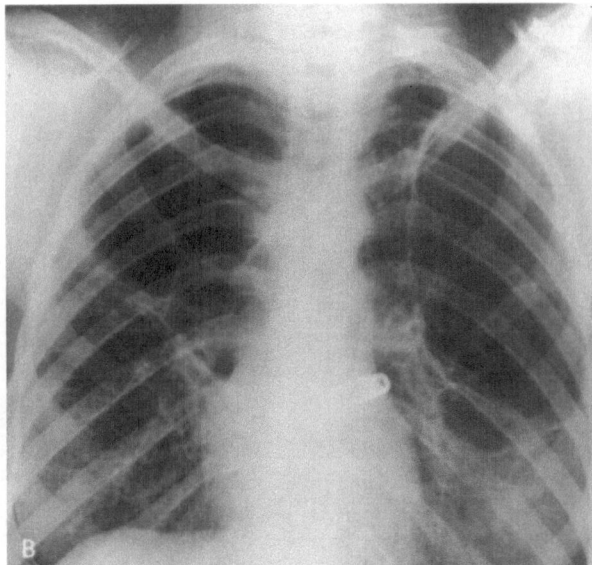

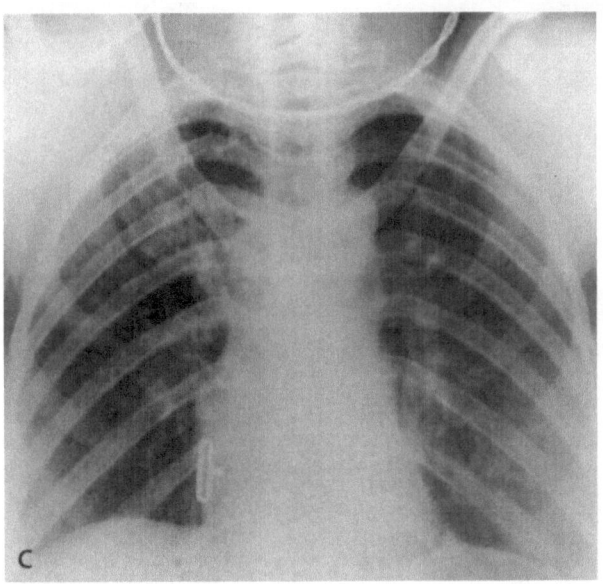

Fig. 28 A–C. In some patients the response to antituberculous treatment can be remarkable. A This African woman presented with severe bilateral cavitating tuberculosis: her sputum was positive for tuberculosis. She was started on treatment and 5 months later (B) there had been improvement in both lower lobes, but there were huge bullae almost replacing both upper lobes (the left particular), and compressing the remainder of both lungs. She was clinically improved but very short of breath. Treatment was continued and (C) 11 months later there was remarkable improvement: the bullae had disappeared and both lungs were clear except for some upper lobe fibrosis. It is difficult to believe that there three chest radiographs are of the same patient. The change in her clinical condition was equally impressive

The significance of cavitation has been overemphasized; it is but one event in the natural history of the disease. It is not even unique to human tuberculosis, air-fluid levels having been reported in the lungs of a tuberculous cockatoo!

When there is hemoptysis, the hemorrhage may come either from the lining of a cavity or from endobronchial disease, as when bronchoscopy shows that the blood comes from a lung segment in which there is no radiologically visible cavity. In tuberculosis, arteriography has not been a very satisfactory method of diagnosing or treating the source of the hemorrhage, probably because there is almost always a great deal of associated fibrosis and scarring. A clot in a tuberculous cavity can mimic a mycetoma or become infected by a fungus. It is only the

reabsorption of the clot after a relatively short period, 1 or 2 weeks, which can exclude a fungus ball. Most cavities heal by obliteration, leaving a scar, but some remain as thin-walled bullae.

Fibrosis is equally unrewarding prognostically (Figs. 27–29). Change is almost always slow, so that comparison with a radiograph taken a few weeks previously may show no alteration. It is essential always to compare findings with the very earliest film available. It is never possible to judge the status of the infection from one radiograph; even extensive pulmonary fibrosis may include active disease. The radiological assessment of activity is very unreliable (Fig. 30), and is often made more difficult by a secondary pyogenic infection in the fibrotic lung: clinically it may be some other concomitant illness,

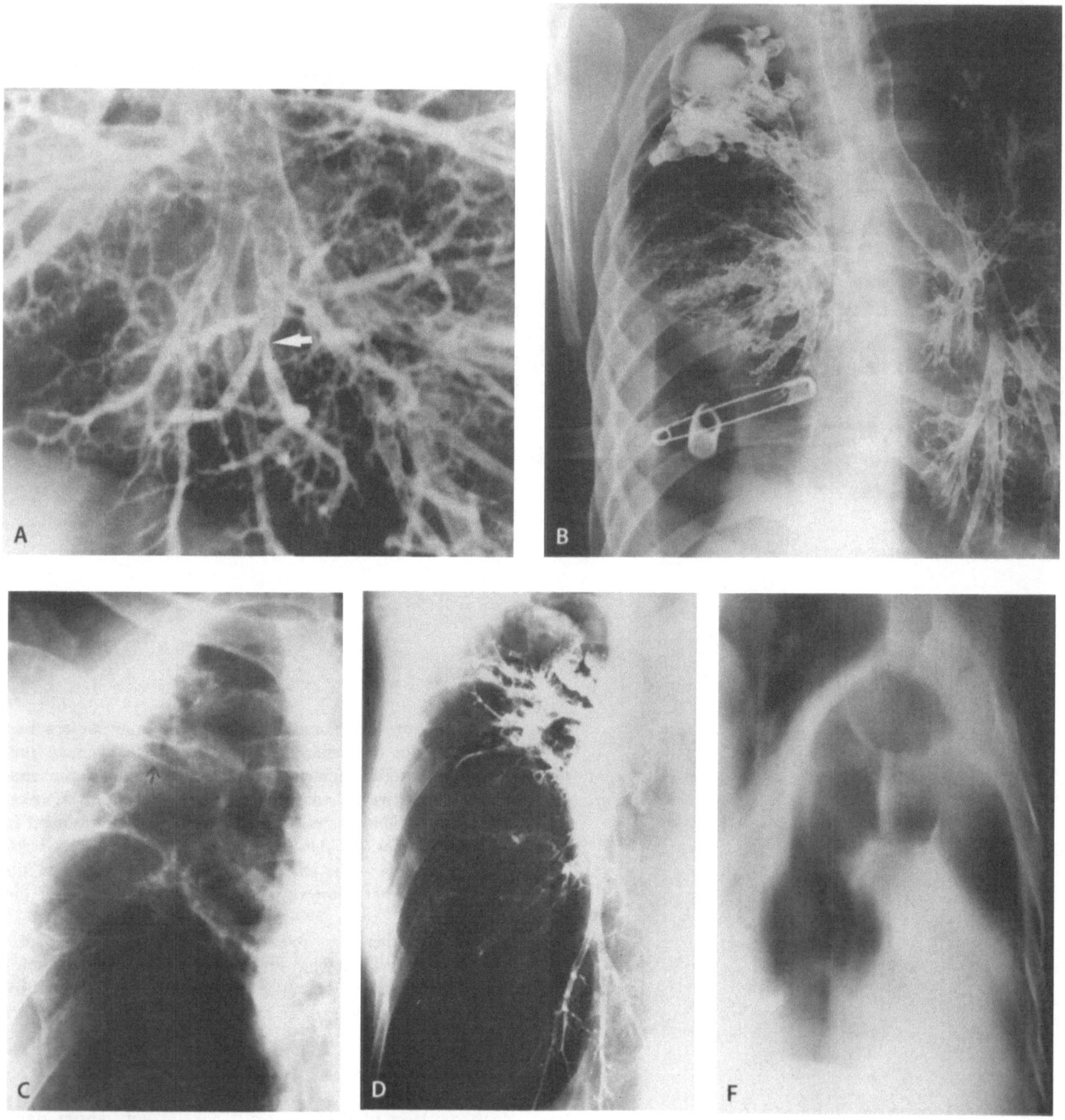

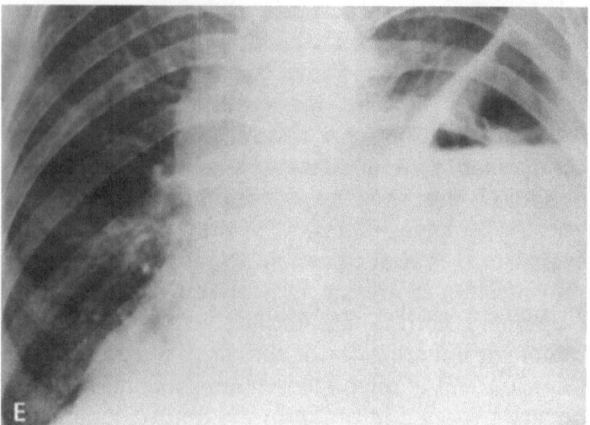

Fig. 29 A–F. Tuberculosis heals by fibrosis. It may cause (A) a localized stricture in a small bronchus (*arrow*) and result in repeated segmental infection or (B) gross bronchiectasis and contraction. The appearance of "apical" fibrosis (C) may be misleading. A bronchogram in the same patient (D) shows that the right upper and middle lobes and the superior segment of the lower lober are contracted upwards, with gross emphysema of the lower lobe. E Fibrosis also affects the pleura and it is often difficult to be sure whether the air-fluid level is in the lung or pleural space. F Standard tomography of the same patient shows the multiple fibrotic septa and thickening of the visceral pleura. It is not necessary to have a secondary pyogenic infection to produce this result

such as amebiasis or dysentery, which has led to deterioration in the patient's general health. When there is lung fibrosis it is impossible by any imaging technique to differentiate pyogenic infection from recurrent tuberculosis, or to exclude contamination by a saprophytic fungus (Fig. 31). The diagnosis has to be made clinically with the help of the bacteriologist; unfortunately, relapsing secondary infections are common.

However, to end this section on a more positive note, the majority of patients eventually heal or fibrose their tuberculosis into a state of inactivity provided they continue with adequate treatment.

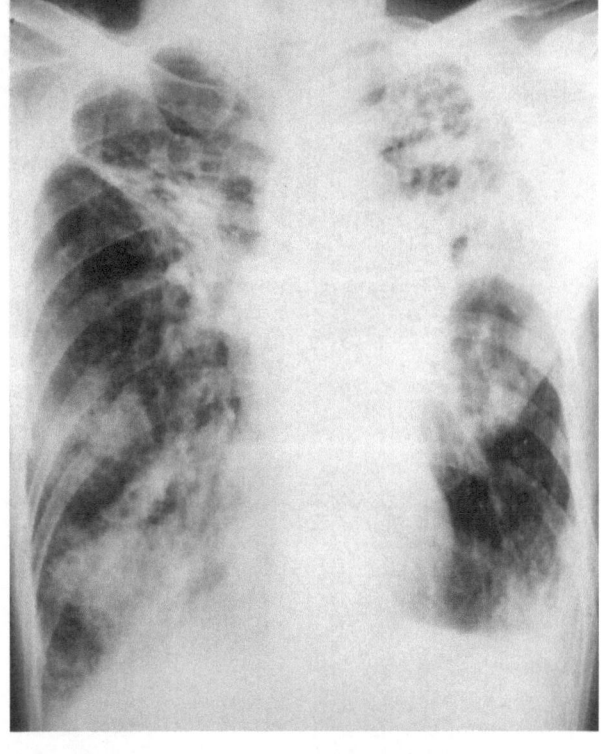

Fig. 30. Severe bilateral chronic tuberculosis will remain unchanged radiographically for many years. In this 70-year-old patient no part of the lung has escaped. He occasionally produced sputum which contained a few *M. tuberculosis* bacilli and so remained a danger to his family, with whom he lived. Surgery was obviously impossible, but the success of continuing antituberculous therapy is doubtful

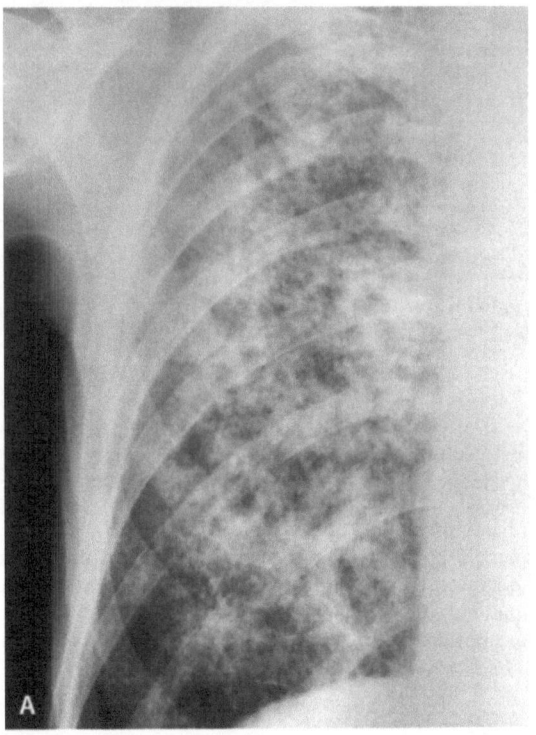

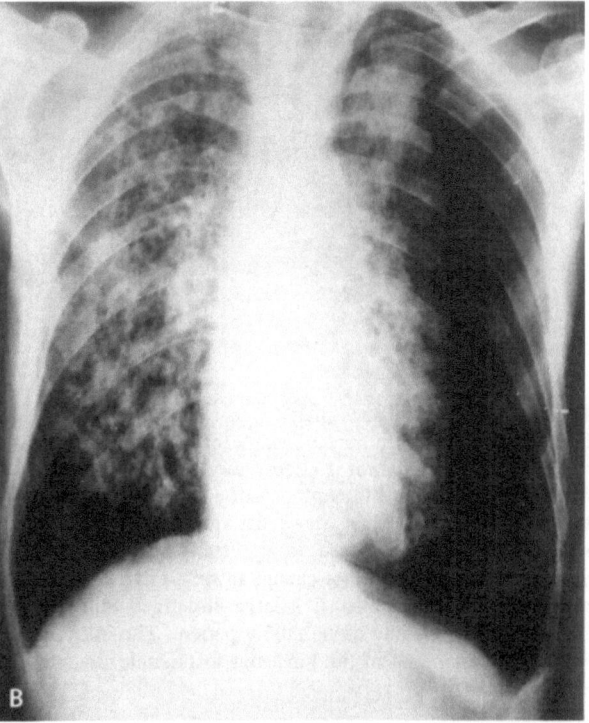

Fig. 31 A, B. Reactivation tuberculosis must be included among the many causes of interstitial fibrosis, particularly in middle-aged patients. The clinical complaint is usually of dyspnea and generall ill health. A Both lungs of this 50-year-old man showed severe interstitial fibrosis and honeycombing. The sputum was consistently negative for *M. tuberculosis* and the diagnosis was established by surgical biopsy. There was no radiographic change after 12 months of antituberculous therapy. B Lungs with interstitial fibrosis are fragile. An acute pneumothorax can develop after paroxysmal coughing, in this case bilaterally but much more severely on the left

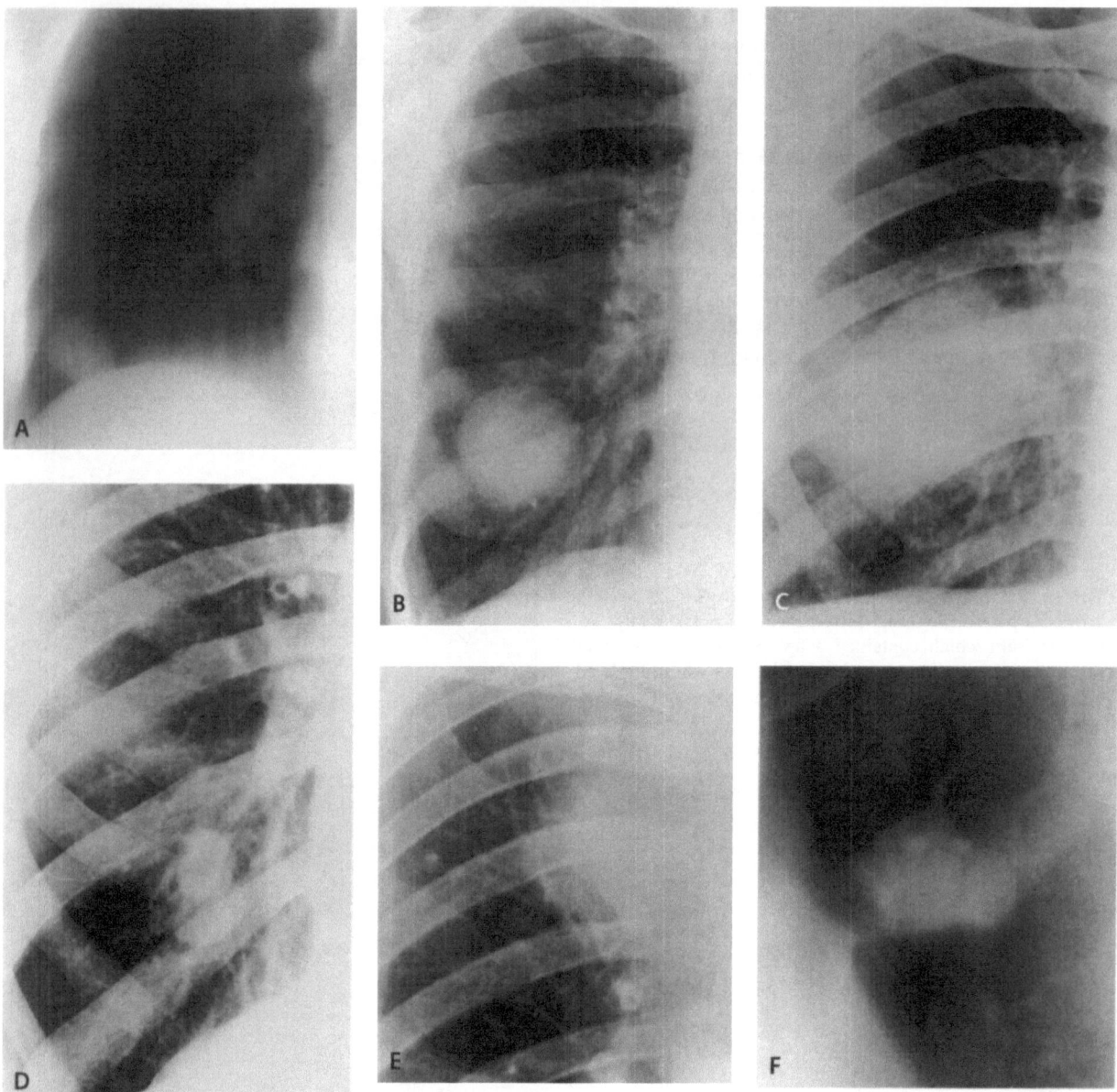

Fig. 32 A–J. Tuberculomas may appear in any part of the lung, at any age; they may be single or multiple and are of varying size. The sputum is almost always negative for *M. tuberculosis*. **A** Standard tomography of a small tuberculoma in the right lower lobe just above the diaphragm. **B** A larger smooth round tuberculoma in the right lower chest of a young woman. This did not change in spite of treatment over a period of many months. **C** A large ill-defined tuberculoma in the right lower lobe of an elderly patient. The initial diagnosis was carcinoma of the lung, but fortunately the sputum was positive for tuberculosis. **D** Some tuberculomas are lobulated or multiple. These two in the right middle lobe are well defined and could be mistaken for metastases. **E** A paramediastinal tuberculoma, clinically and radiologically suspected of being malignant. The diagnosis was made after surgical removal. **F** Standard tomograms of a lobulated tuberculoma in the left upper lobe, which was removed surgically. Even if tomography (including CT) demonstrates calcification, it is seldom possible to completely exclude malignancy by imaging. G–J see p. 433

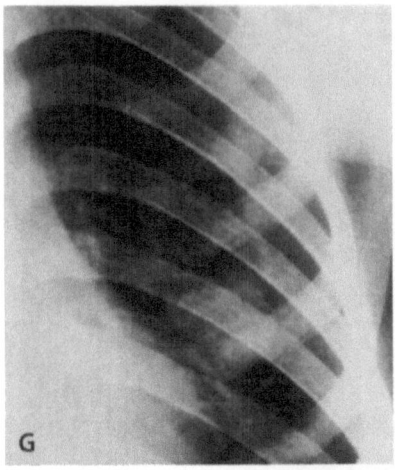 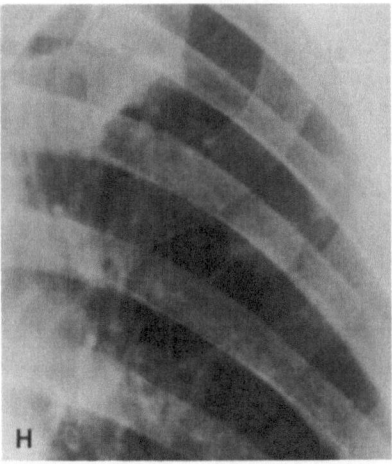 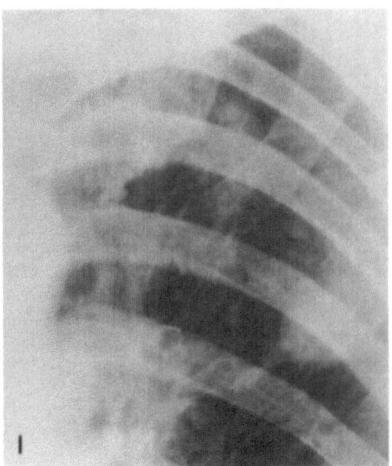

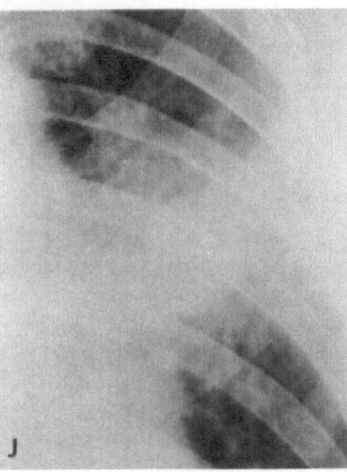

Fig. 32 (*continued*). G Multiple well-defined tuberculomas in the left mid-lung. These closely resemble metastases, but in this young African patient there was no known primary. H Tuberculomas are not always quiescent and harmless. There is a small tuberculoma with satellite nodules in the left upper lobe: it had not changed over 4 months. The patient refused surgery because he was clinically well. I Six months later the outline had become indistinct and further nodules had spread throughout the upper lobe. He remained clinically well but antituberculous therapy was started. J Unfortunately in spite of treatment, the infection continued to spread and coalesced into a caviating area of pneumonia which eventually led to further bronchogenic spread into the left lower lobe

Tuberculoma

Histopathologically, a tuberculoma is a focus of caseating pneumonia, although this is an oversimplified definition using a controversial explanation. Radiologically, a tuberculoma is a nodular, usually circular opacity in the lung, from 1 to 10 cm in diameter. Tuberculomas may be dense or hazy: while most are well defined, a few are more irregular in outline. CT or careful radiography may demonstrate small satellite tubercles. About ten percent of tuberculomas have central cavitation (necrosis). Tuberculomas are usually solitary but may be multiple; sometimes several occur in one lung segment. They may be found in any part of the lung, most often towards the periphery. Calcification, if present, can be central or eccentric, granular or in the form of small scattered flecks best seen with CT. The majority of tuberculomas show no calcification, even on CT or tomography, and there is no hilar lymphadenopathy; there is seldom hilar calcification. Most tuberculomas do not have any surrounding parenchymal reaction, although occasionally there will be ha-

ziness of part of their outline, apparently due to some increased vascularity or lymphatic thickening directed toward the hilum; this may result from continued activity or reactivation.

A tuberculoma may be found in a patient of any age, sometimes by chance on a radiograph taken for some other reason. Other patients present with cough and loss of weight. The appearance of the tuberculoma may remain unchanged for weeks or months, or it may unexpectedly break down as the result of some alteration in the patient's immune status. This can cause bronchogenic infection, or local spread becoming an area of pneumonia. No tuberculoma is "safe," even when unchanged for a long period.

The differential diagnostic problems are considerable: many tuberculomas are only diagnosed after surgical removal because a tumor has been suspected. If the patient is young, malignancy is unlikely and in the tropics, tuberculosis will be the most common etiology. A mycotic infection, such as histoplasmosis or coccidiomycosis, is possible; an early pyogenic or amebic lung abscess must be ex-

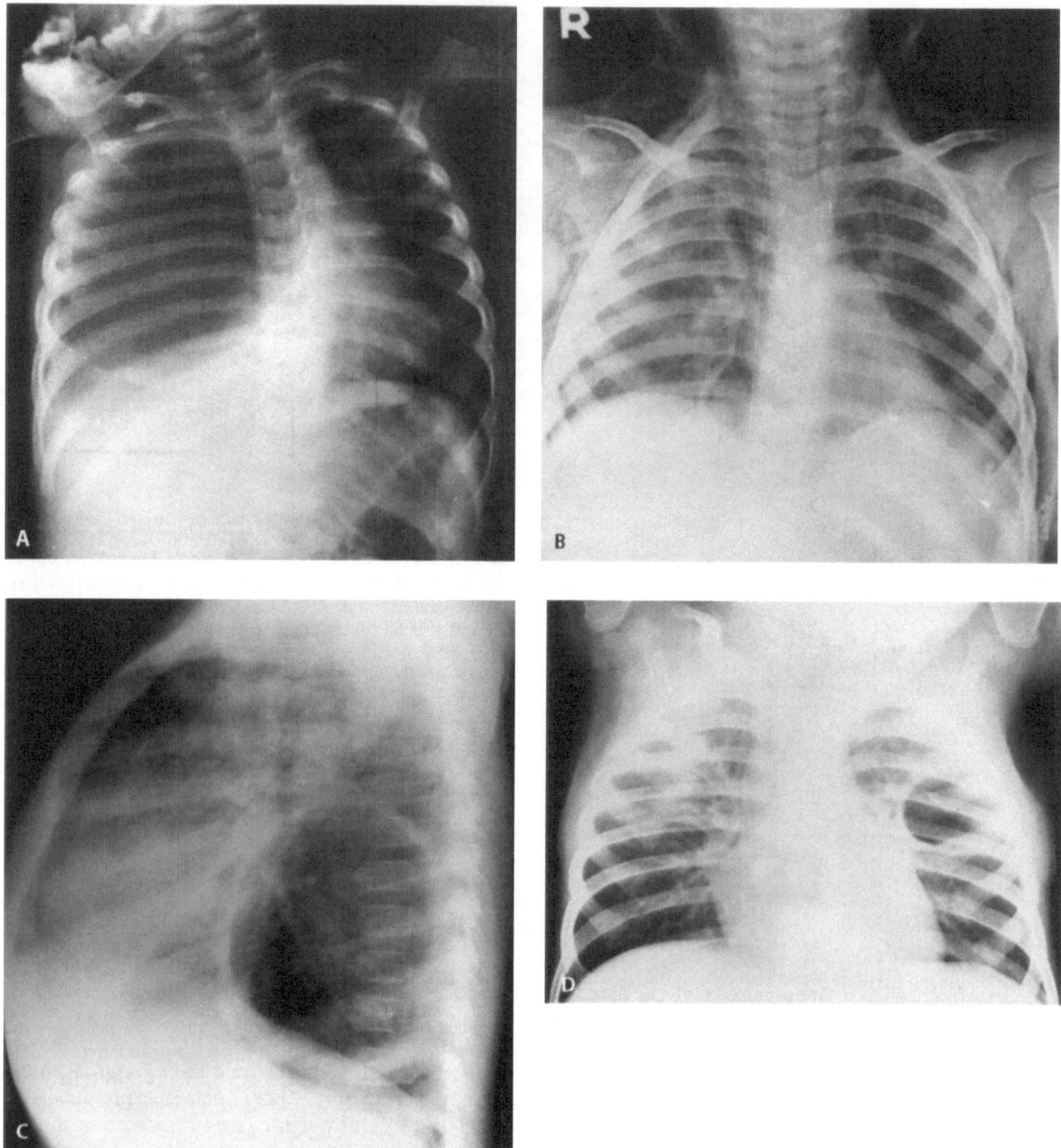

Fig. 33 A–D. Large emphysematous bullae can develop in children and adults, and can occur spontaneously or in association with isoniazed therapy. **A** A large thin-walled air-containing cyst in the right lung, displacing the mediastinum to the left. It is very easy to mistake this for a pneumothorax and attempt to decompress it. This may result in a broncho-pleural fistula, but in some patients the pressure is so severe that the risk has to be taken. **B** Spontaneous rupture of a similar cyst has caused subcutaneous emphysema and a pneumomediastinum. If the pressure increases, this may be fatal. **C** The lateral radiograph showing a similar large cyst in a child. This is quite thick walled, and there is tuberculous lymphadenopathy. However, this is not a tuberculous cavity. **D** Another child with a large bulla in the left mid-lung. The patients illustrated in **C** and **D** were both being treated with isoniazid

cluded. Unlike the majority of tuberculomas, lung abscesses are progressive and seldom remain unchanged. The tuberculin skin test will usually be positive, but this may not be a helpful differentiating feature. A hydatid cyst will have to be excluded, but this is usually not too difficult. In adults, primary carcinoma of the lung must be considered, although it is still uncommon in many parts of the tropics. For the solitary nodule which strongly resembles primary or metastatic lung carcinoma radiographically, and which is too small or inaccessible for CT-guided biopsy, surgical excision may be the only way to establish the diagnosis but, if tuberculosis is considered possible, it is a wise precaution to give the patient adequate antituberculous therapy prior to biopsy or surgery.

Isoniazid Cysts

Isoniazid is used in chemoprophylaxis, particularly to prevent development of tuberculosis in infected persons or those who have been in contact with patients who have sputum which is positive for tuberculous bacilli. It is also used for treatment, together with antibiotics. Prolonged therapy with isoniazid, apart from risking liver damage and peripheral neuropathy, may result in the development of thinwalled cysts (Fig. 33). Such cysts may regress or develop ball-valve obstruction due to blockage of the segmental bronchi. They may enlarge to resemble a single emphysematous bulla and may burst, causing an acute pneumothorax or pneumomediastinum. This can happen in children or in adults.

Chronic Pleural Disease

A pleural effusion during the reactivation stage of tuberculosis almost always means direct spread of the infection into the pleural cavity and is really an empyema. It is a different process to the primary pleural effusion which occurs in the nonimmune patient. If there is a spontaneous air-fluid level in the pleura, then a bronchopleural fistula has developed. Tuberculosis often causes adhesions between the parietal and visceral pleura, causing loculation of the pleural fluid, and, following aspiration, there may be loculated air-fluid levels within thick-walled pleural cavities. On a standard chest radiograph it can be very difficult to differentiate this from pulmonary cavitation; ultrasonography, CT, or standard tomography may be helpful, but intrapleural contrast examinations may be necessary to show the cavity and the connections if decortication is considered. Eventually, many, but not all, tuberculous effusions and pleuritis result in pleural calcification (see p. 27) which may be extreme in some individuals, extending over half or more of the involved hemithorax.

Chest Wall

Direct spread of tuberculosis from the lung or pleura to any part of the chest wall does occur, but is extremely uncommon. It is a little more frequent in atypical mycobacterial infections. When there is a tuberculous empyema the ribs can be directly involved (Fig. 34). Infection of the sternum, ribs or spine occurs in children, particularly during the progressive primary type of infection. Hematogenous dissemination accounts for other cases with skeletal involvement (see Fig. 90, p. 121).

In many cases of spinal tuberculosis, there is no pulmonary infection (see p. 85), but an abscess from infected vertebrae can track around the intercostal space and surface alongside the sternum (see p. 116). The vertebral lesion may not be obvious radiologically and radioisotope bone scans or CT or MR may be advisable, or standard tomography where scanning is not available.

Tuberculosis of the ribs or sternum follows the same pattern as skeletal tuberculosis elsewhere (see Fig. 85, p. 117).

Congenital Tuberculosis

Infection of an infant within the uterus or during parturition is rare. However, many infants become infected from their mother within a few days after birth. There are about 150 recorded cases of undoubted congenital tuberculosis. The skin reaction to tuberculin is not always positive and the diagnosis can be difficult. About half these children will die wihin 3 or 4 weeks. It is not known whether this pattern or frequency of congenital infection will change in mothers who are HIV-positive. A chest radiograph of the infant will show a pattern of either miliary disease or primary tuberculosis, and progressive tuberculosis may develop. Some may be born with or develop primary skin tuberculosis. It is possible that the mother will have no radiological and little clinical evidence of tuberculosis.

Immunization with BCG

Many countries with a high incidence of tuberculosis recommend BCG vaccination (bacille Calmette-Guérin, an attenuated but live strain of *M. tuberculosis*), often during the first few days of life. The protection privided varies in different countries: it

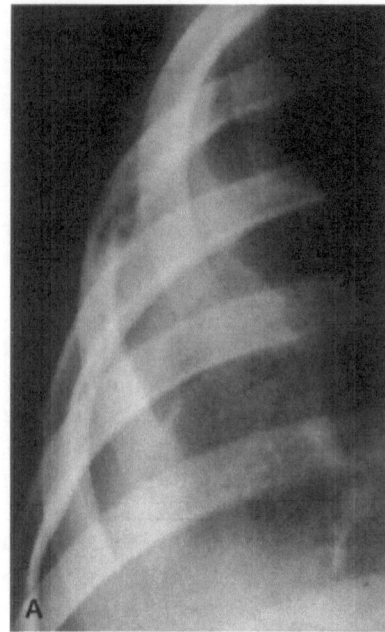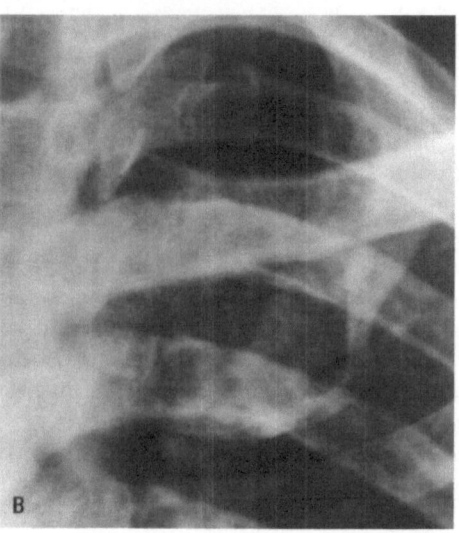

Fig. 34 A, B. Tuberculosis of the chest wall is uncommon. **A** This patient had a tuberculous empyema and the fifth right rib became infected directly from the empyema. There is also some periosteal reaction along the underside of the sixth right rib. **B** There is a lytic lesion in the anterior end of the first left rib which was found on biopsy to be tuberculous. The lung was not infected and no other source of tuberculosis was found. (See also Fig. 85 on p. 117.)

has been excellent in Haiti, but less successful in parts of India and Puerto Rico. The difference in results is not easily explained; sometimes it is due to technical problems with the vaccine and its administration. Another theory suggests that where environmental mycobacteria are common (see p. 4), people are already protected. However, where successful, the main benefit has been to reduce the risk of tuberculous meningitis and acute disseminated infections.

Protection following BCG immunization can last from 10 to 20 years; there are a few complications. Between 3 and 6 months after vaccination, children occasionally develop axillary (or inguinal) lymphadenitis on the same side (Fig. 35). This is probably no more than a classical primary complex, but in some children the nodal swelling is considerable and a very few nodes may suppurate: other children may be generally unwell and pyrexial, but the chest x-ray is almost invariably normal and remains so. Some countries use a more potent BCG vaccine when immunizing neonates, and this has been inadvertently used for older children. Although there was an increased skin reaction, which took longer to heal, there were no other complications and after 6 months the resulting scar was much the same as if the correct strength has been used.

The response to BCG vaccination is very dependent on the state of health and nutrition of the individual. When healthy, a positive reaction will be obtained in 85% of those vaccinated. The BCG response is grossly impaired in children and young adults with kwashiorkor (severe protein deficiency) and very depressed also when the patients are marasmic with kwashiorkor (combined protein and calorie deficiency). However, if the individual is marasmic only (calorie deficient) the successful vaccination rate is the same as in healthy patients. This is important, because the pattern of tuberculosis which may develop in those who have been vaccinated is very dependent on the success or otherwise of the vaccination. After successful BCG vaccination the tuberculin skin reaction should be positive and, if active tuberculosis does develops some years later, it will follow the hyperergic (secondary immune) pattern.

Although BCG osteomyelitis has been recorded, it is rare. The lesion is usually eccentric in the metaphysis of a long bone, breaking the cortex but seldom causing a periosteal reaction. Generalized BCG tuberculosis and tuberculous meningitis are even more rare; these complications may have an increased incidence in HIV+ patients. Intravesical BCG has been used successfully in the treatment of bladder carcinoma: subsequent miliary tuberculosis (recognized on a chest radiograph) has been recorded in two patients. The response to appropriate antituberculous therapy has been good.

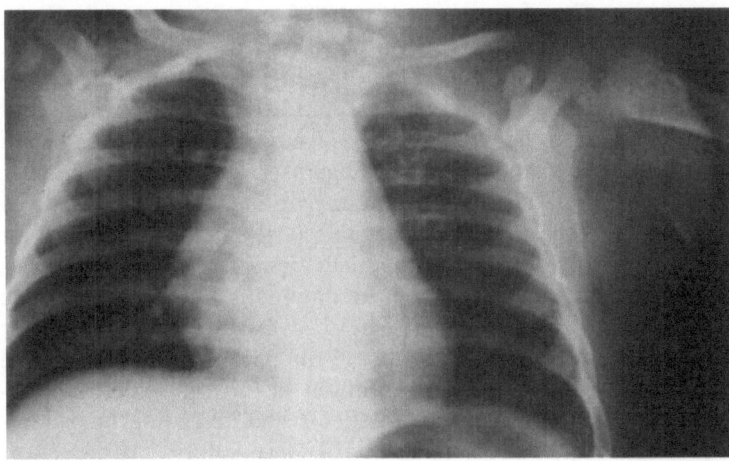

Fig. 35. The chest radiograph of a child who had had BCG vaccination 4 months previously. The chest is normal but the left axillary lymph nodes are considerably enlarged. It is very unusual for there to be any pulmonary abnormality following BCG vaccination

PPD Conversion

When an individual known to have a negative tuberculin skin test (PPD, Heaf or Mantoux) converts "spontaneously" to a positive reaction it is presumptive evidence of a tuberculous infection, although the site of the focus may never be found. A year of isoniazid or other therapy is often recommended. Several large series have shown that routine chest x-rays during this treatment are unnecessary, provided the patient takes the prescribed drugs correctly and continuously for at least 9 of the required 12 months. Only if treatment is discontinued or is inadequate should chest radiographs be taken regularly. Not all authorities agree with this regimen, and in the tropics the conversion may be due not to tuberculosis but to other mycobacteria (see p. 4). There is also some health risk from the chemotherapy, statistically about the same as for an otherwise healthy individual developing active tuberculosis after the tuberculin skin reaction has become positive. The risks of isoniazid-induced hepatitis are age dependent, and many authorities do not recommend its use in patients older than 40 years. Even if patients are not given active prophylaxis, repeated chest radiographs should not be routine but be taken only when indicated by the clinical condition of the individual.

Clubbing

Hypertrophic pulmonary osteoarthropathy, "clubbing," is not now usually associated with pulmonary tuberculosis (Fig. 36). It used to be more common and was a frequent finding in America and elsewhere before antituberculuous therapy was available. However, it may still occur in up to one-third of African men with debilitating pulmonary tuberculosis. It seems to be more common in those who

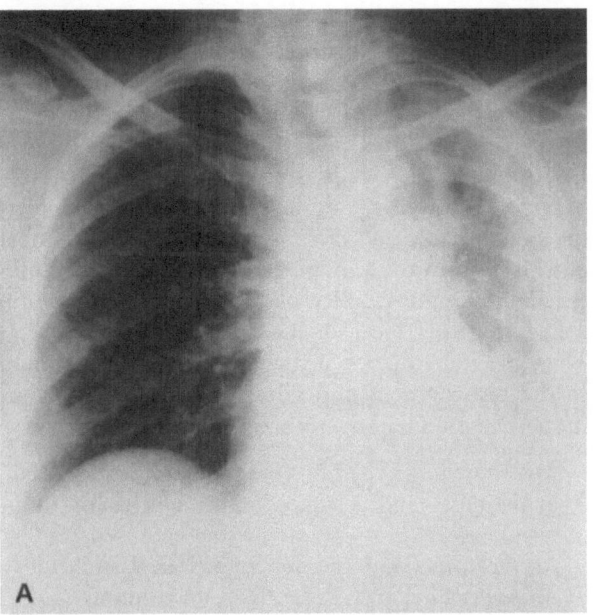

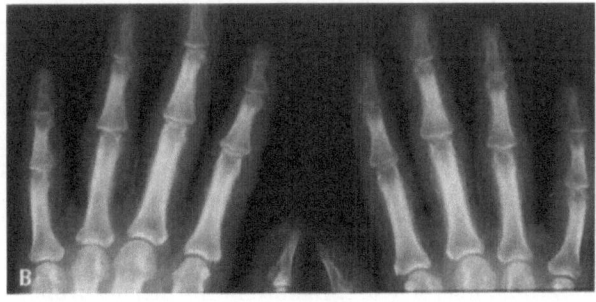

Fig. 36 A, B. Clubbing is much less common in patients with pulmonary tuberculosis than in the past. A This middle-aged patient with chronic tuberculosis, which has caused fibrosis of his left lung and pleura with flattening of the left chest wall, has (B) well-marked hypertrophic pulmonary osteoarthropathy. There is some apical fibrosis in his right lung, but he was not more dyspneic than would be expected, and there was no other reason for his clubbing. (There was no evidence of any bone foci.)

are undernourished but cannot be correlated with the extent of the findings on the chest x-ray or any other likely cause. Whether clubbing is increasing again in pulmonary tuberculosis in other regions is not yet recorded. It is important as a possible explanation for periosteal reaction on the clavicles, femora, digits, or elsewhere.

The Other Mycobacterioses

Nontuberculous mycobacterial infections (MOTT or "mycobacteria other than tuberculosis") cannot be distinguished clinically, radiologically, or pathologically from tuberculosis in the lungs or any other organ system, including bones. Some authorities state that there are about 40 different species; others recognize over 60 species. All agree that about half are potentially able to cause disease in humans. The distinction between *M. tuberculosis* and nontuberculous mycobacteria can only be made in the laboratory and requires selective culture, biochemical, DNA/polymerase chain reactions, and RNA sequencing, as well as serology. Differential skin testing is unreliable and cross-sensitivity occurs; almost all MOTT produce hypersensitivity to tuberculin and confuse the validity of the standard tuberculin skin tests (PPD, Heaf, or Mantoux reactions).

These nontuberculous mycobacteria are all common in the environment and all are acid-fast. The commonest human pathogens are:

- *Mycobacterium leprae* (with reservoirs in man, armadillos, and perhaps some species of monkeys)
- *M. ulcerans* (skin and subcutaneous tissue)
- *M. kansasii* (lungs, bones; can disseminate)
- *M. marinum* (*M. balnei*) (skin granuloma)
- *M. avium-intracellulare* (lungs), *M. scrofulaceum* (lymph nodes) complex: the MAIS complex, formerly the Battey bacillus, can disseminate (see also discussion of *M. paratuberculosis* below)
- *M. xenopi*
- *M. fortuitum* complex (can disseminate rapidly)
- *M. chelonei* (can disseminate rapidly)
- *M. simiae* and *M. szulgai* (both of which are very rare)
- *M. malmoense* (rare: mainly in northern Europe)

Apart from *M. ulcerans*, all have low virulence and poor infectivity and seem to lack person-to-person transmission, but there is evidence that this may be altered by immunosuppression, since all the nontuberculous mycobacterial infections are being seen much more frequently because of the AIDS epidemic. The majority are resistant to most antituberculous drugs, but may be treated by multidrug regimens, often aided by surgery.

The frequency of MOTT in Europe and North America is 1%–3% of all new cases of tuberculosis. In Western Australia the frequency has increased to 16%. The prevalence of tuberculosis is not affected by other mycobacteria, nor do the MOTT bacilli give any protection against tuberculosis. Repeated isolation of the same mycobacterium is required before suggesting that this is the causal organism for any illness. If *M. tuberculosis* is also present, then this is likely to be the dominant organism. Although the incidence of MOTT disease is now showing a marked increase due to the AIDS epidemic, it has in the past been uncommon in most developing countries except when patients had preceding lung damage; for example, MOTT were cultured twice as often in African gold miners who had silicosis compared with those who did not. Despite AIDS, MOTT is still less common in the tropics than elsewhere. The MOTT bacteria may be more important as subliminal infections.

Many MOTT are low-grade pathogens and others exist as saprophytes. The known precipitating factors in infections are chronic lung damage, such as emphysema and bronchitis, industrial lung disease (both mining and agricultural), urban air pollution, and immunosuppression. All MOTT (except *M. ulcerans*) can rarely cause pulmonary lesions. *M. scrofulaceum* may cause cervical lymphadenitits in children under 12 years old. *M. kansasii* and *M. avium-intracellulare* often affect males over the age of 45. *M. ulcerans* infects mainly the panniculus; *M. marinum* (balnei) also affects the skin, as do *M. fortuitum* and *M. chelonei* (see also p. 104).

There have been many descriptions of the radiological appearances of different MOTT pulmonary infections, and there are some small differences, but as already noted, it is not possible to distinguish radiologically between the mycobacteria. When an otherwise healthy patient with an apparently straightforward tuberculous infection fails to respond to adequate therapy, the commonest cause is drug resistance and the alternative is infection with a nontuberculous mycobacterium.

Mycobacterium paratuberculosis (Johne's bacillus). Within the *M. avium-intracellulare* complex is *M. paratuberculosis*, which is very common in the intestine of many animals, including cattle, sheep, and primates. It has been cultured from humans with chronic intestinal inflammation, resembling Crohn's disease. It is very difficult to recognize by culture and in some instances cannot be cultured at all. It is very resistant to most antituberculous drugs.

Its incidence in the tropics is not documented, but it has caused cervical lymphadenitis and chronic granulomas of the small intestine, closely resembling those found in the animal infection. Radiologically, a cobblestone appearance of the mucosa, with segmental narrowing of the terminal ileum, has been reported. Human infection is acquired from drinking infected milk from animals which may be apparently well and are subclinically infected. The AIDS epidemic, lowering the human immunity, opens the way to more widespread infection by this bacillus.

Tuberculosis of the Alimentary Tract

Despite the overall prevalence of tuberculosis in many parts of the tropics, there is no part of the world where tuberculosis commonly affects the gastrointestinal system. In the past, before drug therapy became available and patients died of advanced tuberculosis, about 80% were found to have abdominal tuberculosis at autopsy; the more advanced the pulmonary tuberculosis, the more likely there was to be bowel infection. However, now that treatment has become possible, the correlation of pulmonary tuberculosis with abdominal tuberculosis has altered, and less than 50% (in some series 25%) of patients with abdominal tuberculosis have pulmonary tuberculosis also. This ratio may again alter significantly with AIDS; for example, already in Haiti patients with AIDS are 3.5 times more likely to have extrapulmonary tuberculosis than patients who are HIV-negative.

Tuberculosis in the bowel starts as a localized inflammation of the lymphoid tissue and progresses to necrosis. The reaction can be ulcerating or hypertrophic or both. Tuberculous ulcers in the bowel are often transverse and linear, rather than round. Throughout the alimentary tract, tuberculosis forms granulomas which may be demonstrated as a mass, as distortion due to fibrosis, or as a stricture. When the infection has healed, there may be a residual scar, adhesions, or an ulcer on the damaged tissue.

Except in tuberculous peritonitis (p. 67) there are no clinical or laboratory criteria which are of any significance other than the histology. Stool culture for tuberculosis is unreliable. The symptomatology is vague and cannot be related satisfactorily to the abdominal lesions in the majority of patients. Ill health, vomiting, diarrhea (30%), and constipation (20%) are all generalized symptoms. Malabsorption is not uncommon. Analysis of the blood is normal except for anemia, which is very common in tropical countries and is more likely to indicate the presence of parasites rather than tuberculous bowel. There is no close relationship between pulmonary and alimentary tuberculosis. Many patients will have active lung disease, but a normal chest x-ray does not exclude tuberculosis in the alimentary tract.

The development of imaging by scanning (ultrasonography, CT, or MRI) has provided a better understanding of the pathophysiology and clinical pattern of tuberculosis in the abdomen. The diagnostic index of suspicion for tuberculosis has been raised, but unfortunately this is all that imaging can do; there are no specific changes for any form of intra-abdominal tuberculosis. Nevertheless, all current methods of imaging provide very useful guidance and are invaluable in follow-up during or after treatment.

The two most common alimentary forms are peritoneal and cecal tuberculosis; other sites are less frequently affected. But again, it is probable that the incidence and pattern will change in AIDS patients, particularly late in their disease, adding to the already difficult problem of differential diagnosis.

Tuberculosis of the Esophagus

There are three ways in which tuberculosis may present in the esophagus: (1) as a long stricture; (2) as gross dilatation above a narrow stricture, and (3) as ulceration, with or without a diverticulum. A tuberculous granuloma presenting as a mass in the esophagus does not appear to have been reported, but is a theoretical possibility.

The common site of infection is in the upper half of the esophagus (Fig. 37). The patient may present with difficulty in swallowing or with sudden impaction of food and acute dysphagia. Intrinsic tuberculosis of the upper third of the esophagus must be differentiated from the traction and fusiform pseudodiverticula which may be caused by fibrotic tuberculosis in the lung apices. In these patients the trachea is always deviated. Many texts state that traction diverticula in the mid-third of the esophagus are the result of adhesions from tuberculous mediastinal lymphadenopathy. None of the authors of "The Imaging of Tropical Diseases", who together have considerable tropical and tuberculous experience, have ever seen this occur; it may happen, but it is not the cause of most esophageal diverticula. There are many thousands of patients with tuberculous mediastinal lymphadenopathy and very few with esophageal diverticula.

Tuberculosis may cause a hard nodular swelling in the thyroid, which often grows quite rapidly; it is only later in the disease that an abscess develops, either in the neck or, rarely, behind the sternum. A mediastinal tuberculous abscess spreading from nodal or vertebral infection also may cause pressure displacement of the esophagus but a pyogenic abscess or an intrathoracic goiter must also be considered. Tuberculous infection of the esophagus has been known to cause both bronchoesophageal fistu-

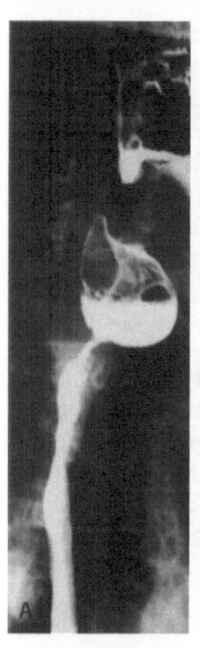
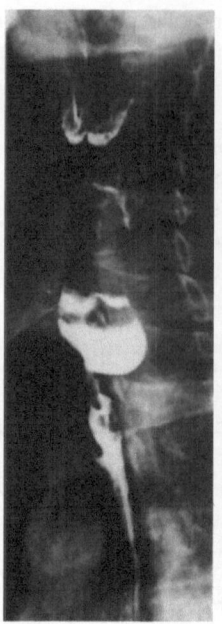
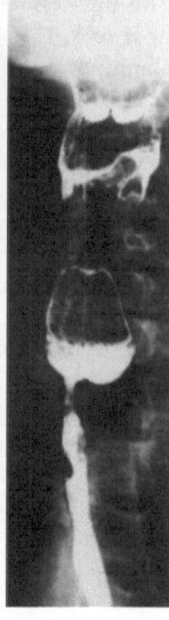

Fig. 37. Tuberculosis of the esophagus. **A** A stricture in the upper third of the esophagus which has all the characteristics of a carcinoma: only the histology can provide the diagnosis. (Courtesy of Dr. Jack Farman, Great Neck) **B** Mucosal edema and ulceration in the upper third of the esophagus of an Indian: he suffered from pain and dysphagia. This type of tuberculosis may eventually perforate (**C**) into the mediastinum but will still heal satisfactorily with the correct treatment. **D** A similar infection which has formed a pseudodiverticulum posteriorly in the lower third of the esophagus

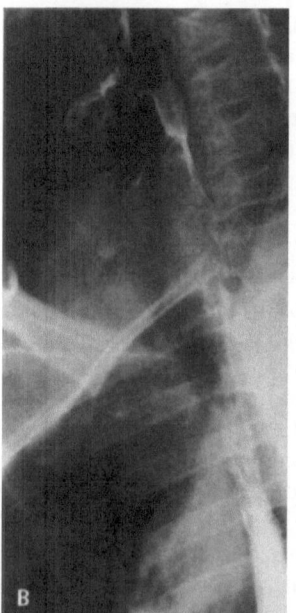
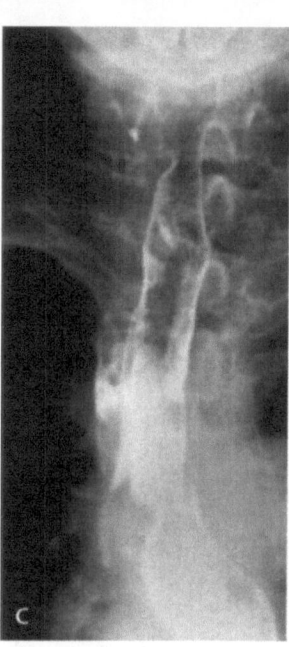
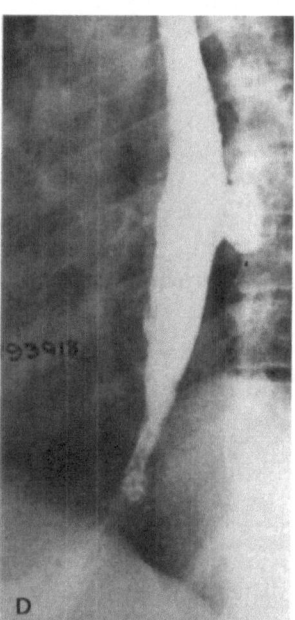

lae and erosion into an aortic aneurysm. These complications may become more common with AIDS.

Long-segment esophageal strictures with irregular granulations on the surface have to be differentiated from caustic burns, inflammatory or fungal esophagitis, and malignancy (Fig. 38). Carcinoma of the esophagus in the tropics may be multifocal and extend over 15 cm (6 inches) or more. Short-segment tuberculous strictures involving the lower third of the esophagus have to be differentiated from achalasia, peptic esophagitis and neoplasm; tuberculosis is a rare cause, and such differentiation is often impossible radiologically.

Tuberculosis of the Stomach

Tuberculosis of the Stomach is a very uncommon form of tuberculosis, although it may become more frequent in AIDS patients. However, it is unlikely to become easier to diagnose either clinically or radiologically. The clinical history and the imaging appearance may suggest simple peptic ulceration, other granulomatous diseases, lymphoma, or carcinoma of the stomach. Even at surgery the macroscopic appearances can resemble malignancy and the differential diagnosis may only be made histologically.

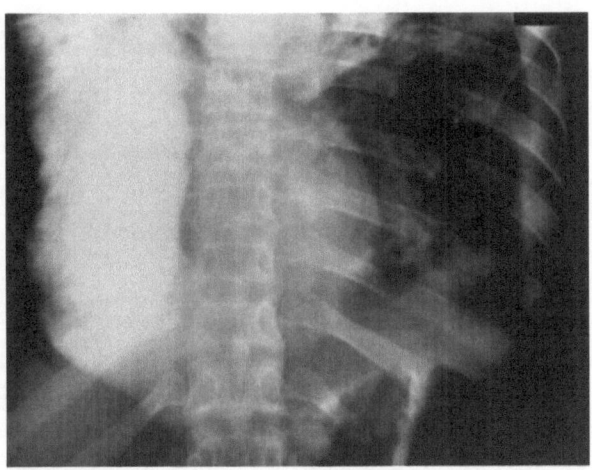

Fig. 38. Tuberculosis of the lower end of the esophagus causing a stricture just above the cardia. The huge dilated esophagus lies on the right side of the mediastinum and closely resembles the esophagus in achalasia or Chagas' disease. The patient had active pulmonary tuberculosis and a left pleural effusion, but histological examination of the esophagus provided the correct diagnosis. (Courtesy of Ms. A. A. Whitemore, Harare)

Radiologically, tuberculosis of the stomach may present as shallow ulceration, (which can be extensive), a granulomatous mass or as fibrosis (Fig. 39). The most common of these is ulceration at the pylorus, which causes gastric outlet obstruction. It resembles and often cannot be distinguished from fibrosis following peptic ulceration or from malignancy. Tuberculosis is yet another cause of gastric dilatation, but is sufficiently rare as to be a surprise to both the radiologist and the surgeon when histology is obtained. Tuberculous gastric ulceration has been reported on both the greater and lesser curvatures off the stomach, usually surrounded by induration and therefore reducing peristalsis.

These ulcers are usually shallow but can be quite large. The likely mistaken radiological diagnosis will be lymphoma or carcinoma.

The tuberculous granulomatous mass in the stomach is the least common presentation of this uncommon form of tuberculosis. With such a mass, there is always associated mucosal ulceration and surrounding fibrosis. The ulcers are chronic and will be mistaken for malignancy.

The shape of the stomach (and duodenum) can be distorted by extrinsic pressure from enlarged tuberculous lymph nodes (Fig. 40, 41). These usually compress on the distal third of the stomach and in some cases are adherent. In one such patient there was a very large gastroesophageal mass at the cardia causing distortion, with an overlying gastric ulcer. Although ultrasonography, CT or MRI can de-

monstrate the lymphadenopathy, the correct diagnosis is not likely to be achieved by imaging.

Fibrotic tuberculosis can mimic linitis plastica, scirrhous carcinoma, infiltrating lymphoma, or even syphilis. To this differential diagnosis must be added Crohn's disease, which is rare in the tropics.

Tuberculosis of the Duodenum and Small Intestine (Fig. 41, 42)

The clinical presentation in most cases is of upper small bowel obstruction, but occasionally the obstruction is lower, in the ileum. There is usually a history of intermittent attacks of abdominal pain, with nausea and sometimes vomiting; when the small bowel is involved, malabsorption and malnutrition occur more frequently. Eventually the obstruction becomes persistent.

Abdominal radiographs in the supine and erect positions will show dilated loops of small bowel and fluid levels. In one series from India, there were dilated loops of bowel in 54% of patients with abdominal tuberculosis and, in over half of them, there were fluid levels on the erect radiographs. Tuberculosis causes different patterns of small bowel disease. Infiltrating tuberculous granulomas around the duodenum result in loss of normal mucosal pattern, rigidity, and narrow strictures; a stricture of the third part of the duodenum (particularly in India) should raise the suspicion of tuberculosis. In the ileum, the Peyer's patches are affected and transverse ulcers develop with typical undermined edges. In the early stages, when there is minimal dilatation of the ileal loops with excess fluid in the lumen and no significant stricture, the barium forms bubbles with ill-defined edges; later as the strictures become well established, the barium column is trapped proximal to the stricture, and outlines the stricture and dilated lumen. The dilated ileum can reach mammoth proportions. It may be necessary to introduce a peroral small intestinal tube for a selective small-bowel enema to demonstrate the stricture. These strictures are most common in the ileum but may be seen in the jejunum and even in the duodenum. Tuberculous strictures are not as long as those of regional enteritis, and in addition, as noted earlier, Crohn's disease is extremely rare in India and many other tropical countries. Occasionally in tuberculosis there are fistulae in addition to the strictures, and the differential diagnosis then becomes impossible without histological examination. Calcified enteroliths are commonly seen within dilated loops of small bowel proximal to tuberculous strictures, but are uncommon in any other chronic enteritis (see Fig. 42 F and p. 67).

(Text continues on p. 58)

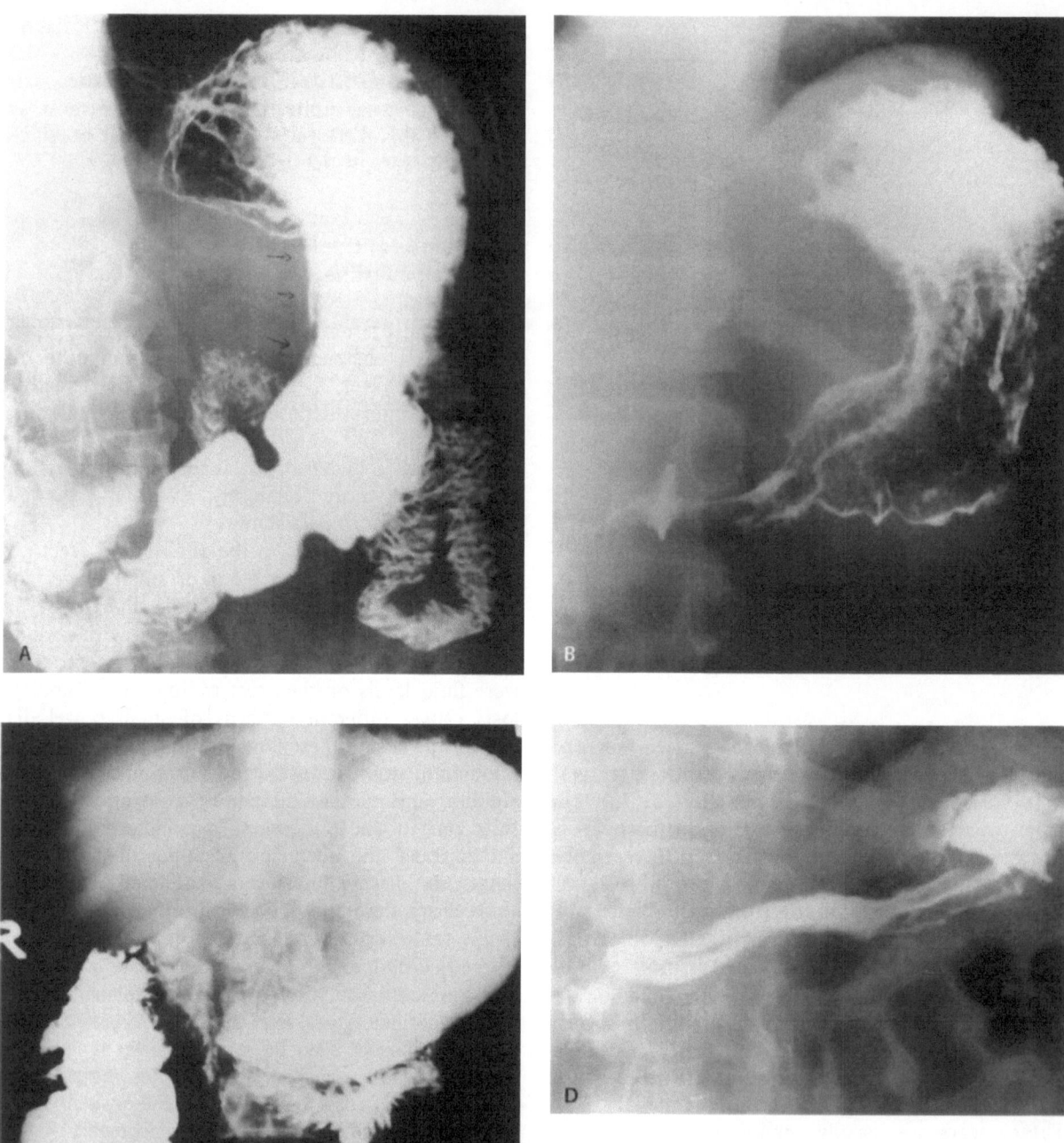

▲ **Fig. 39 A–D.** Tuberculosis of the stomach. **A** Mucosal ulceration and edema along the lesser curvature of the stomach of a patient from India. **B** Tuberculous gastritis causing thickening of the wall of the stomach and narrowing of the pyloric antrum of an African patient. **C** Granulomatous tuberculosis thickening and immobilizing the lesser curvature of the stomach from below the cardia down to the pylorus. In each of these patients the radiological diagnosis is likely to be carcinoma or lymphoma because tuberculosis of the stomach is rare. **D** Tuberculosis of the stomach resembling linitis plastica. The stomach would not distend beyond this size and peristalsis was absent. The radiological diagnosis was malignancy: it may be impossible even at surgery to distinguish tuberculosis without histological confirmation

Fig. 40 A–F. Pressure on the stomach and duodenum from ▶ tuberculous lymphadenopathy. **A** Displacement of the second part of the duodenum downwards and the third portion upwards. An African patient. **B** Distortion of both the lesser and greater curvature of the stomach and the pyloric antrum, with stretching of the pylorus and partial obstruction of the second and third parts of the duodenum. **C** Persistent pressure on the great curvature of the stomach which was thought to be due to carcinoma of the pancreas. The patient complained of abdominal swelling and pain. At surgery there was a large inoperable "tumor" which was found on histology to be due to active tuberculosis. Where it had originated remained undecided, but the patient responded to appropriate treatment. (Courtesy of the University of Capetown, Radiology Library) Continued on p. 53

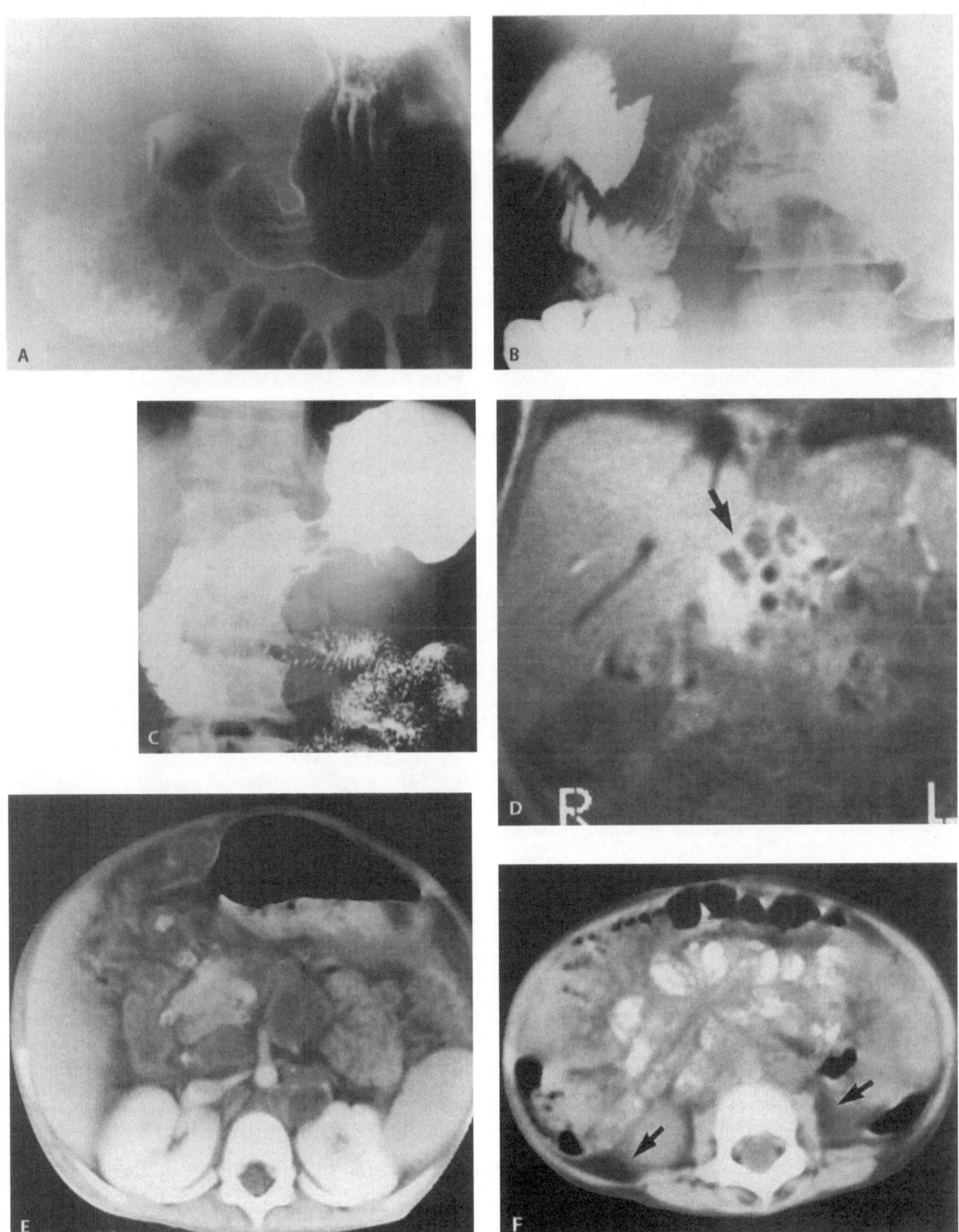

Fig. 40 (*continued*). **D** Intravenous gadolinium-enhanced T1-weighted MR scan showing an enhanced nodal inflammatory mass (*arrow*) with areas of nonenhancing necrosis. **E** A contrast-enhanced CT scan showing multiple rim-enhancing nodes in the periaortic region. **F** A CT scan of a child from South Africa shows diffuse calcification in enlarged mesenteric lymph nodes. There is also some ascites (*arrows*). (**D–F** from Cremin and Jamieson 1995)

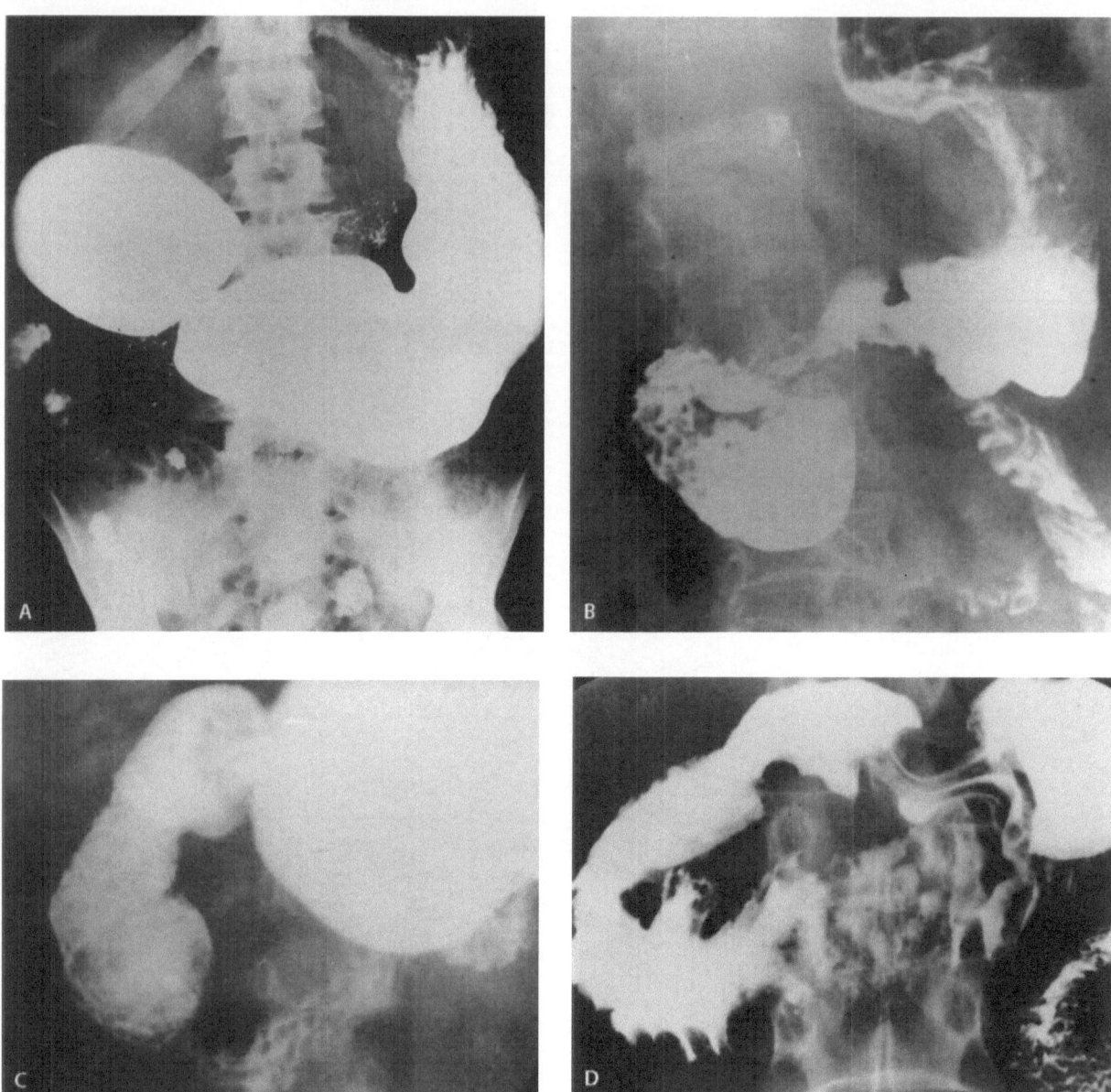

Fig. 41 A–H. The pressure from tuberculous lymph nodes may eventually cause obstruction in the duodenum. **A** A large duodenal bulb due to obstruction of the second part of the duodenum, caused by fibrosis following tuberculous lymphadenitis. **B** A stricture at the junction of the second and third parts of the duodenum of a 6-year-old child from India, which caused partial obstruction. This was the result of adhesions from previous tuberculous lymphadenitis. Both A and B could be misdiagnosed as the result of congenital bands (Ladd's syndrome). **C** Pressure on the second part of the duodenum and obstruction at the junction of the third and fourth parts by enlarged lymph nodes and tuberculous duodenitis. **D** Tuberculous mucosal ulceration and edema of the second, third, and fourth parts of the duodenum. This could be mistaken for lymphoma or even a severe parasitic infection. With proper treatment this condition may heal completely, or result in strictures. E–H see p. 55

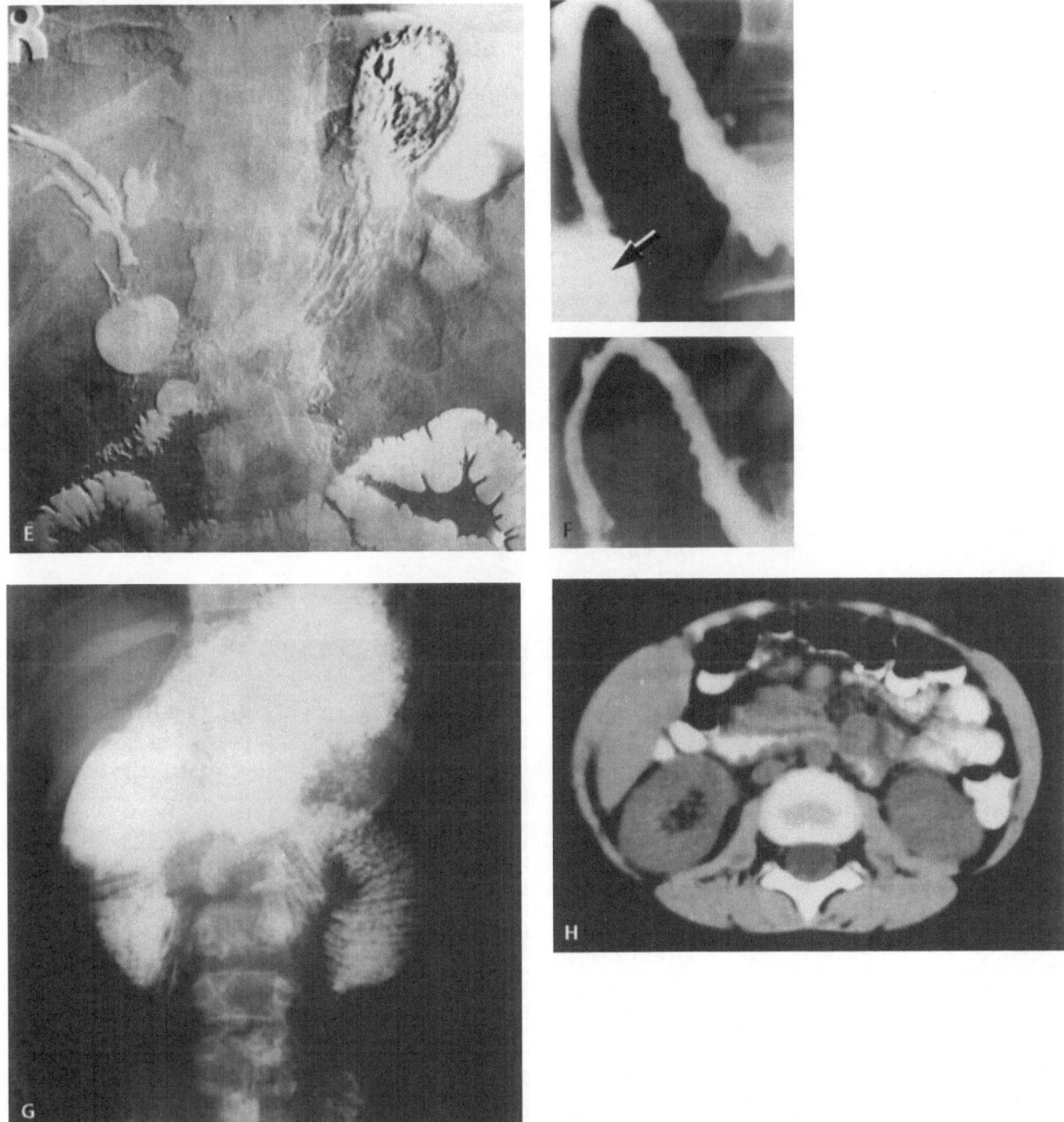

Fig. 41 (*continued*). **E** Perforation of a tuberculous ulcer of the second part of the duodenum into the biliary tract, with reflux of bile. (There are two incidental duodenal diverticula.) **F** Intense spasm due to tuberculous ulceration of the first and second parts of the duodenum of an Arab patient. Tu-berculosis was confirmed histologically. (The *arrow* points to an artefact.) **G** A funnel-shaped stricture in the proximal jejunum of a 12-year-old Arab female. **H** A CT scan shows pressure on the transverse duodenum from lymphadenopathy. There was also an ileal stricture which is not demonstrated

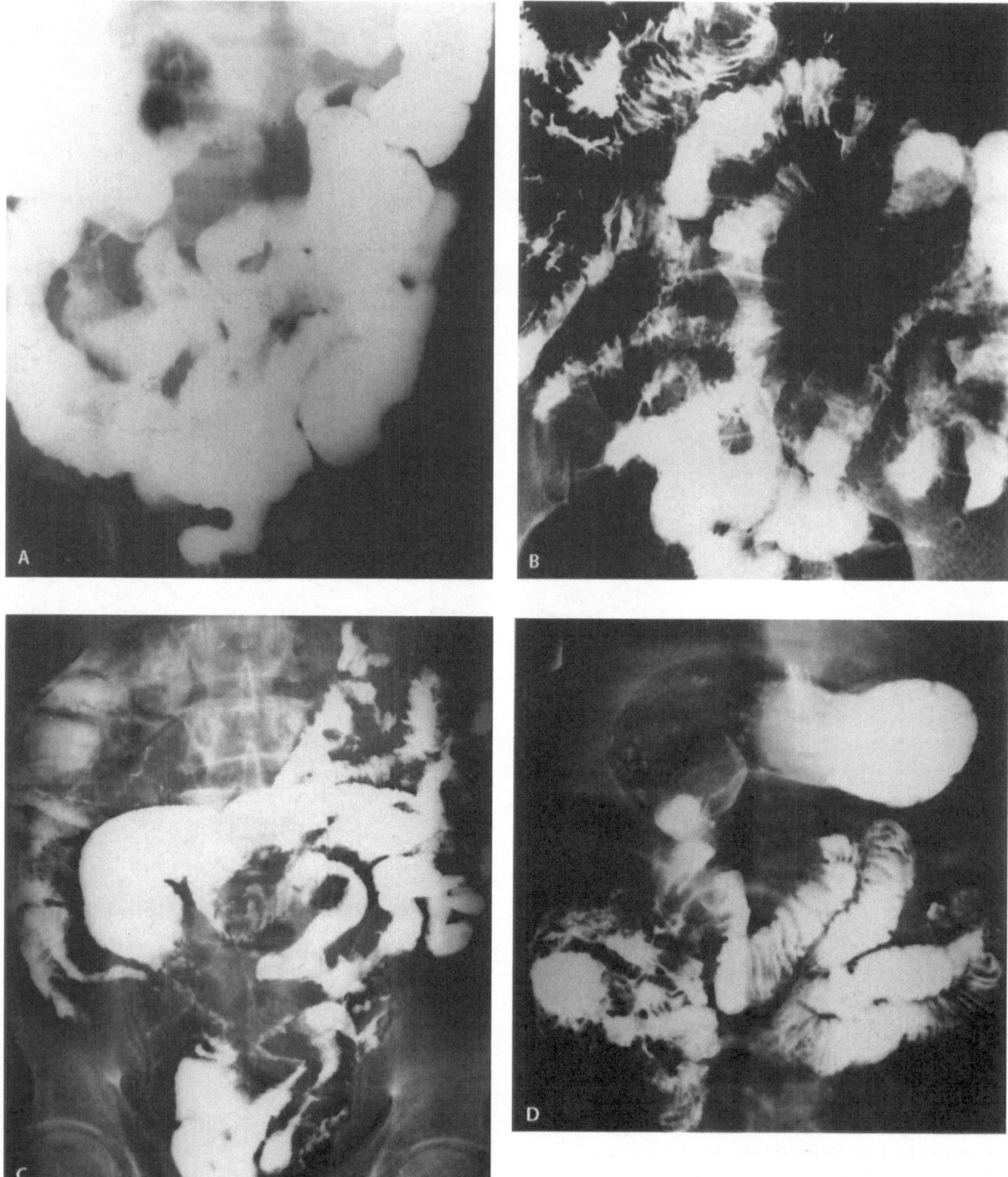

Fig. 42 A–H. Tuberculosis of the small intestine. **A** The acute phase with loss of normal bowel pattern, excess fluid within the gut, and rapid transit of barium. **B** Multiple granulomatous strictures and dilated lengths of bowel. **C** Multiple lengths of dilatation and strictures with fistula formation, indistinguishable from regional enteritis except by histology. **D** A tuberculous duodenal-colic fistula in a 7-year-old child with malabsorption (incidental stomach rotation). **E–H** see p. 57

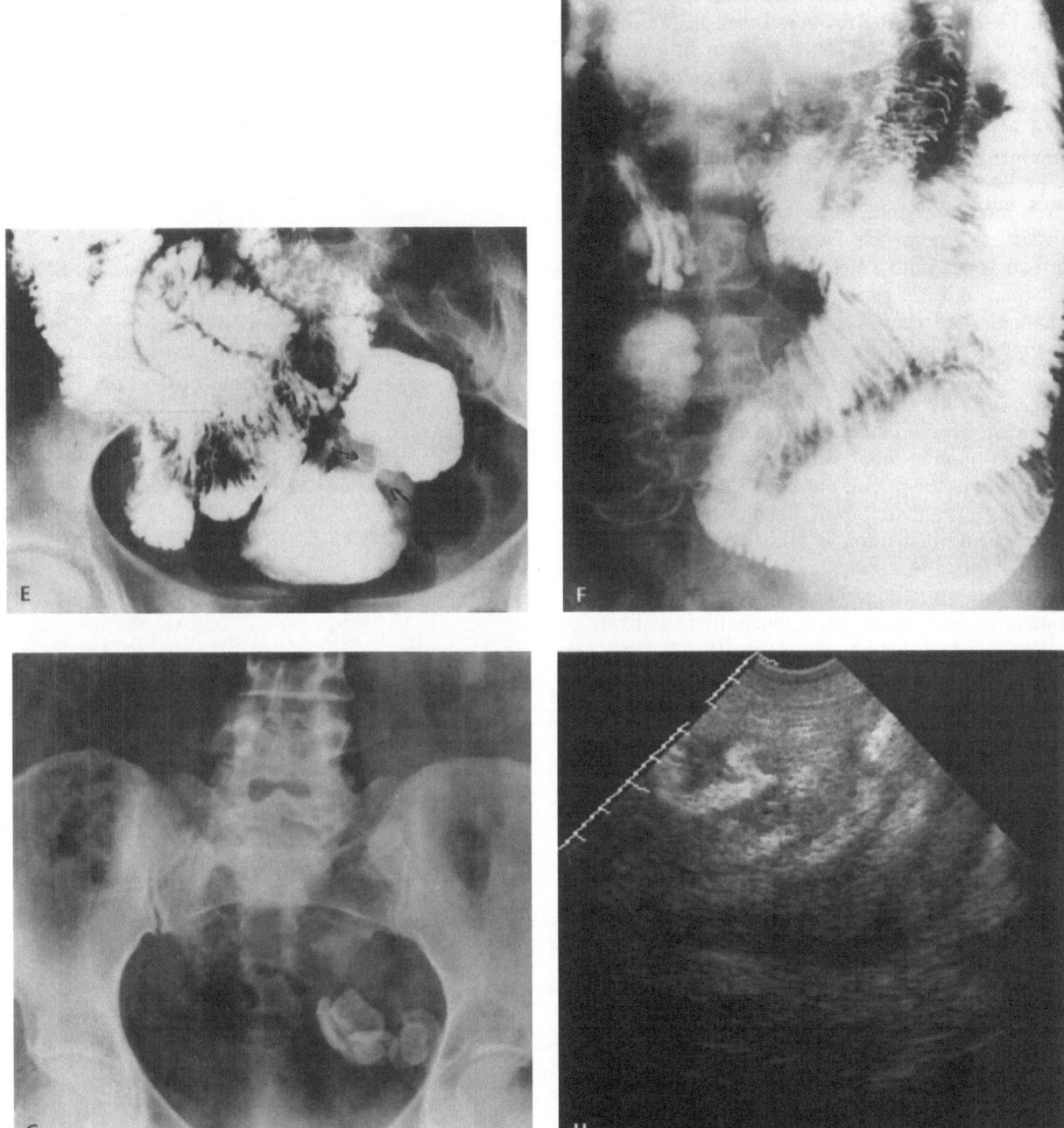

Fig. 42 (*continued*). **E** A narrow "apple core" stricture in the ileum (*arrows*). **F** Partial obstruction of the mid-ileum due to tuberculous granulation tissue. **G** Enteroliths in the small bowel of a patient from India, with multiple strictures in the small bowel from tuberculosis. **H** Ultrasonography can be used to show the thick wall of tuberculous small bowel and will often demonstrate that the bowel loops are adherent of clumped together. This patient was an African male, aged 43 years, from Zimbabwe. (**B** courtesy of the University of Cape-town Radiology Library; **C** AFIP 229499; **D** courtesy of Dr. D. Makanjuola and *Eur J Radiol*, 1998; **H** courtesy of Dr. Sam Mindel)

When scanned by ultrasonography, CT, or MRI, the bowel lesions demonstrate the same imaging pattern. There is concentric bowel wall thickening, often with ulceration; this may be seen in any region of the small bowel or even into the colon. There is also narrowing of the lumen, which can be correlated with barium studies. It is not always easy to demonstrate such strictures by scanning. The bowel loops become matted together and adhesions, exudates, inter-loop fluid, or abscess may form complex masses. A "sandwich" appearance has been described when fluid collects close to the bowel wall and there is focal ascites; this results in alternating echogenic and echo-free layers through the wall of the bowel, showing the serosa, the fluid, and then the pattern of the adjacent bowel wall. Omental thickening (described as "caking") can also be demonstrated; a tuberculoma of the omentum has been recorded. Some loops of bowel will be dilated, others more normal or narrowed. Bowel wall thickness of more than 5 mm during contraction and 3 mm during distention is said to be abnormal, but whether this also applies in patients who have malabsorption or added parasitic infection is not known. In abdominal tuberculosis there is likely to be concomitant lymphadenopathy (p. 67), and any part of the small bowel, but especially the duodenum and jejunum, can be distorted and displaced by enlarged tuberculous lymph nodes. Adhesions are frequent and the mucosal folds may be stretched over the nodes. It is unusual for complete obstruction to result from this type of extrinsic lesion.

Tuberculosis of the Cecum
(Hyperplastic Tuberculosis of the GI Tract)

Tuberculosis of the cecum is one of the most common types of intra-abdominal tuberculosis, and in some series it accounts for up to 60% of all tuberculous bowel infections. (The other common sites are the peritoneum and lymph nodes.) The clinical symptoms are vague. There may be alternating diarrhea and constipation, changes in bowel habit, a palpable mass and vague crampy pain in the right iliac fossa, general ill health, and anemia, but all are nonspecific. When tuberculosis affects the cecum and distal ileum (Fig. 43), initially there will be intense ileocecal spasm seen on barium studies or small bowel examination (Stierlin's sign). Later, the lumen and size of the cecum are compromised as the granulomas form into a mass and undergo fibrosis. The mesocolon of the cecum contracts; the cecum is pulled up out of the right iliac fossa and shrinks (Figs. 44, 45), often drastically, culminating in the classical conical, pyramidal, or at times, pear-shaped configuration. The associated fibrosis causes

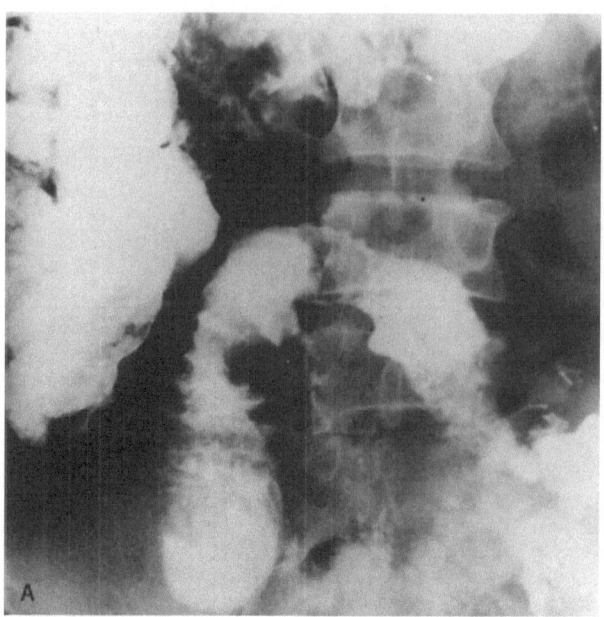

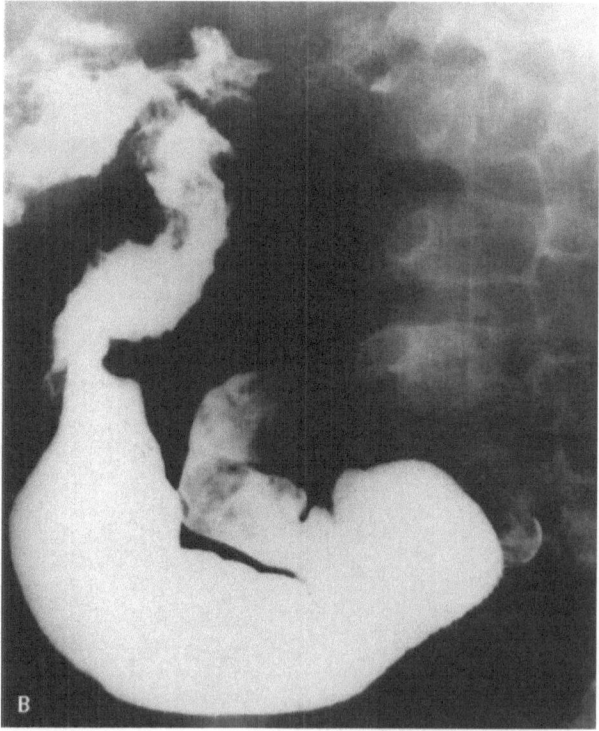

Fig. 43 A–E. Tuberculosis of the terminal ileum and cecum. A Dilatation, edema, and ulceration of the terminal ileum. B The contracted cecum which is the result of tuberculosis in a five-year-old child from India. There is marked adynamic dilatation of the terminal ileum, mucosal ulceration, and loss of the normal ileal pattern. C–E see p. 59

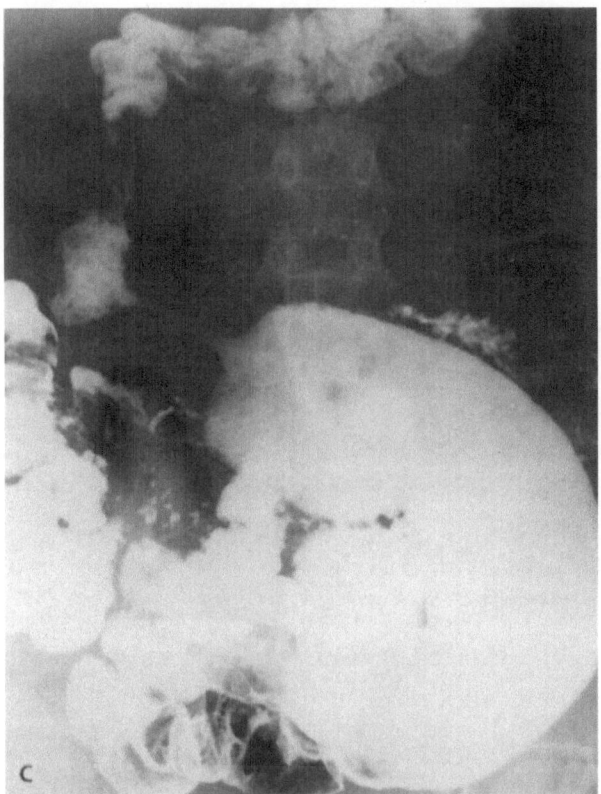

a partial stricture at the ileocecal region and the dilated terminal ileum appears to be suspended from the cecum like a pendulum. There is restricted movement of the cecum on palpation, not only transversely but also in the craniocaudal axis. Sinus formation is uncommon unless there has been inadequate surgical interference; sinus formation and osteomyelitis of the ilium suggest a mycotic infection, usually actinomycosis, not tuberculosis.

Scanning by any method, but particularly ultrasonography will show thickening of the bowel wall in the ileocecal region and in the ascending colon (Fig. 46, p. 62). In many patients the right iliac fossa may be comparatively empty because the cecum has been drawn up by fibrosis. When thickening occurs, it is nearly always concentric but may be distorted by the peritoneal adhesions. Ulceration in the ileocecal area is always associated with bowel wall thickening and may be superficial or deep. The correlation between scanning and barium studies is good, but the typical appearance as shown by barium enema may be more diagnostically accurate. The irregular pattern of spicules and progressive narrowing, together with displacement of the cecum, is likely to be more marked than in amebiasis.

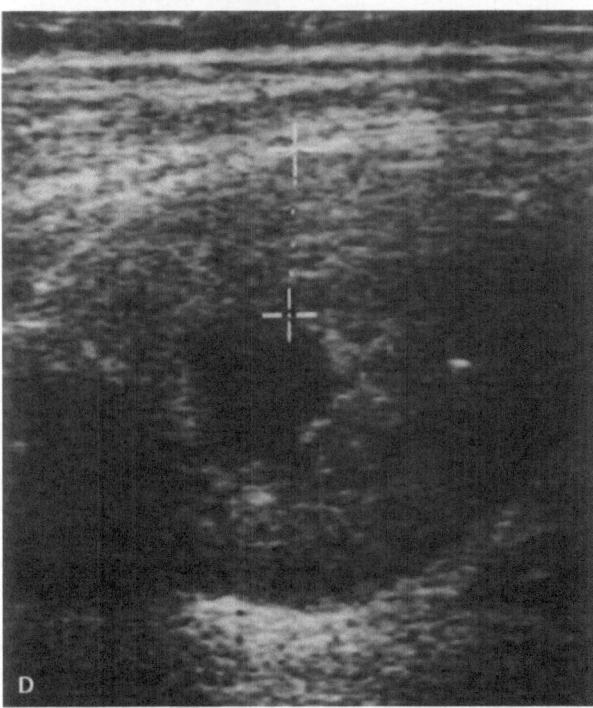

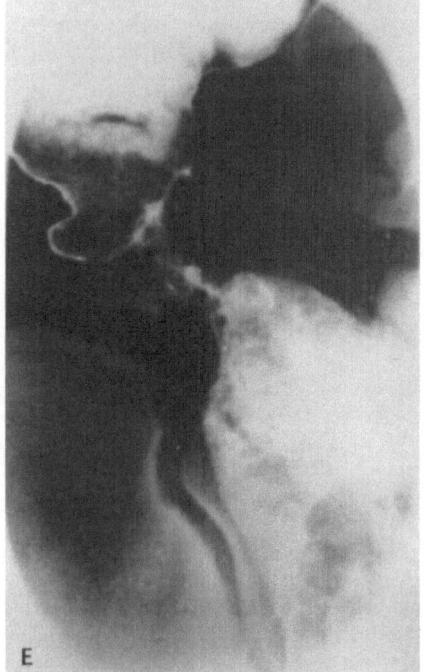

Fig. 43 (*continued*). C Gross dilatation of the terminal ileum above a long stricture proximal to a contracted tuberculous cecum. D The sonogram of the thickened cecum of a child from the Pacific Islands with ileocecal tuberculosis. E A tuberculous ulcer at the ileocecal junction, causing marked constriction and distortion, with resulting obstruction and dilatation of the ileum above it. (D courtesy of Dr. Cheryl Sisler, Hawaii)

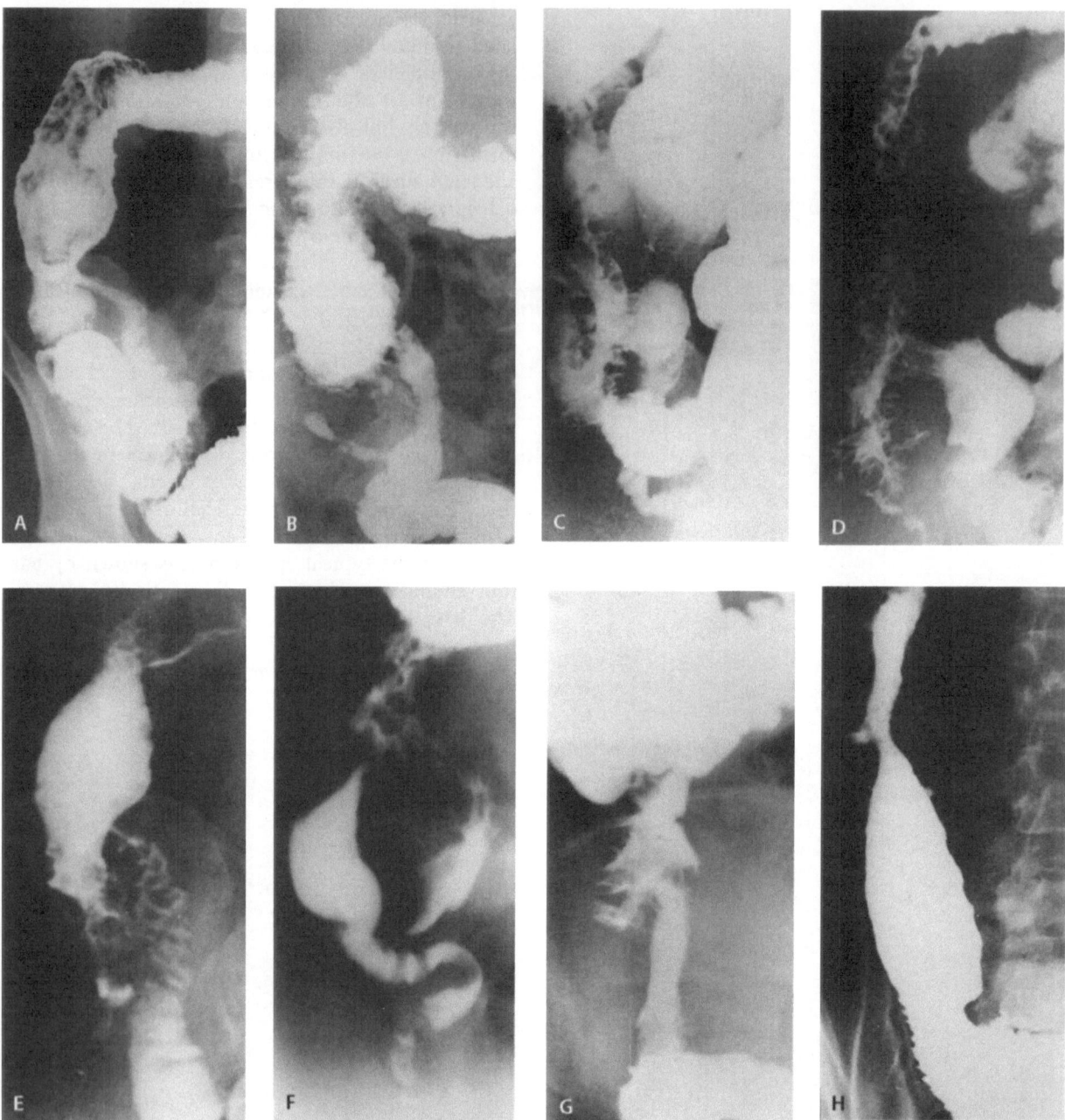

Fig. 44 A–H. Tuberculosis affects the cecum in different ways, but the end result is almost always contraction and considerable distortion. These figures illustrate eight different cases from different parts of the world. **A** Early mucosal ulceration and edema, with shrinkage. The infection has caused ulceration of the proximal half of the transverse colon. **B** Mucosal edema and ulceration of the right side of the colon and of the terminal ileum, which is beginning to dilate. The appendix is also infected. **C** The cecum is contracted, grossly edematous, and ulcerated, and the terminal ileum is dilated. **D** Gross and extensive mucosal edema with nodulation and spiculation of the cecum and ascending colon. The ileocecal valve is patent and the terminal ileum is also edematous and nodular. **E–H** Four different examples of shrinkage of the tuberculous cecum. In **E** and **F** the terminal ileum is not affected or even obstructed, but the cecum is much reduced in size. In **G** there is some ulceration of the terminal ileum, but only minimal obstruction and mild dilatation. In **H** the cecum is very small, and the ileum is dilated and almost seems to hang from the small narrow cecum

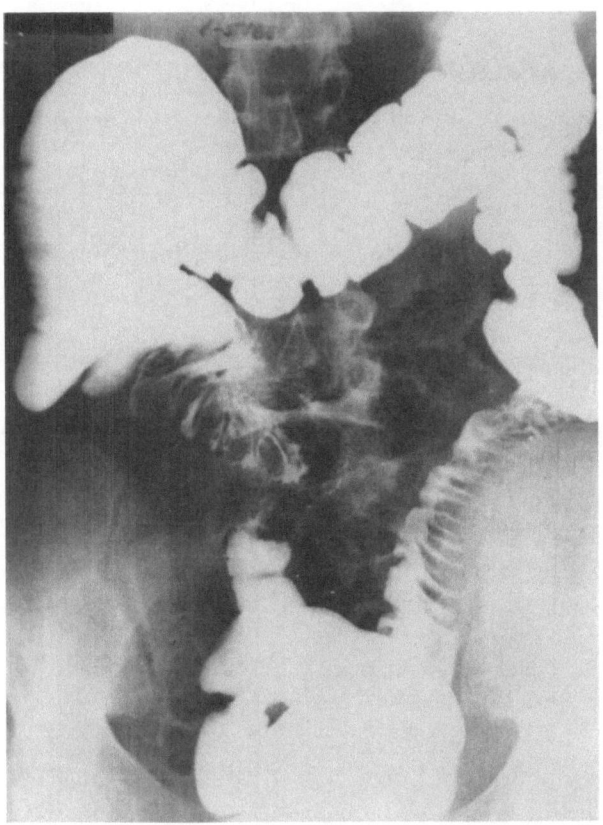

Fig. 45. Tuberculosis often affects more than one part of the bowel. This patient had a palpable mass in the right iliac fossa which clinically suggested malignancy. The barium enema showed that the cecum was displaced upwards and both the cecum and terminal ileum had edematous, swollen mucosa. However, there were marked mucosal changes in the sigmoid colon also, which led to the alternative diagnosis of amebiasis. Tuberculosis was not considered but the combination of fixation of the ileocecal junction, the contracted cecum, and small bowel adhesions is typical of tuberculosis. The palpable mass was due to enlarged lymph nodes

In children in particular, the ileocecal valve may be so thick that it mimics a mass in the cecum; it remains patent because of the fibrosis. Oral contrast is helpful with CT scanning to show the thickness of the bowel wall and narrowing of the lumen. However, in many cases only histology will provide the correct diagnosis.

The differential diagnosis includes an appendiceal abscess, amebiasis, actinomycosis, granulomatous colitis, and malignancy. Tenderness on palpation is maximal with appendiceal abscess, tethering of the cecum is maximal in tuberculosis, and sinus formation and adjacent osteomyelitis are most common in actinomycosis. Reflux into the ileum due to loss of ileocecal sphincter action is easily demonstrated in amebiasis and malignancy, in spite of the tumor; reflux is less common in tuberculosis in the early

stages because of spasm, and later because of fibrosis. Amebic infection can present as granulomatous disease also, but the fibrosis accompanying tuberculosis is more severe than that seen in amebiasis. Multiple granulomas and strictures may also occur in amebic infection, but in over 90% of patients with amebiasis only the last 6 inches of small bowel are involved, whereas in tuberculosis and Crohn's disease the terminal ileum is involved in 45%–50% or more of patients with cecal disease. Lymphoma will have to be considered, but carcinoma, granulomatous colitis, and ischemic colitis are all rare in most tropical countries.

Tuberculosis of the Large Intestine

When tuberculosis affects the large bowel, the clinical symptoms vary from constipation to diarrhea. The presentation may be incomplete obstruction; there is seldom pain, but the majority of patients are in general ill health (though a small number are surprisingly well).

Radiologically and endoscopically, tuberculous granulomas present in the colon either as colitis or stricture; both can be multifocal. In the early stages the mucosa may be ulcerated (Fig. 47). There may be aphthoid ulcers (with central flecks of barium, some surrounded by a translucent halo on a normal mucosa). Short lengths are most commonly affected; tuberculosis does not involve the colon extensively, but in some cases it is associated with small bowel infection.

The radiological differential diagnosis can be very difficult and tuberculous colitis cannot be distinguished from amebiasis (Fig. 48); however, aphthoid ulcers are more commonly associated with Crohn's disease or amebiasis. Tuberculous granulomas cause strictures quite indistinguishable from an ameboma and both tuberculosis and amebiasis can also mimic the apple-core stricture of carcinoma of the colon, producing the same chronic, progressive obstruction. Radiological differentiation may be impossible; only histology can be accurate. Fortunately, malignant disease of the large bowel is uncommon in much of the tropics.

(Text continues on p. 65)

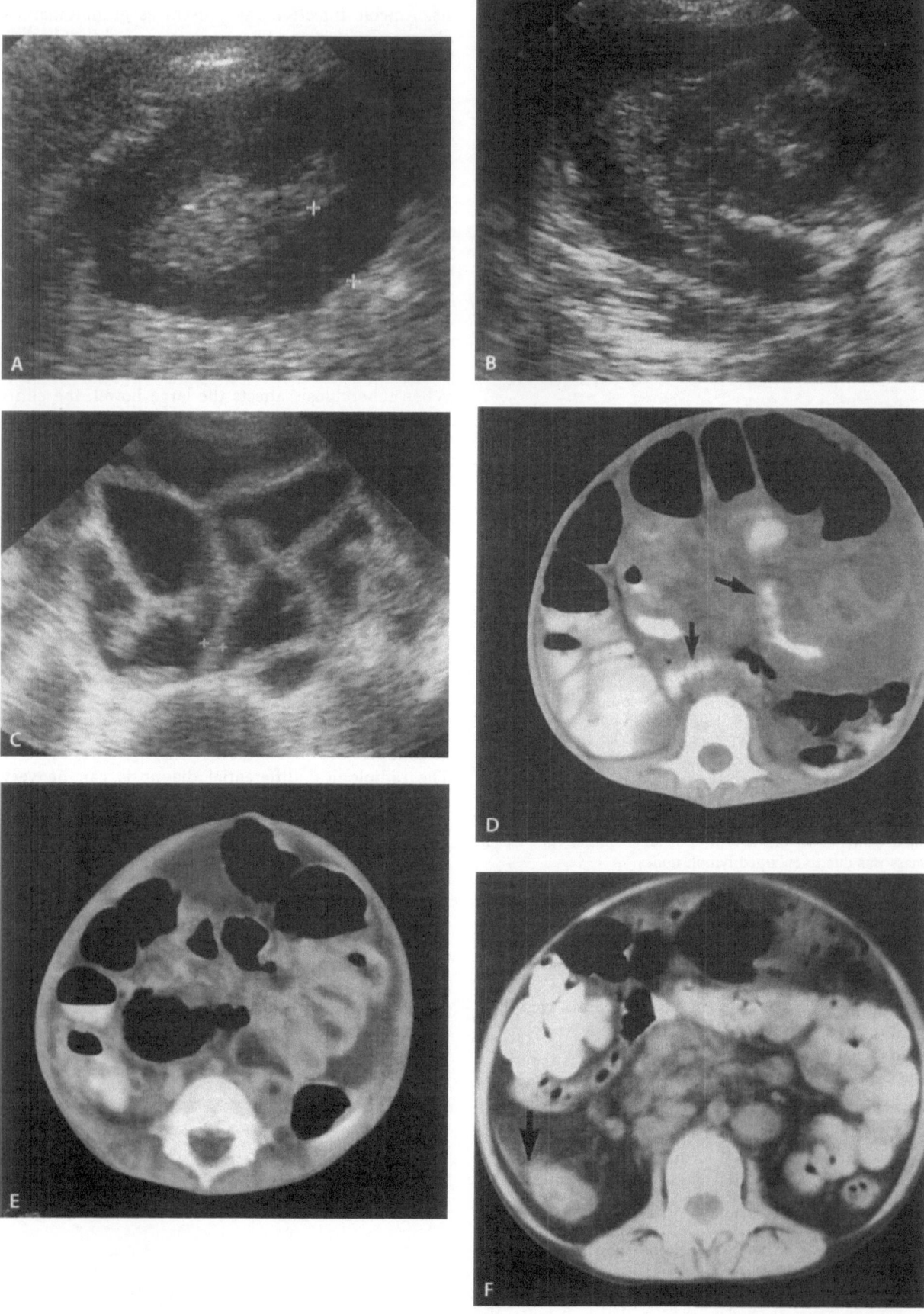

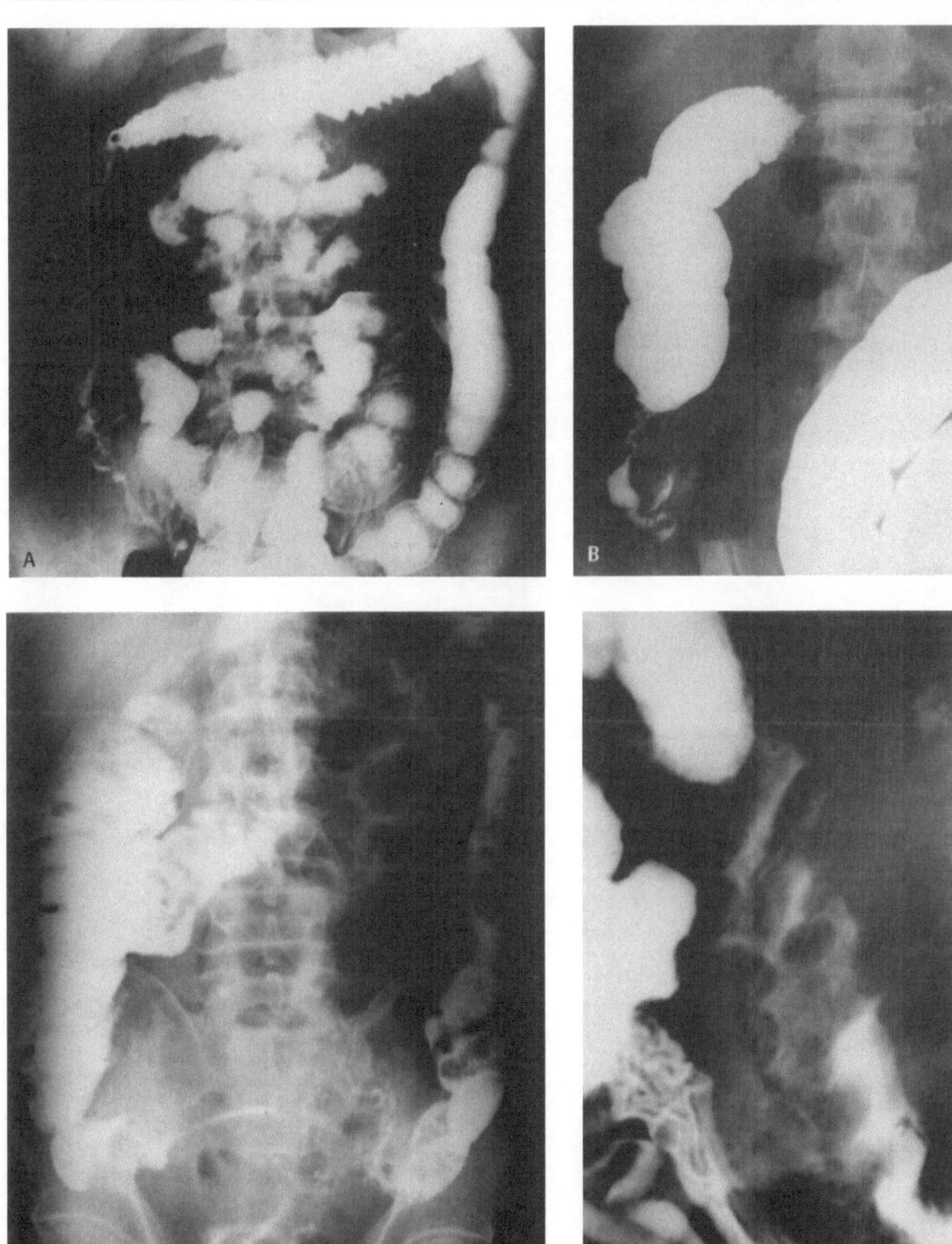

◀ **Fig. 46 A–F.** Scanning, both ultrasonography and CT, can demonstrate the type of mass shown in Fig. 45 but can also be used to show thickening of the bowel wall and ascites. **A–C** Sonograms showing thick-walled bowel, groups of thick-walled bowel matted together, and distended small bowel with echogenic walls and ascites. **D–F** scans of children showing the thick walls of ileum (*arrows*) filled with contrast, contrast-enhanced mural thickening, and a thickened irregular contrast-filled cecum (*arrow*). The patients shown in **D–F** all had ascites. (From Cremin and Jamieson 1995)

Fig. 47 A–H. Tuberculous colitis can affect any part of the large bowel. **A** Tuberculosis affecting the cecum, ascending colon and much of the transverse colon. **B** Tuberculosis affecting the cecum, apparently sparing the ascending colon but involving the transverse colon. **C** Tuberculosis affecting the left half of the colon. In each of these patients there is spasm, saw-tooth distortion of the bowel wall, mucusal ulceration, and edema. In **A** and **C** there is excess mucus which is better seen in **B**, in which the edema is so acute that there is thumbprinting. This patient had active pulmonary tuberculosis. **D** Mucosal ulceration in the sigmoid colon of a child: the lower part of the colon is straight because of the peritoneal thickening. It would be difficult by any imaging method to distinguish any of these cases from amebiasis. **E–H** see p. 64

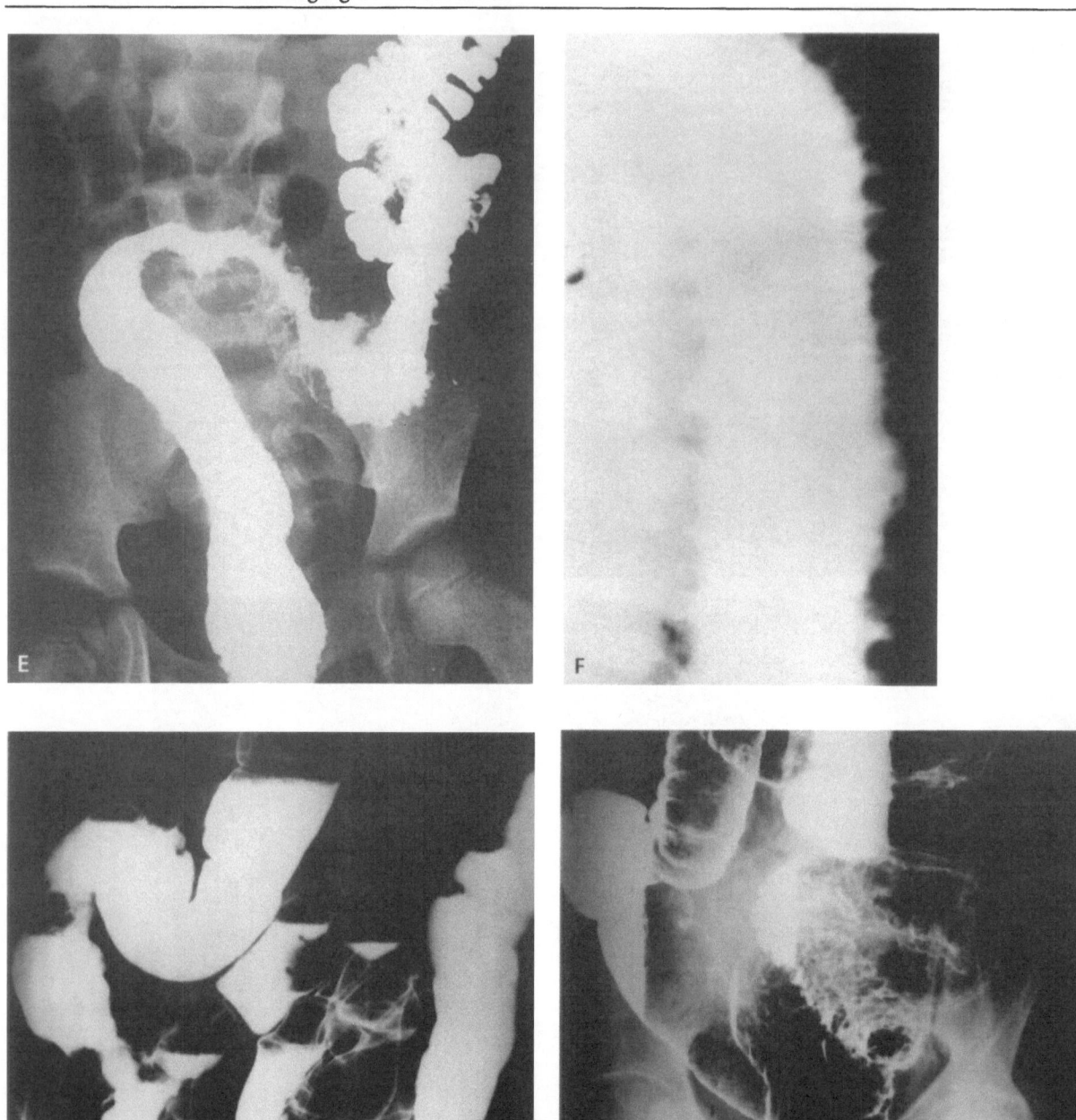

Fig. 47 (continued). E The irregular saw-tooth appearance of tuberculous mucosal ulceration (descending colon). F Spasm and mucosal edema affecting the whole of the descending and proximal sigmoid colon of a 53-year-old African patient. The cecum and ascending colon were also affected, but could not be filled with the barium enema because of the spasm and discomfort. G Skip ulcers encircling and constricting the ascending colon of a 42-year-old Arab female. H Extensive transverse ulceration without any narrowing in the sigmoid colon of a 38-year-old Arab female. (C AFIP 229499; E from Cremin and Jamieson 1995)

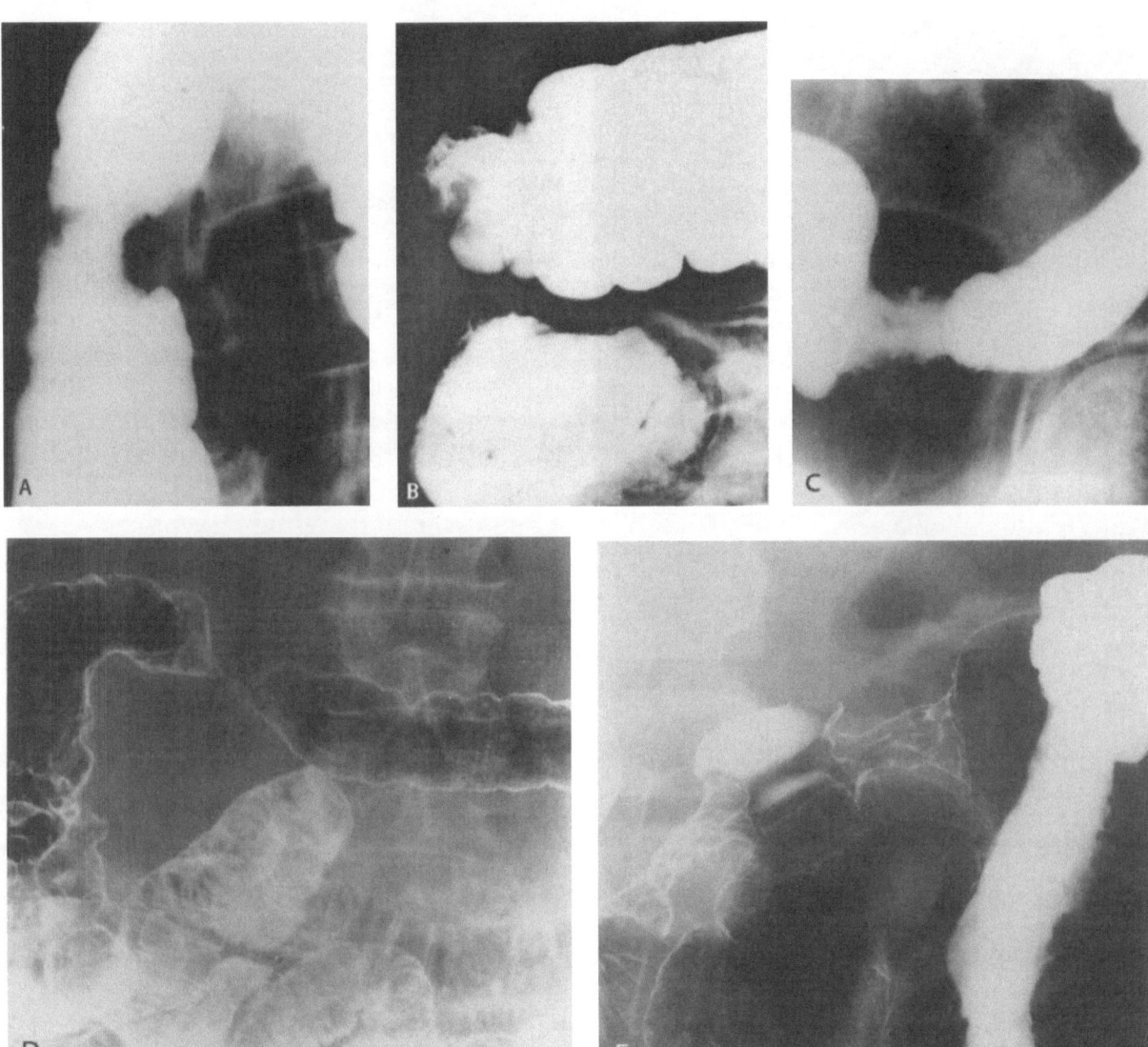

Fig. 48 A–E. Tuberculosis can cause local narrowing in any part of the bowel at any age. The differential diagnosis is then very difficult. **A** Stricture in the ascending colon of a 40 year old man. **B** A very tight stricture just proximal to the hepatic flexure in a young woman. **C** A constriction with ulceration in the sigmoid colon of a middle-aged man; these tuberculous strictures resmble localized amebomas, and carcinoma would have to be considered, although it is rare in some parts of the tropics. **D** A tight stricture in the proximal transverse colon. **E** Two elongated strictures in the transverse colon of a child: there is also ulceration and thickening of the descending colon. These patients come from three different continents: all had normal chest radiographs and in each the diagnosis of tuberculosis was confirmed histologically. (**E** from Cremin and Jamieson 1995)

Tuberculosis of the Rectum

The rectum is the least common site for tuberculosis in the gastrointestinal tract. There may be ulcerating procitis, fistulas and even stricture (Fig. 49). Any chronic ischiorectal abscess in the tropics should raise the possibility of underlying tuberculosis. A chest x-ray is often requested in patients who have a perianal or ischiorectal abscess but who are otherwise in good health. The majority of these will be negative. Imaging with barium, ultrasonography, or CT may demonstrate the abscess but it can be very difficult to demonstrate any fistula or connection between the abscess and deeper tissues. The differential diagnosis will include lymphogranuloma venereum, schistosomiasis, amebiasis, and rarely, actinomycosis or Crohn's disease.

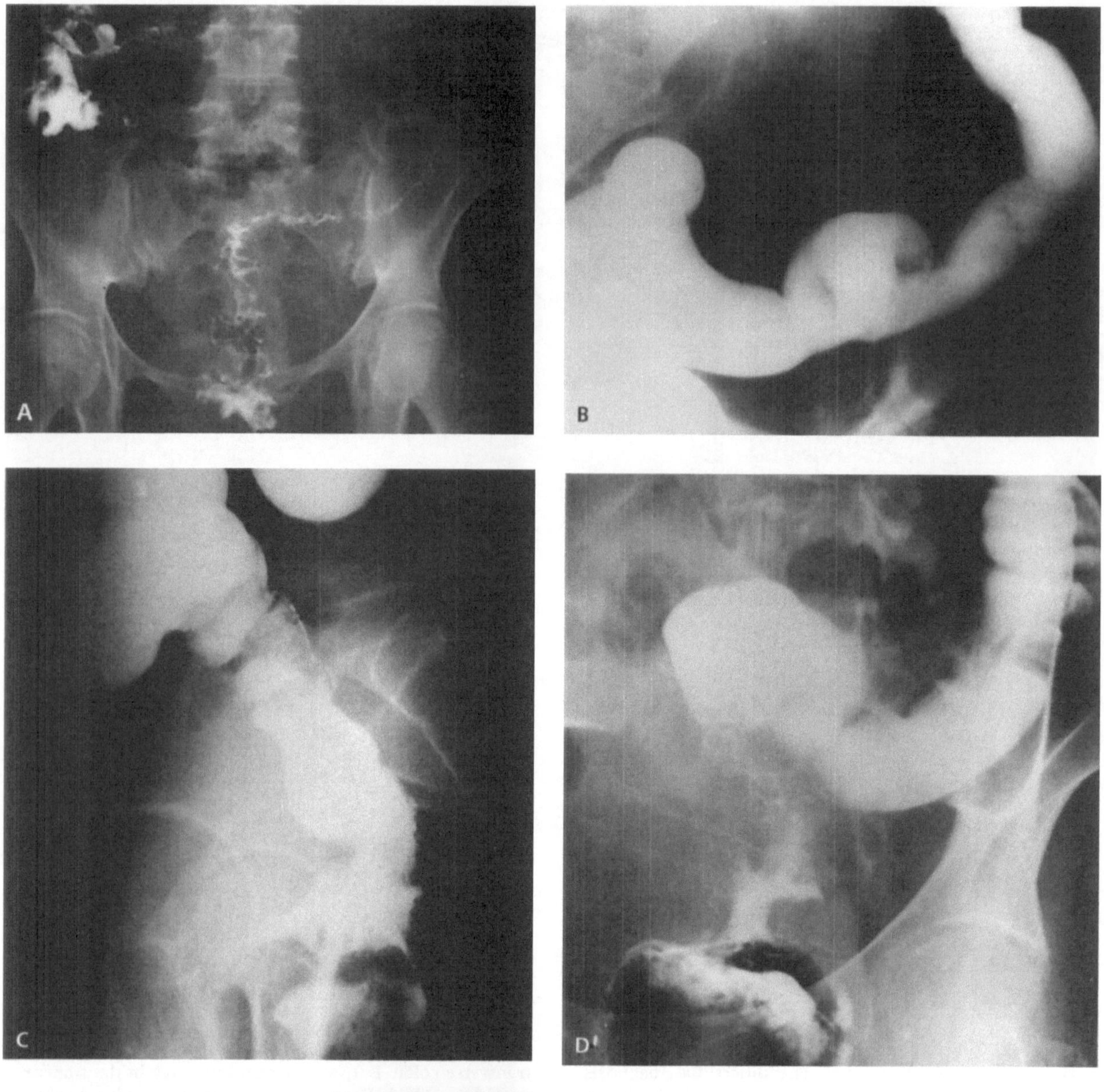

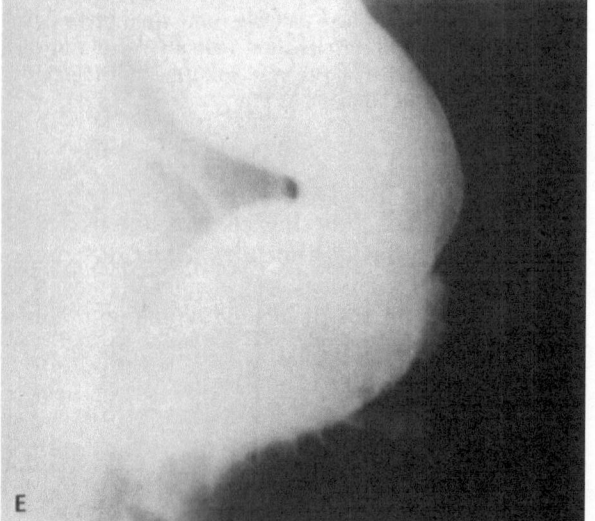

Fig. 49 A–E. Sigmoid and rectal tuberculosis. **A** Sigmoid colitis extending into the upper part of the rectum, with edema, ulceration, and mural thickening. **B** The possible end result: the sigmoid is smooth, lacking normal haustration or mucosal pattern. In both the acute and chronic stages, tuberculosis resembles amebiasis or any other ulcerating colitis. **C, D** The rectum is narrowed, ulcerating, and thickened. There is minimal obstruction at this stage, but if untreated the rectum will narrow and the sigmoid will dilate. **E** A different lateral projection of the same patient shows two fistulae which may connect with the skin in the perineum. Only biopsy will differentiate tuberculosis from lymphogranuloma venereum or amebiasis. Schistosomiasis can cause the same mucosal changes but fistulae are much less common. (**A** AFIP 68-2419-4; **B** AFIP 29499-185)

Enteroliths

Intestinal calculi are not seen very frequently, but may occur above any chronic bowel stricture. They seem to be more common in India than elsewhere, but isolated cases have been noted from many countries where tuberculosis is common. When the stricture is high within the small bowel, the enteroliths are composed of choleic acid and are usually radiographically nonopaque but may be seen as a dense mass on ultrasonography or CT. In the lower bowel, where there is a greater alkaline content and a higher concentration of calcium salts, enteroliths often become radiographically opaque (see Fig. 42 G, p. 57). Some are completely opacified, but others have translucent centers with a ring of calcification. These calcified enteroliths may be found in the lower ileum or colon in up to 3%–4% of cases of intestinal tuberculosis. They vary from multiple small stones to a single large laminated calculus. Massive nontuberculous enterolithiasis has been associated with ileal dysgenesis: the enteroliths were located in the distal ileum and there was an ileo-transverse-colonic fistula. Enteroliths must also be differentiated from calcified granulomatous lymph nodes, renal stones, gallstones, or less commonly, vesical stones. Cross-sectional imaging with ultrasonography or CT is an ideal way to localize them accurately. Barium studies may not always be successful because of the bowel stricture which is the underlying cause.

Tuberculosis of the Peritoneum and Abdominal Lymph Nodes

Tuberculous peritonitis occurs at any age and in both sexes, and in some series accounts for 30% of all nonpulmonary tuberculosis and for at least 20% of all cases of ascites. In most countries about half the cases of abdominal tuberculosis will be due to peritoneal infection, but in children and young adults the frequency is even higher. In almost every case there is associated abdominal lymphadenopathy; the infection may originate from a primary intestinal or gynecological source, or be blood borne; it is seldom possible to locate the primary focus.

The major clinical symptoms are abdominal distention and abdominal pain, vomiting and diarrhea are less common, occurring in under 30% of patients. Nonspecific findings, such as weight loss and tiredness, are difficult to evaluate, but occur consistently in tuberculous peritonitis.

Clinical examination shows that 30% of the patients are afebrile, but almost all have ascites, which is clinically detectable in only about one-third. The tuberculin skin reaction varies geographically, but is often negative (Nigeria 70%, Ethiopia 23%), and the chest radiograph is often normal (India 60%, Ethiopia 50%, Iran 40%). Over 30% of the patients will have lymphadenopathy elsewhere, most commonly cervical. On palpation the abdomen has an ill-defined "doughy" feeling, and in 15% or more of patients there will be palpable abdominal masses. The liver and spleen are often enlarged. There is a very reliable laboratory test for tuberculous peritonitis. If the fluid is examined and the adenosine deaminase (ADA) level is over 32.3 µ/l, there is a 98% sensitivity and 95% specificity for tuberculosis.

Laparoscopy is one way to obtain a tissue biopsy because even culture of the ascitic fluid may be negative. It is useful to exclude other causes of ascites, such as carcinomatosis, lymphoma, or even worms. However, laparoscopy is not without risk, and in tuberculous patients carries a mortality of 3%–12%.

The course of the disease may be relatively benign (except in AIDS patients), and the response to antituberculous therapy quite rapid. The ascites may persist, even when the patient is improving clinically. Histological examination shows that the majority of the infected abdominal lymph nodes will be caseating and necrotic. These nodes may rupture into the peritoneum, disseminating small tubercles all over the peritoneal cavity and occasionally causing an acute clinical reaction and hemorrhagic exudate. When this happens, the protein content of the ascitic fluid will be high (over 45 g/l) and the fluid/blood glucose ratio will be below 0.96, with marked lymphocytosis. The tubercle bacilli can be seen on direct staining of the fluid or subsequently on culture. Another initially rare complication is intestinal obstruction, which increases in frequency as the fluid is absorbed and adhesions form. At autopsy, cecal or ileal ulceration may occasionally be found.

Tuberculous Peritonitis

Ultrasonography is the method of choice for imaging tuberculous peritonitis. There are three imaging patterns. There may be ascites (the "wet" form), there may be multiple caseous nodules and adhesions (the plastic or "dry" form), or there may be a combination in which loops of bowel, omentum, or mesentery have clumped together, often becoming palpable and associated with ascites.

When there is ascites, it can be free, localized or loculated (Figs. 50, 51). When free, it is either clear fluid or contains multiple thin strands, septa, or floating debris. These strands may be mobile and quite delicate or relatively thick so that adhesions occur. Fluid may be trapped between the thickened loops of bowel, producing alternating echoic and echo-free bands (the "sandwich" appearance). Both on ultrasonography and CT the density of tubercu-

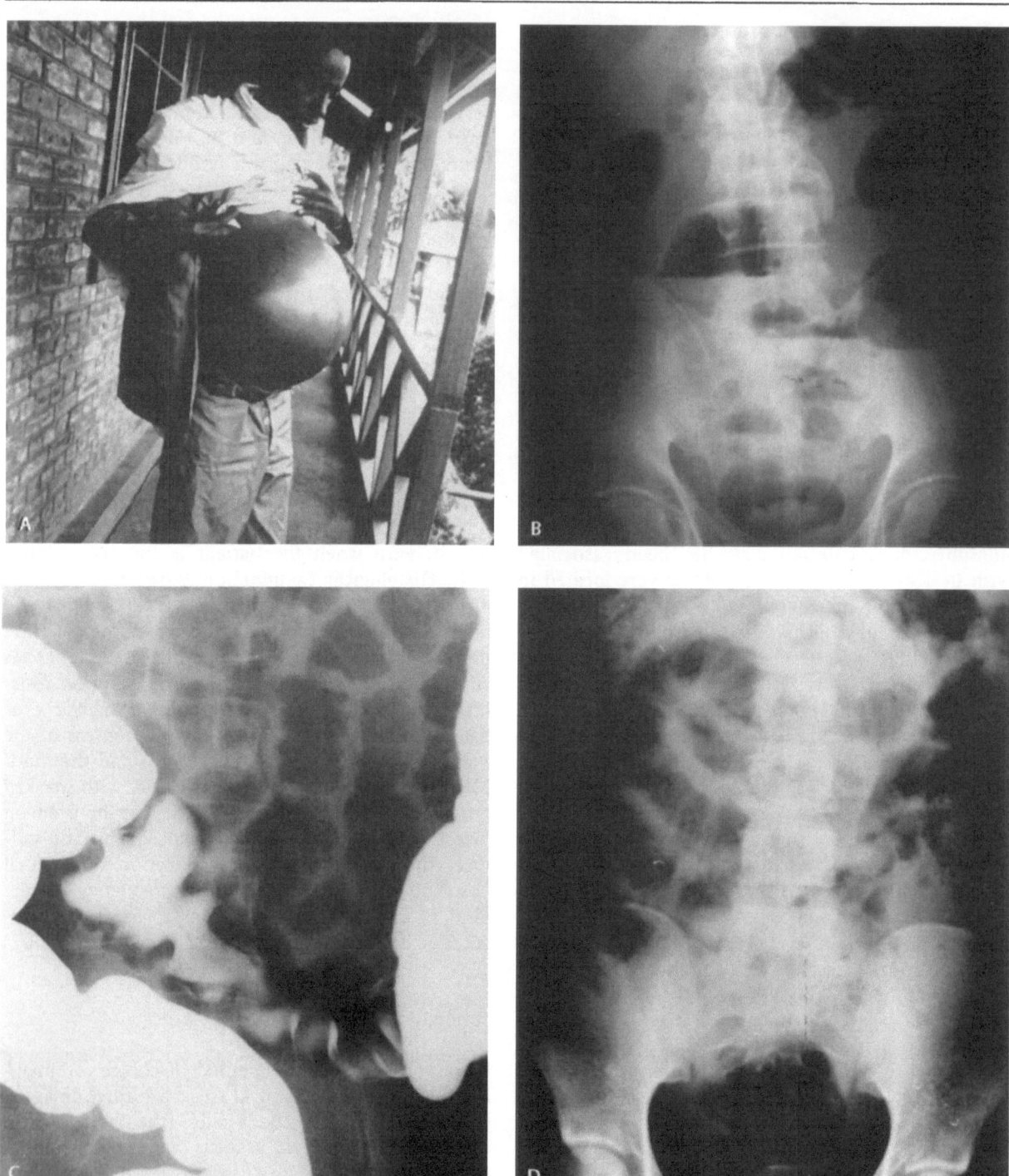

Fig. 50 A–F. Tuberculous peritonitis **A** An Ethiopian patient with tense ascites. (Courtesy of Dr. Richter) **B** Multiple loops of dilated bowel, with fluid levels, floating in ascites. It can be very difficult to decide whether this is due to obstruction or ileus. **C, D** Dilated loops of small bowel with thick edematous walls. This is still an ileus, but obstruction may develop later because of adhesions. **C** is an African and **D** a patient from India. **E, F** see p. 69

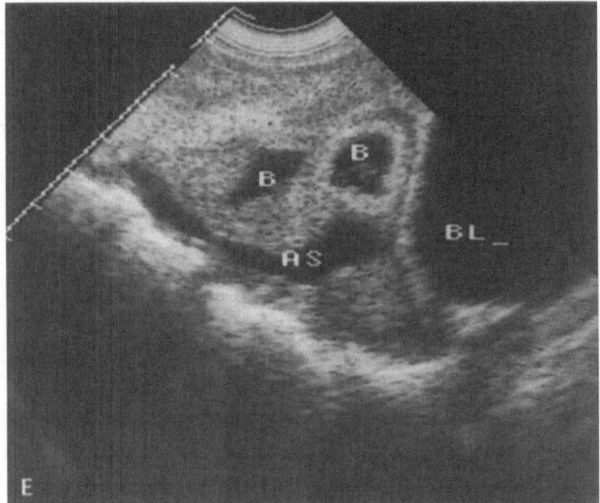

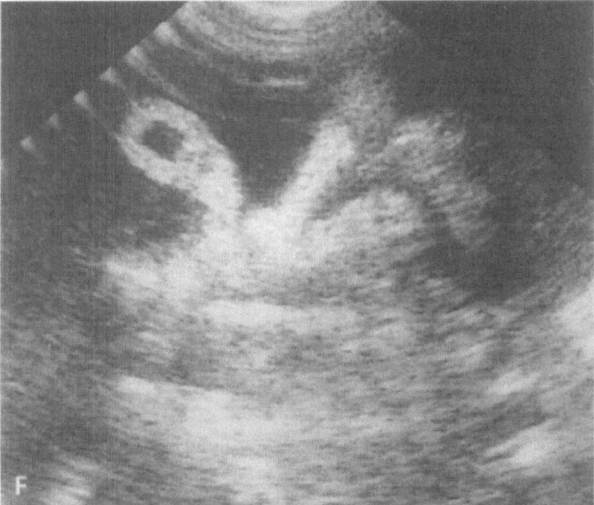

Fig. 50 (*continued*). **E** A sonogram showing thickened bowel walls and ascites due to tuberculosis in a 44-year-old African from Zimbabwe. *B* bowel; *AS* ascites; *BL* urinary bladder. (Courtesy of Dr. Sam Mindel) **F** A sonogram of a South African child showing echogenic bowel loops radiating from the mesenteric root, with ascites. (From Cremin and Jamieson 1995)

lymph nodes can be differentiated. CT may be better than ultrasonography for the anatomical localization of the fluids and to show where it is loculated or has been walled off into a thick abscess. In patients with AIDS the progress is similar but exaggerated.

In the dry form of peritonitis, ultrasonography will demonstrate irregular echo-free or echo-poor, nodular or laminar thickening of the peritoneum. The nodules are poorly echogenic and occur almost anywhere on the peritoneal surface. Histologically they are caseating granulomas. The mesenteric thickening may lead to fixation of the bowel and, when there is fluid, these loops of bowel and mesentery have been described as "radiating from the mesenteric root in a stellate configuration." This is more clearly seen with CT. On MRI performed with gadolinium enhancement, lymphadenopathy can be demonstrated. However, there is seldom any great advantage in using either CT or MRI compared with ultrasonography.

Routine supine and erect radiographs of the abdomen will show free fluid and multiple distended loops of small bowel, often with thickened intestinal walls. There may be scattered fluid levels in the erect position, but the findings are those of an ileus rather than obstruction. The amount of fluid in the abdomen may make the details of the bowel somewhat hazy. The proximal colon may be involved and also dilated to the same extent.

As healing occurs, the degree of dilatation lessens and the wall of the gut is less thickened. Adhesions may form and there may be subacute obstruction with fluid levels. At this stage, barium contrast gastrointestinal studies may be helpful and careful fluoroscopy with palpation will make it possible to decide whether the loops of bowel are fixed or mobile. However, apart from adhesions, the majority of cases will have a normal radiological examination of the small intestine without any obvious bowel involvement.

Tuberculous Abdominal Lymph Nodes

At least one-third of patients with tuberculous peritonitis will have lymphadenopathy, and in some countries, e.g., India, the frequency of lymph node involvement may be as high as 70%. Lymphadenopathy may occur without detectable bowel involvement. All groups of nodes within the abdomen can be infected, particularly those around the pancreas, portal region, aorta, and vena cava. On ultrasonography the enlarged nodes are usually hypoechoic (Fig. 51 A, B), and some display central echogenic areas where caseation has started. Some nodes will be discrete, while others adhere together into large

lous ascites is variable; when clear, it is a transudate in the early stage and becomes thickened later. There may be progression to an abscess, seen on ultrasonography as well-defined localized fluid collections, septate and with internal echoes. Aspiration of this thick fluid can be difficult or even impossible due to its consistency and the multiple septate divisions. But even at this stage, there can be good response to antituberculous therapy.

Computed tomography can demonstrate the ascites and the plastic changes, but preliminary oral contrast should be used within the bowel so that the thickening of the bowel wall, the omentum, and the

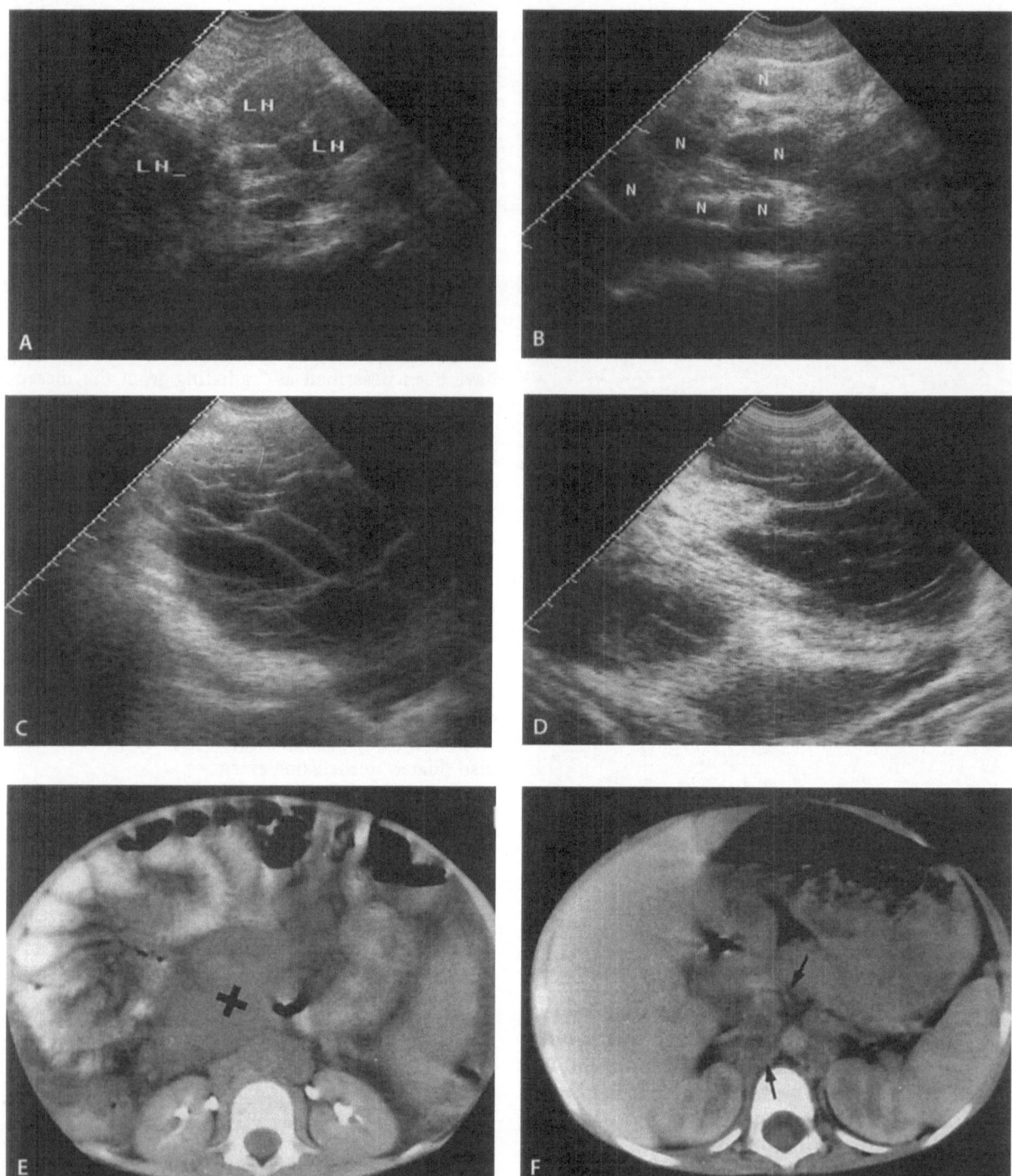

Fig. 51 A–F. With scanning it is possible to identify lymphadenopathy and peritoneal fluid: it is not always possible to be sure of the etiology. **A, B** Ultrasonography shows multile enlarged tuberculous lymph nodes (*LN/N*) and ascites. **C, D** Loculated peritoneal fluid, a very common finding in tuberculous peritonitis. (**A–D** are all patients from Zimbabwe, courtesy of Dr. Sam Mindel) **E** A contrast-enhanced CT scan (with contrast also in the bowel) shows the thickened intestinal loops and a large omental mass (*X*). There is loculated low-density ascites. **F** Contrast-enhanced CT showing rim-enhancing para-aortic lymph nodes (*arrows*). (**E, F** are children from South Africa, from Cremin and Jamieson 1995)

masses. Both CT and ultrasonography can demonstrate nodal calcification before it can be seen on a plain radiograph. On CT, most nodes are of low density with some peripheral rim enhancement (Fig. 51 F). It is probable that some of the abscesses (cold abscesses) within the peritoneum result from extensive caseation of lymph nodes; they have a very variable ultrasound and CT appearance depending on the degree of caseation, central necrosis, and septation.

Enlargement of the abdominal lymph nodes may cause direct pressure on various parts of the gut; there may be distortion of the pyloric antrum, duodenal loop, and upper jejunum in particular (see pp. 52–54). If the nodes are very large and swollen, the bowel may become adherent to them and involved in the tuberculous process. This results in a spiky irregular outline of the intestinal mucosa, localized edema, and ileus. Rupture of enlarged nodes into the intestine has been reported, but only happens when this adhesive process has occurred. It can present as a communicating diverticulum.

Calcification of the mesenteric nodes may be seen on healing, but is not as common as might be expected. It will be detected earlier by CT scanning (see p. 53). Although the presence of caseation and calcification suggest tuberculosis, metastatic malignancy and histoplasmosis can occasionally also calcify. When healing, tuberculous calcification may be seen as discrete lines on both CT and ultrasonography. Radiographically visible abdominal nodal calcification due to tuberculosis is uncommon in Africa and many parts of the tropics despite the frequency of the infection. (However, calcification in nodes elsewhere, such as the neck, mediastinum, and inguinal region, is quite common.)

Lymphangiography has been used for the evaluation of lymph nodes in abdominal tuberculosis, using pedal injections. Sharply outlined central filling defects, extending in some cases to the periphery, have been described as a feature of tuberculous lymphadenitis. As would be expected, the bunching of the lymph nodes together in an adherent mass can also be seen. The central filling defects are probably tuberculous caseation. In some lymph channels there is obstruction to the flow. Most of the changes described on lymphangiography can also be seen in lymphoma and metastatic disease. Ultrasonography and CT scanning have replaced lymphangiography for diagnosis in most countries (which is probably appreciated by the patient).

A rare tuberculous involvement of a large giant cystic lymphangioma has occurred in an Ethiopian male immigrant to Israel. His father had tuberculosis. The patient had a draining right axillary sinus, a large fluctuating abdominal mass, and an equally large reducible right inguinal hernia. There was no palpable lymphadenopathy. Imaging showed this to be one cystic mass extending from the posterior mediastinum into the pelvis, and aspiration of the fluid was positive for *M. tuberculosis*. The mass responded to antituberculous therapy and fluid aspiration. Two similar infections have been reported in cystic hygromas of the hand.

Tuberculosis of the Liver, Spleen, and Pancreas

Hematogenous, disseminated tuberculosis can result in small (miliary) tuberculous nodules in any organ. Clinically, generalised hepatomegaly and less often splenomegaly may be palpated when there is abdominal tuberculosis. Isolated tuberculomas of the spleen have also been reported.

There may be multiple small tuberculous granulomas (tubercles) in the liver (Fig. 52 A–F) or spleen (Fig. 53) which on ultrasonography have a granular echoic or hypoechoic appearance. The granulomas can become macronodular, depending on the stage of development. Some will have central, more echogenic areas. If a tuberculous abscess develops, the hypoechoic center will be surrounded by a hyperechoic rim. One or more areas of caseous necrosis may develop. This is less common in the spleen. In some tuberculous granulomas there will be calcification, seen on plain radiographs or causing acoustic shadowing on ultrasonography.

Large tuberculous masses are unusual, but have been reported. One such mass was 3.5×5.5 cm, and situated in the left lobe of the liver (Fig. 52 G–I). On CT the mass was hypodense and showed peripheral enhancement with contrast. Celiac arteriography in this case showed stretched hepatic vessels and some neovascularity, which suggested that the mass was inflammatory rather than neoplastic in origin.

Only very rarely will a solitary hepatic, splenic, or pancreatic abscess or tuberculoma be the presenting evidence of tuberculosis. Almost always there will be marked tuberculous lymphadenopathy within the abdomen, and often in peripheral nodes also.

Pancreatic tuberculosis may be nodular or (less commonly) form an abscess, usually of complex echogenicity but with less surrounding pancreatitis than a pyogenic abscess. The nodular pattern is less easily recognized. Pancreatic lesions are unlikely to be seen in children. In adults they are probably going to become more common in patients with AIDS.

All tuberculous granulomas may calcify and then be demonstrated by any method of imaging (Fig. 53 A, B). Many cases of splenic or, less often, hepatic calcified tuberculous granulomas are chance findings.

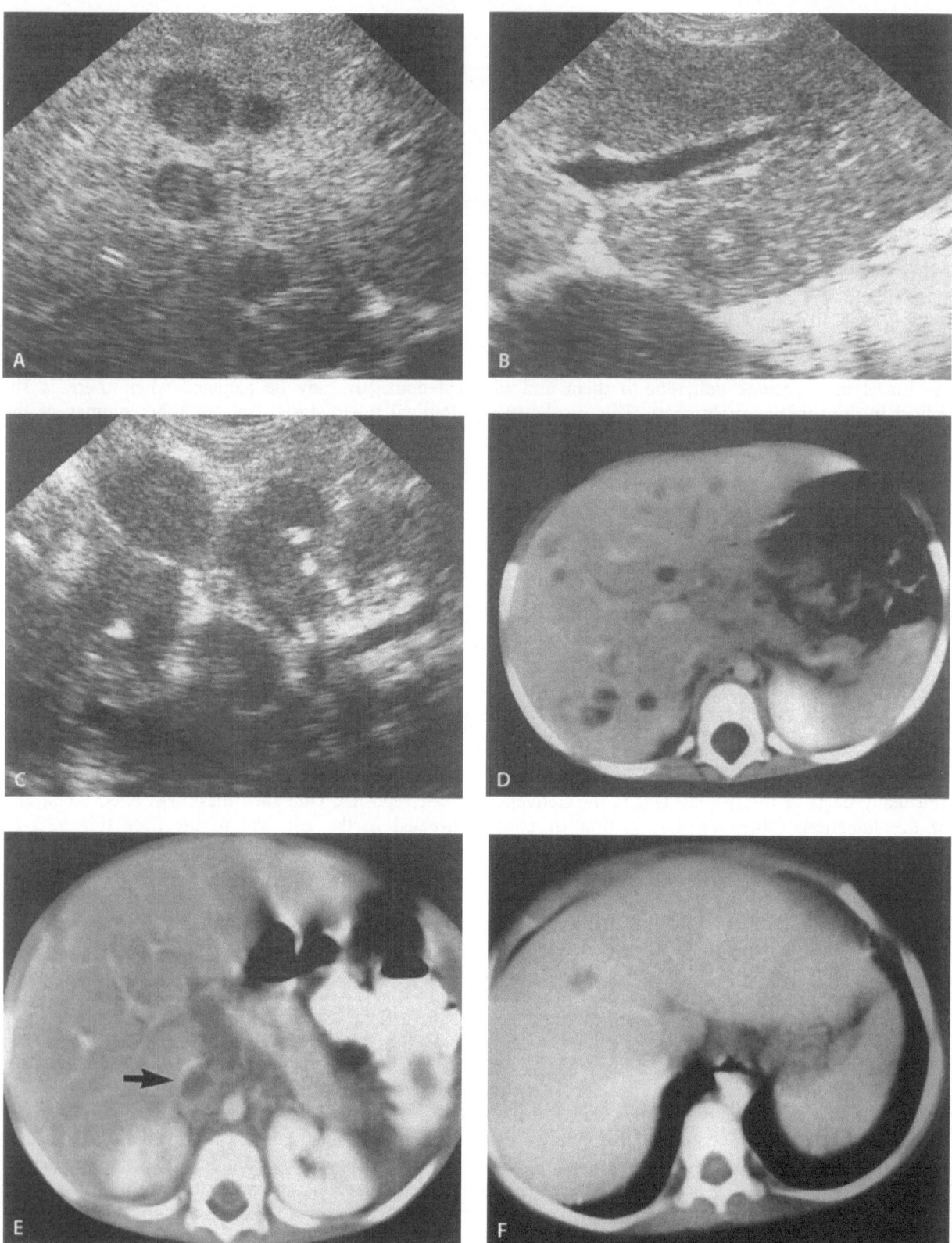

Fig. 52 A–I. Tuberculosis of the liver. **A, B** Ultrasonography showing hypoechoic granulomas and a granuloma with an echogenic center in the livers of two children. **C** In another child there are hypoechoic lymph nodes with echoic calcification at the porta hepatis. **D** A contrast-enhanced CT shows multiple nonenhancing granulomas in a child's liver. **E** In a different child there are rim-enhancing nodes (*arrow*) at the porta hepatis. **F** A child from a Pacific Island has a similar granuloma in the liver. **G–I** see p. 73

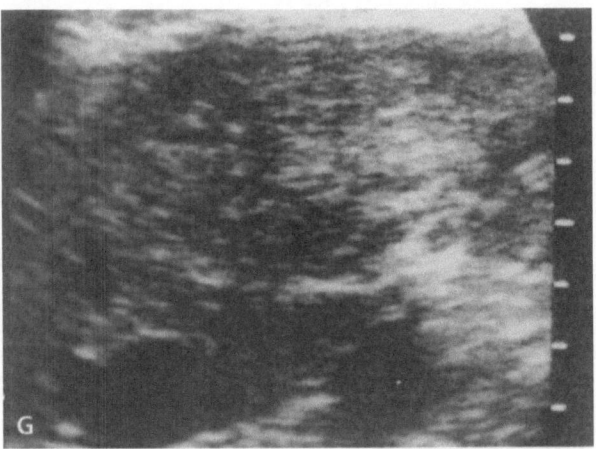

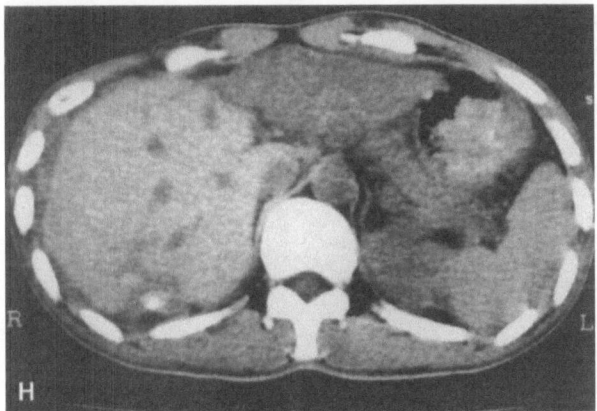

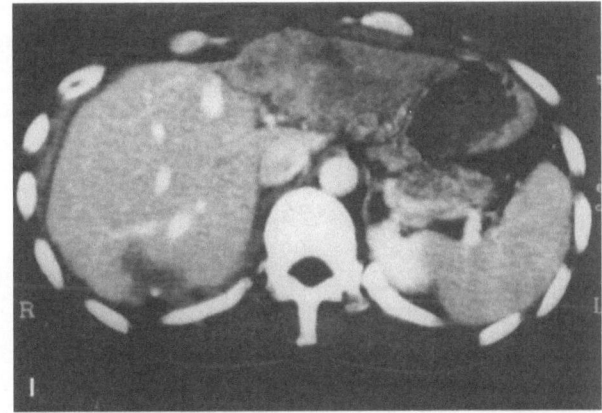

Fig. 52 (*continued*). G Tuberculoma of the liver: A 3.5×5.5 cm hyperechoic mass in the lateral segment of the left lobe (sonogram). An isolated hyperechoic nodule with calcification was also found in the right lobe. H, I CT scans of the same patient show a large hypodense lesion almost filling the lateral segment of the left lobe of the liver. The other nodule in the right lobe, is hypodense with a central hyperdense focus. I After intravenous contrast, the lesion in the left lobe shows peripheral enhancement and central hypodensity; the nodule in the right lobe remains hyperdense. The patient was a 30-year-old male, a known hepatitis B carrier, complaining of right upper quadrant abdominal pain. His chest radiograph showed diffuse reticulonodulation and pleural thickening due to pulmonary tuberculosis and pleurisy. Ultrasound-guided liver biopsy showed acid-fast bacilli. After 9 months of antituberculous treatment the tuberculoma had almost disappeared. (A–E from Cremin and Jamieson 1995; F courtesy of Dr. Cheryl Sisler, Hawaii; G–I courtesy of Dr. T.C.F. Tan et al. and *Br J Radiol*, 1997)

The differential diagnosis can be difficult because of the rarity of active infection in the liver, pancreas, or spleen. A chest x-ray may be normal, but the tuberculin skin test is usually positive. Most of these tuberculous lesions in solid organs will be non-enhancing, although there may be rim enhancement at some stage. If noncalcified, they can be mistaken for hydatid disease, amebic abscesses, or, more commonly, malignancy. Image-guided aspiration can be used to make the diagnosis and, in some cases, for instillation of drugs. When lesions are calcified their etiology can be more difficult to identify, since they may resemble a collapsed hydatid cyst, an old pyogenic or mycotic infection, or even a calcified hematoma following trauma (especially in the spleen). For noncalcified, active granulomas or abscesses, imaging can be used to follow treatment, and prolonged follow-up is advisable because recurrence of tuberculosis from these isolated abscesses has been recorded.

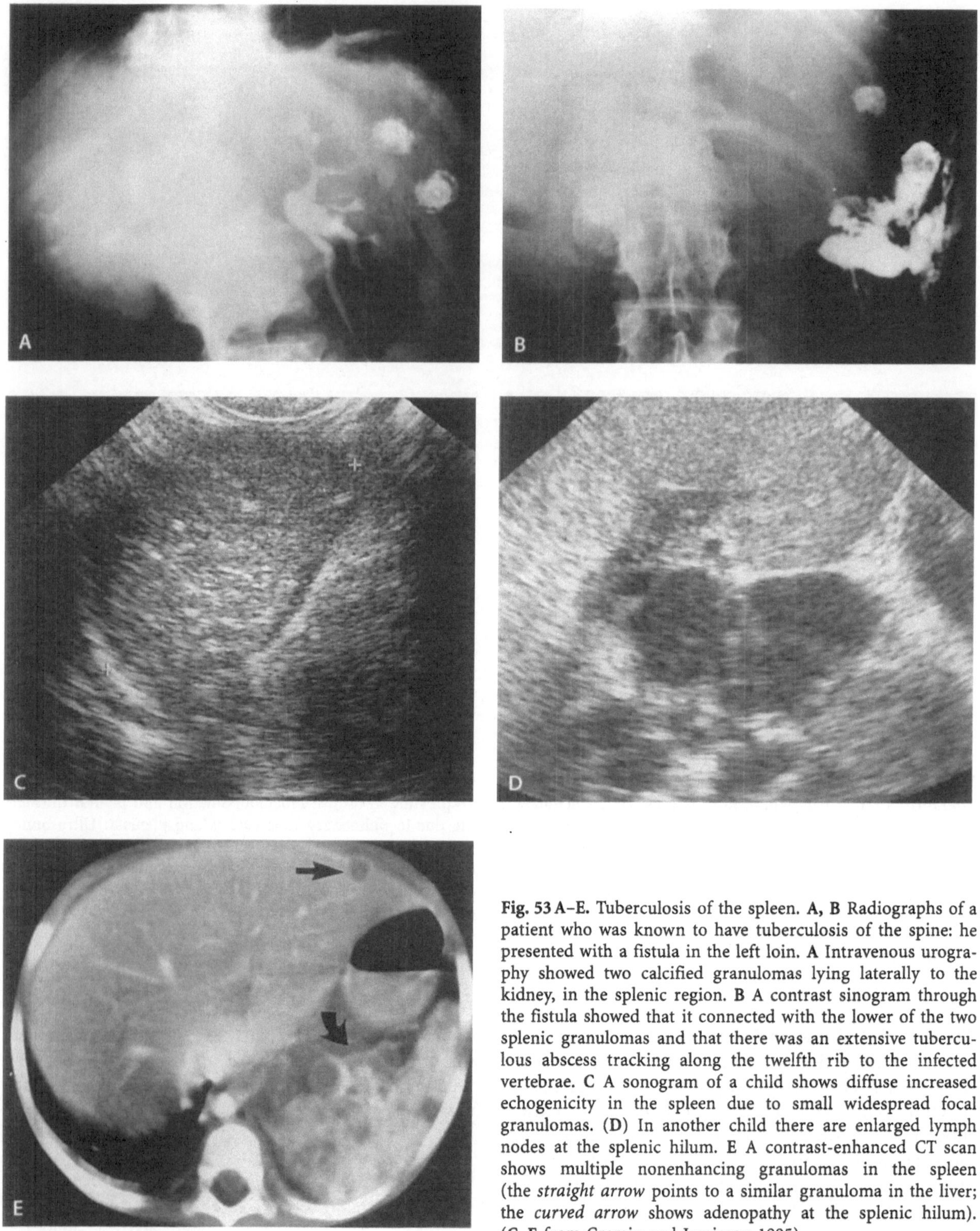

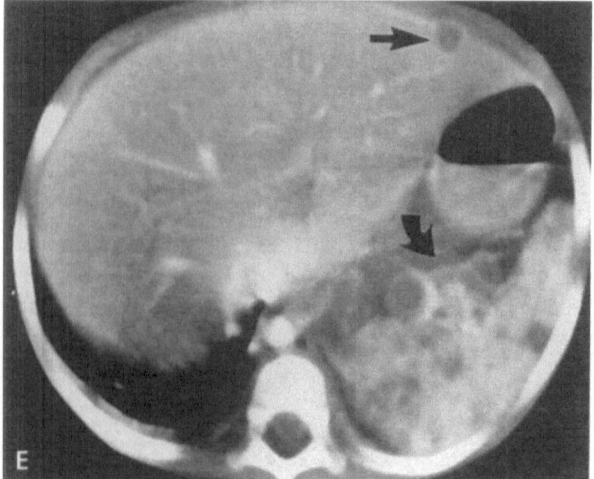

Fig. 53 A–E. Tuberculosis of the spleen. **A, B** Radiographs of a patient who was known to have tuberculosis of the spine: he presented with a fistula in the left loin. **A** Intravenous urography showed two calcified granulomas lying laterally to the kidney, in the splenic region. **B** A contrast sinogram through the fistula showed that it connected with the lower of the two splenic granulomas and that there was an extensive tuberculous abscess tracking along the twelfth rib to the infected vertebrae. **C** A sonogram of a child shows diffuse increased echogenicity in the spleen due to small widespread focal granulomas. **(D)** In another child there are enlarged lymph nodes at the splenic hilum. **E** A contrast-enhanced CT scan shows multiple nonenhancing granulomas in the spleen (the *straight arrow* points to a similar granuloma in the liver; the *curved arrow* shows adenopathy at the splenic hilum). (C–E from Cremin and Jamieson 1995)

Tuberculosis of the Urinary Tract

Kidneys and Ureters

The incidence of tuberculosis of the kidney varies throughout the tropics; wherever it occurs, it is more common in the higher socioeconomic groups, repeating the pattern of the disease in Europe. It is uncommon in tropical Africa, despite the fact that tuberculosis is prevalent in most other tissues. It occurs frequently in much of Asia and India, particularly in association with diabetes. There are often tuberculous foci in the chest and skeleton also.

Renal tuberculosis in the tropics is probably almost always bilateral, although it may be first identified on one side only, especially if contrast urography is used to make the diagnosis. It usually starts as a localized caseating lesion, most commonly in the upper pole of either kidney, although it may arise anywhere. Such foci are caused by hematogenous spread. Alternatively, it may present as pyelonephritis as the result of reflux from an infected bladder.

The nidus of infection in the renal parenchyma enlarges and ruptures into a neighboring calyx, discharging necrotic caseous material and distorting the calyx (Fig. 54). This can be demonstrated by intravenous or retrograde urography, ultrasonography, CT, or MRI. If there is a communication into the tuberculous cavity, it may fill during an intervenous contrast examination; the affected calyx becomes an ulcerated cavernous lesion. The infection spreads to involve the draining calyceal infundibulum, which may then develop a stricture and seal off the infected calyx. If the ulcer and stricture are located in the renal pelvis, there will be obstruction to the outflow of urine and the calyces will become clubbed. Later, a stricture of the renal pelvis can seal off the kidney, and fibrosis and calcification may follow. This may result in autoamputation of the kidney. If there is direct extension of the tuberculous infection into the rest of the kidney, the entire kidney becomes a bag of caseous necrotic pus.

Ultrasonography is a very satisfactory way to image each step of this pathological process (Fig. 55). Although it is possible to demonstrate the irregular and dilated calyces and the connecting cavities by urography, ultrasonography gives a better image of the renal parenchyma and will show increased echogenicity and mixed echogenicity around the calyceal lesions. Focal parenchymal granulomas which cannot be seen radiographically may be identified by ultrasonography. (Scanning should always include both kidneys even when there is an obvious lesion on one side only). On ultrasonography a tuberculous abscess will appear as an echogenic, irregular mass, often containing debris. Local or general hydronephrosis can be imaged and in the end stage the kidney will become irregular in outline, with varying thickness of the cortex. Eventually there may be calcification in the granulomas or abscess, showing as bright areas on an ultrasound scan (Fig. 56). Punctute calcification of the upper pole and curvilinear calcification outlining the entire kidney are the two extremes of renal involvement. CT and MRI are alternative ways to scan.

Autoamputation, the end result of advanced tuberculosis, can cause remarkably little clinical disturbance, especially in patients who are malnourished and in ill health for other reasons. There can be complete nonfunction on contrast urography or scintigraphy, a finding noted in half the patients in one pediatric series. On ultrasonography or CT, the distorted renal mass can be accurately demonstrated, showing the ultrasonically echogenic fibrosis and contraction and the bright echoes of calcification.

In tuberculosis, the ureters are dilated proximally, with irregular granulomas which result in one or more strictures and which ultimately lead to hydronephrosis (Fig. 57). Eventually, spotty ureteric calcification develops; in extreme cases this may merge into extensive pipe calcification along the length of the ureters. Intravenous, or even better, retrograde pyelography is the most accurate way to delineate the full length of the ureters to demonstrate strictures or calcification. Although the echogenic irregularity and granulomatous masses, as well as the dilatation of the ureters, can be demonstrated by ultrasonography, it is not so easy to assess the peristalsis. CT and MRI are less reliable (and MRI particularly is not a good way to demonstrate calcification) and are both a more costly way of evaluating the ureters.

Differentiation from schistosomiasis is usually fairly reliable. In schistosomiasis calcification is first seen in the lower end of the ureters and the bladder and then extends up the ureters; it is most unusual to see much ureteric calcification without bladder calcification, and the nodular irregularity of the ureters as seen on contrast studies will be useful in making the differentiation. In tuberculosis, the calcification extends down the ureter and the bladder is very seldom calcified to the same extend as in schistosomiasis. The multiple strictures and nodules of ureteritis cystica are rare in the pattern of urinary tuberculosis in the tropics. Transitional cell carcinoma of the ureter is rare in much of the tropics but does produce irregular strictures with mucosal destruction and nodular masses, often with the characteristic goblet or champagne glass appearance.

(Text continues on p. 81)

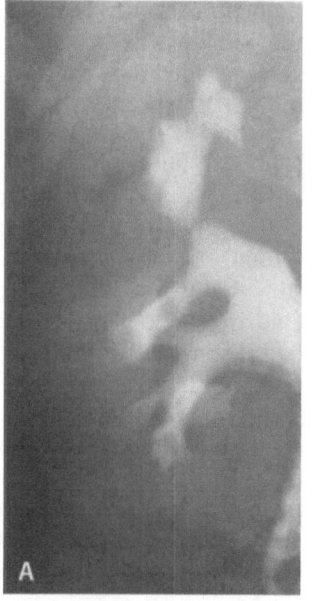

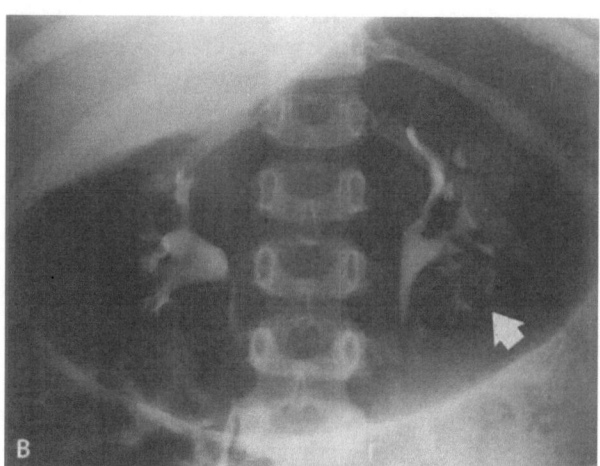

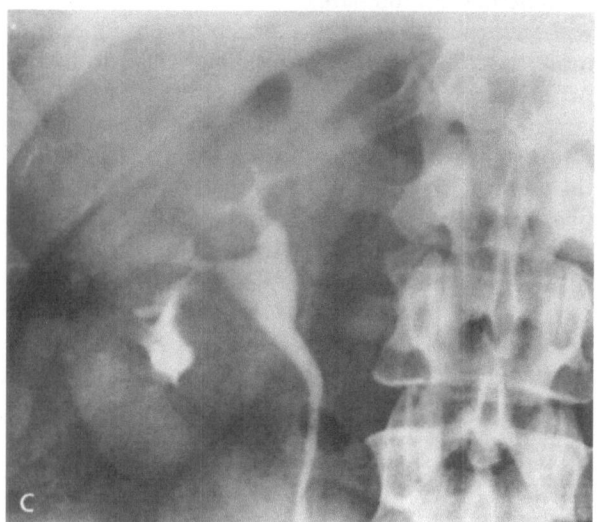

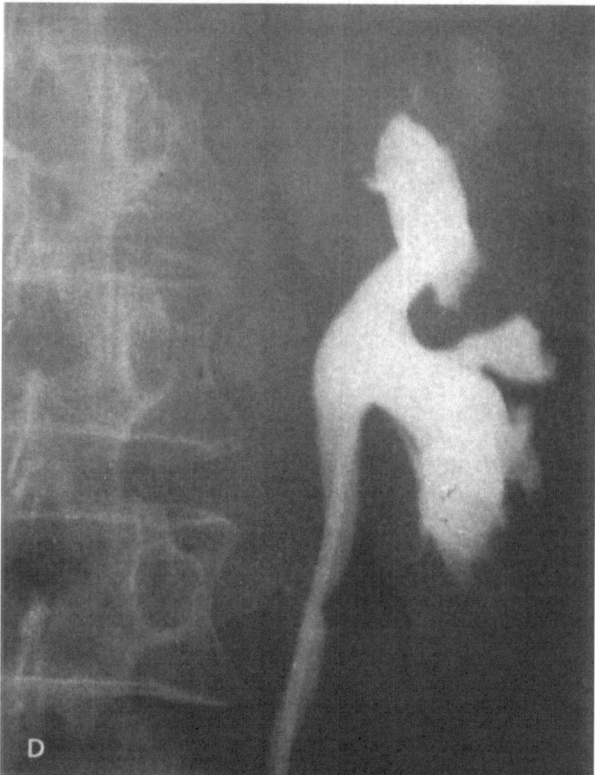

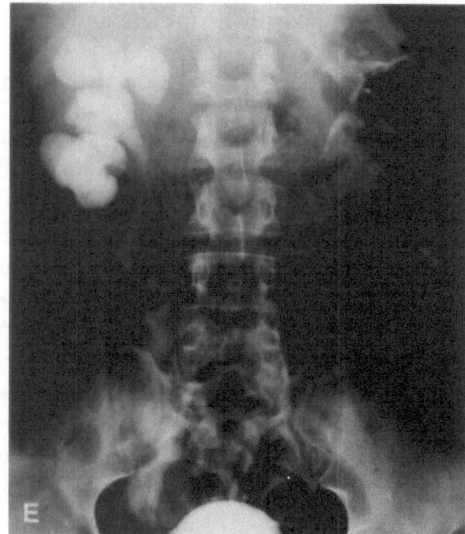

Fig. 54 A–E. Tuberculosis of the kidney; appearance on excretory urograms. **A** Widening and distortion of the upper calyces of the right kidney, with a stricture just above the renal pelvis. **B** Destruction of the lower calyces in the left kidney (*arrow*) with the formation of crescents outlining the edges fo the tuberculous abscess (cavity) **C** A tuberculous cavity in the lower pole of the kidney. **D** All the calyces in this left kidney are clubbed and distorted, and there is an abscess in the upper pole. **E** A more advanced infection in the right kidney with hydronephrosis and dilatation of the lower half of the right ureter. The left kidney and ureter are normal so far

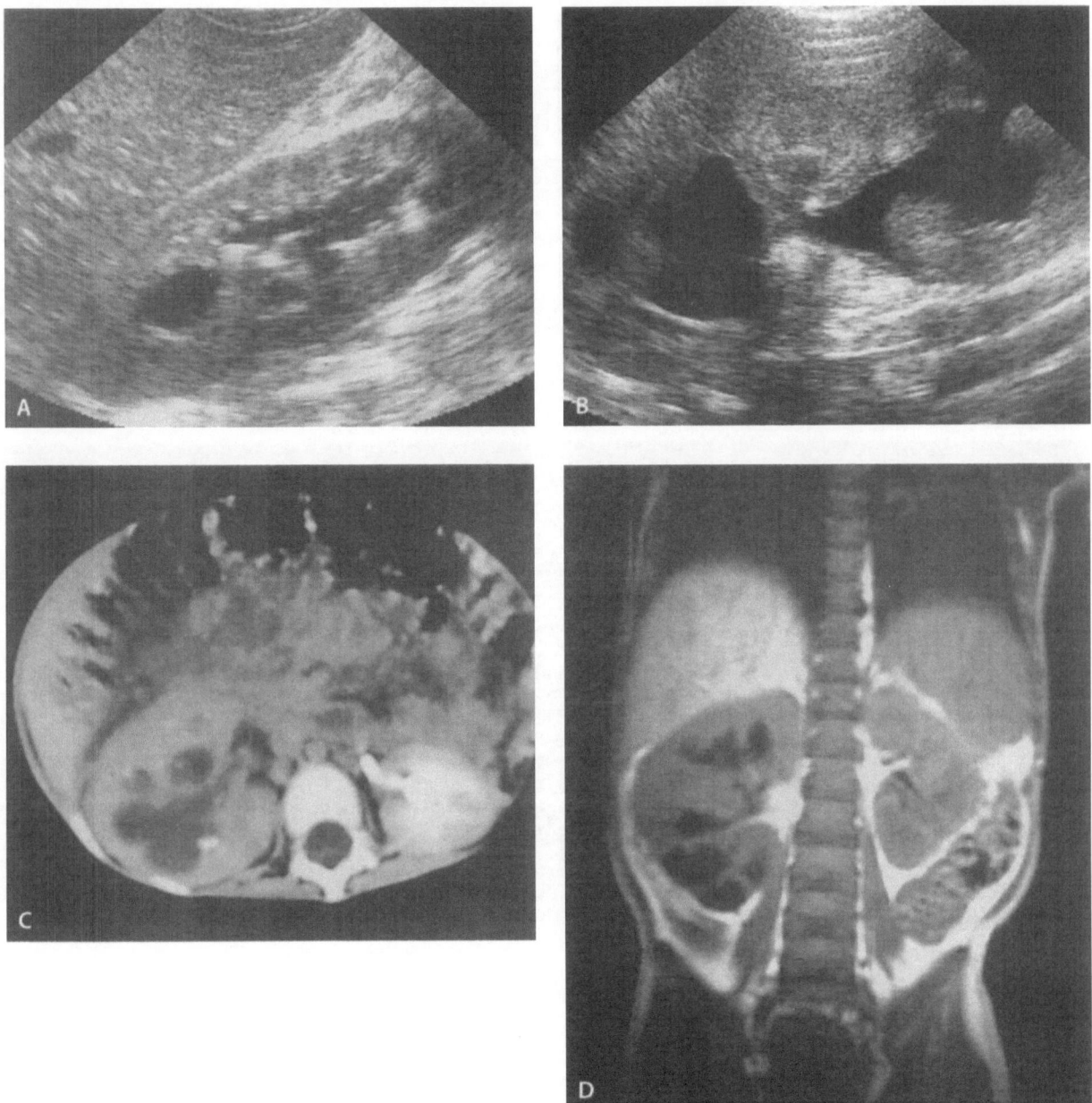

Fig. 55 A–F. Scanning can show parenchymal tuberculosis and its progress more accurately than urography. **A, B** Ultrasonography shows echogenic foci in the renal parenchyma as well as irregular caliectasis. **C** More advanced destruction of the calyceal system in the right kidney, with some parenchymal foci. **D** A coronal T1-weighed MR scan of the same patient showing the extent of the renal infection. **E, F** see p. 78

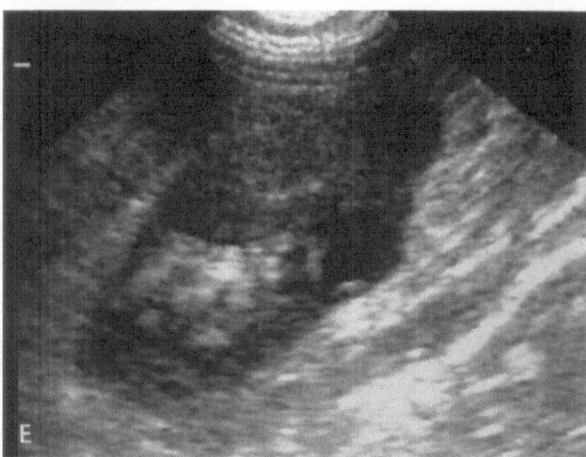

Fig. 55 (*continued*). E Ultrasonography of a different patient showing an abscess in the right kidney. F Angiography is useful in the differential diagnosis between a tumor and infection. The extent of the renal tuberculosis is well shown in this right renal series. (A–D from Cremin and Jamieson 1995; E courtesy of WHO: *The manual of diagnostic ultrasound*, Geneva, WHO, 1995; F courtesy of the University of Capetown Radiology Library)

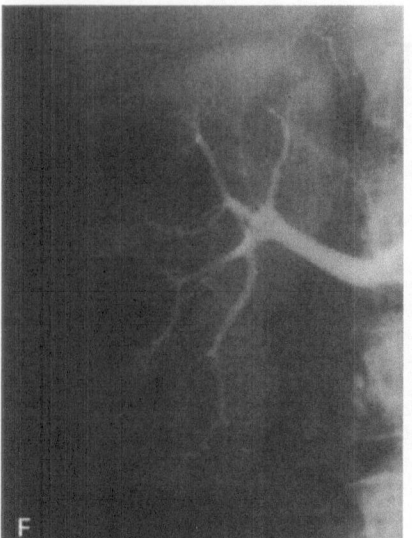
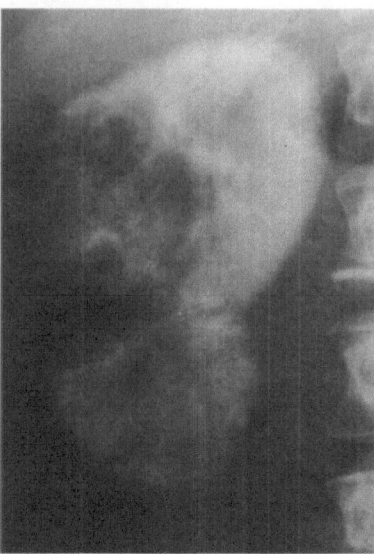
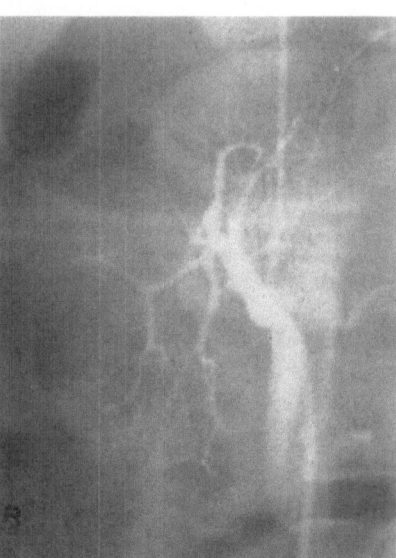

Fig. 56 A–E. When renal tuberculosis heals, it usually calcifies. A A large parenchymal granuloma in the lower pole of the left kidney. B Fine scattered calcification in the upper pole of the left kidney and the lower pole of the right kidney with scarring and shrinkage of the renal parenchyma. C Almost complete calcification of the right kidney. (There is no contrast medium.) An African from Kenya. This is the end result of renal tuberculosis. As well as heavy calcification within the right kidney, linear calcification can be seen in the right ureter, but not in the bladder. The ureteric calcification ▶ is unlike the intermittent, spotty calcification seen in schistosomiasis and, if that were the cause, there would almost certainly by bladder calcification also and a comparatively normal kidney. D, E Auto-amputation shown by ultrasonography (D) and by contrast CT (E). In D there is also echogenic material blocking the renal pelvis. (C courtesy of Dr. S. Malik, Nairobi; D, E from Cremin and Jamieson 1995)

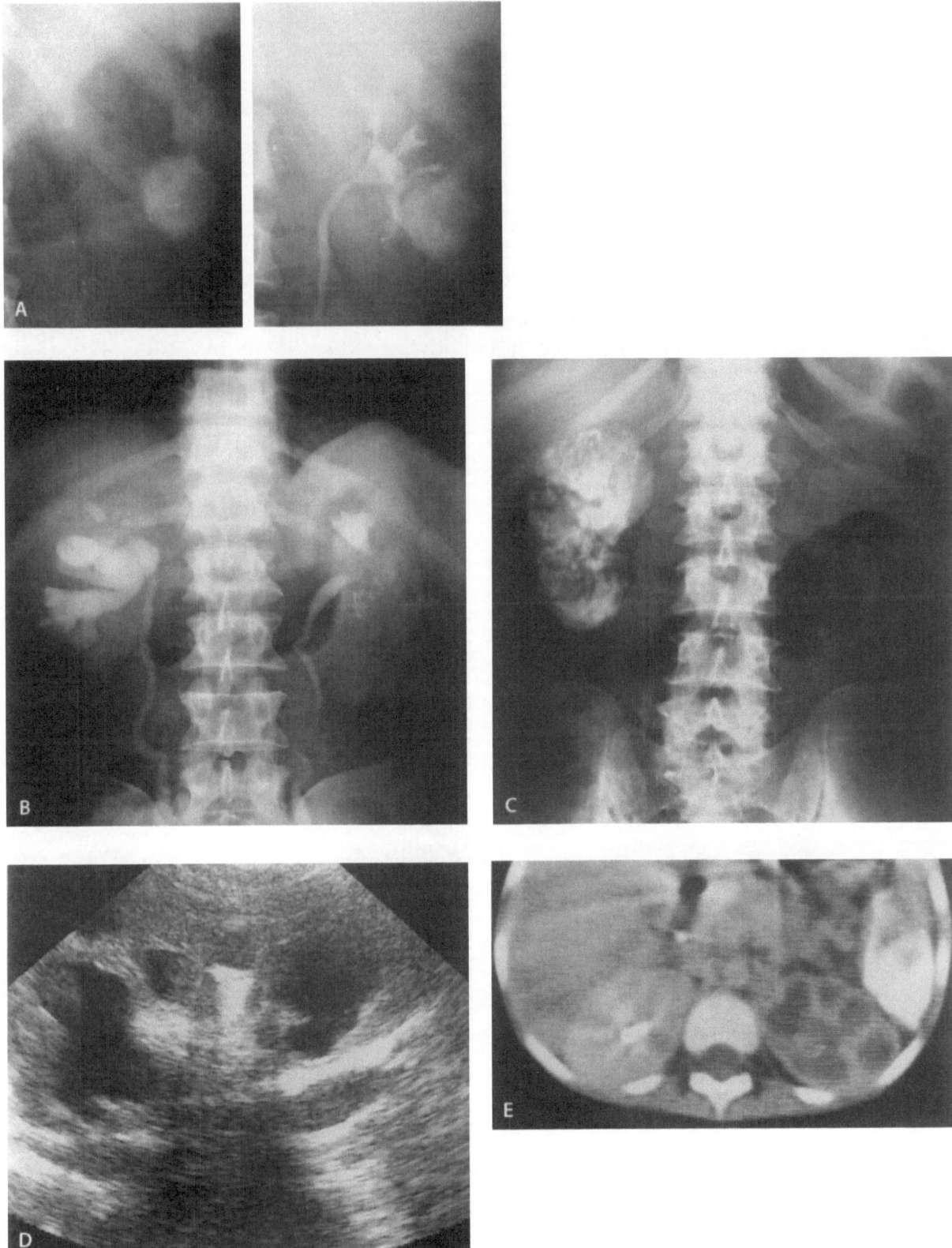

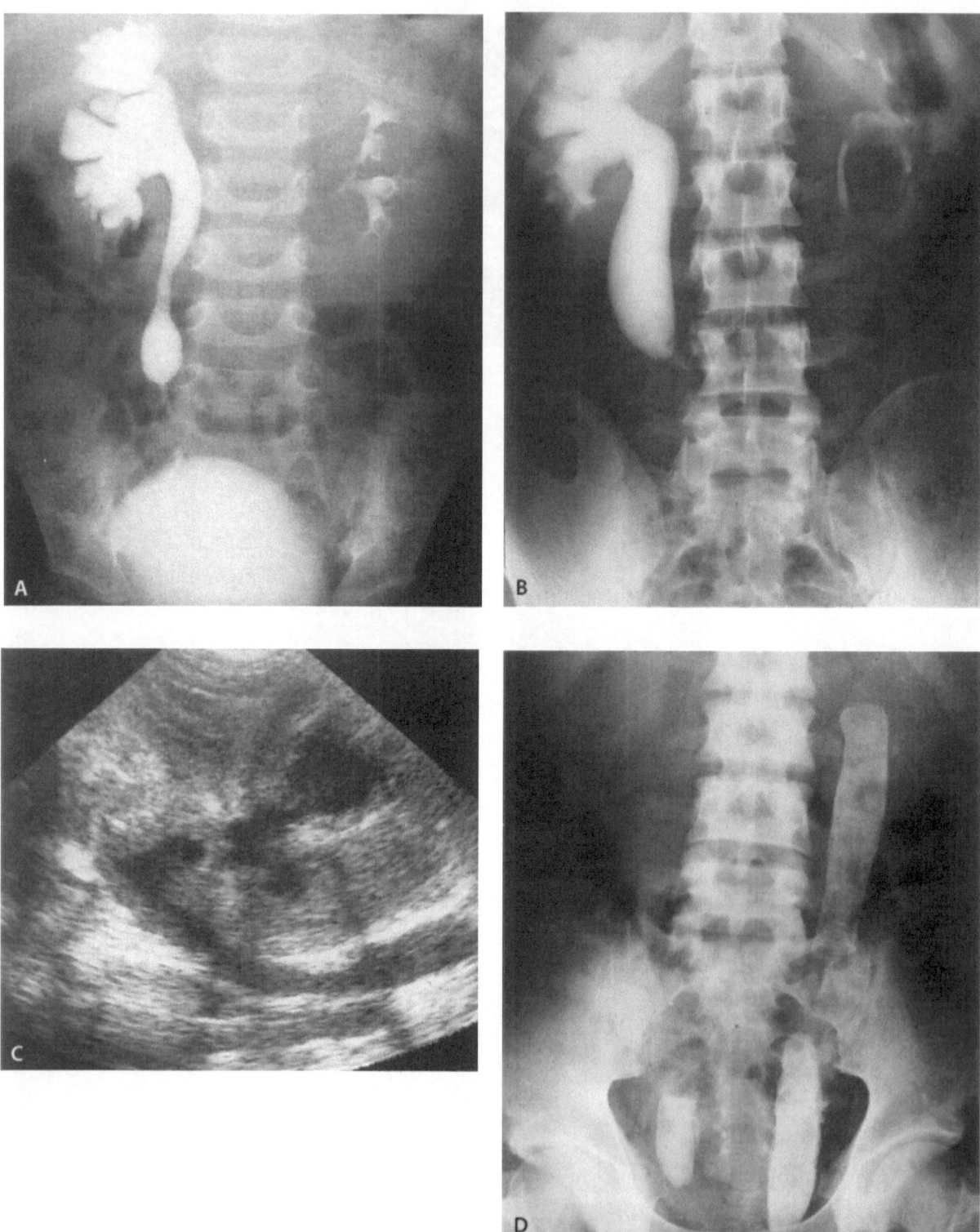

Fig. 57 A–D. Tuberculosis of the ureters. Strictures and dilatation can occur at almost any level along the ureter. **A** Hydronephrosis of the right kidney resulting from a stricture of the distal ureter in a child from South Africa. **B** A similar almost complete ureteric obstruction in an adult Kenyan African. It is difficult to know whether the kidney is also infected, or only obstructed. **C** Ultrasonography showing dilatation of the upper ureter and thickening of the ureteric epithelium, as well as parenchymal cavitation. **D** Advanced bilateral ureteric calcification. Unlike schistosomiasis, the bladder is not calcified. (**A, C** from Cremin and Jamieson 1995)

Bladder

When tuberculosis affects the bladder there is gradual and usually localized thickening of the bladder wall, with increasing diminution of the bladder volume. Trabeculation of the mucosa may develop. The vesicoureteric orifices are affected by this progressive fibrosis and there will then be bilateral, often asymmetrical hydroureter and hydronephrosis. Both CT and ultrasonography will demonstrate the irregular and thick bladder wall caused by the tuberculous granulomas (Fig. 58). To show this clearly, the bladder must be full which, in some patients with tuberculosis, may cause considerable discomfort because chronic contraction is usual. Contrast radiography or cystography is likely to demonstrate reflux up to the dilated ureters because the orifices will usually be rigid and held open. Bladder calcification due to tuberculosis is very uncommon and, when it occurs, is patchy. Calcification in small granulomas may be seen more easily on ultrasonography or CT. When bladder calcification is seen on a radiograph, schistosomiasis is the diagnosis until proven otherwise. Spontaneous tumor calcification can occur in bladder neoplasm (particularly in India), and needs to be differentiated from tuberculosis (but only after schistosomiasis has been excluded).

Adrenal Tuberculosis

Ultrasonography and CT (and MRI) have shown that adrenal tuberculosis is a little more common than previously suspected. It is usually bilateral, which is important in the differential diagnosis from adrenal hemorrhage in adults or from tumors (although metastases, especially from a lung primary, can also be bilateral).

Both ultrasonography and CT will shown enlargement of the adrenal glands: if it is easy to see the adrenals with ultrasonography they are probably enlarged (except in infants). On CT, there can be enhancement after contrast injection. When healing occurs, there will be calcification in the tuberculous granuloma and it can be diffuse, localized, or punctate: it is not always easy to recognize on plain radiograph. Tuberculous abscesses in the adrenal are initially well defined but later become complex.

The differential diagnosis is not difficult, particularly when there is calcification. Adrenal hemorrhage is not uncommon bilaterally during infancy but usually occurs on one side only in adults. Tuberculosis often causes bilateral enlargement and the enlargement may decrease but still persist after healing and calcification.

Genital Tuberculosis

Tuberculosis can infect any part of the genital tract but most commonly it is the fallopian tubes which are affected (Fig. 59 B–E). Hydrosalpinx and pyosalpinx are fairly common and may be large. There may be calcification of the pyosalpinx.

Hysterosalpingography shows flask-shaped dilatation of the fallopian tubes due to obstruction at the fimbriae. Sometimes, the block is at the uterine opening, and then the tubes are not visualized. Tuberculous endometritis and cervicitis may show as synechiae, irregular uterine mucosal lining, or an elongated cervix with loss of differentiation at the uterocervical junction. Surprisingly, tuberculous endometritis is not a significant cause of sterility (2% or less), whereas fallopian tube blockage will be, whatever the cause. Both ultrasonography and CT can demonstrate these changes and show the anatomical relationship clearly. However, the overall picture provided by hysterosalpingography has some advantages, particularly in the investigation of sterility.

Tuberculous masses within the pelvis can be the result of tuberculous peritonitis (see p. 67), with matting of the omentum, mesentery, and bowel, or they can be due to tuberculous tubo-ovarian abscesses. Both types of mass can be large and have been mistaken for carcinoma of the ovary or, in males, for the large nodal mass of lymphoma. On ultrasonography, such masses will have mixed echogenicity and there will be mixed density on CT because almost all will have both solid and cystic components. At least one tuberculous abscess has involved the bladder and vagina. Tuberculous tubo-ovarian abscesses may calcify and be seen in either side of the pelvis (together or singly) as well-defined homogeneous circular masses, sometimes with areas of increased density, probably due to the granuloma. Serpiginous or linear calcification can occur in the fallopian tubes.

Tuberculosis can affect the testes, presenting clinically as a swollen testes with a hydrocele. Ultrasonography will show a mass of mixed echogenicity, often with a central cavity containing debris (Fig. 59 A). It can be very difficult to differentiate tuberculosis from a tumor or other infection. Ultrasonography can also demonstrate the thickening due to edema caused by tuberculosis of the epididymis and vas deferens, both of which may eventually calcify and be seen on radiographs of the male pelvis. This must be differentiated from schistosomiasis and diabetes. Contrast studies of the vas and spermatic cord have not proven useful.

(Text continues on p. 84)

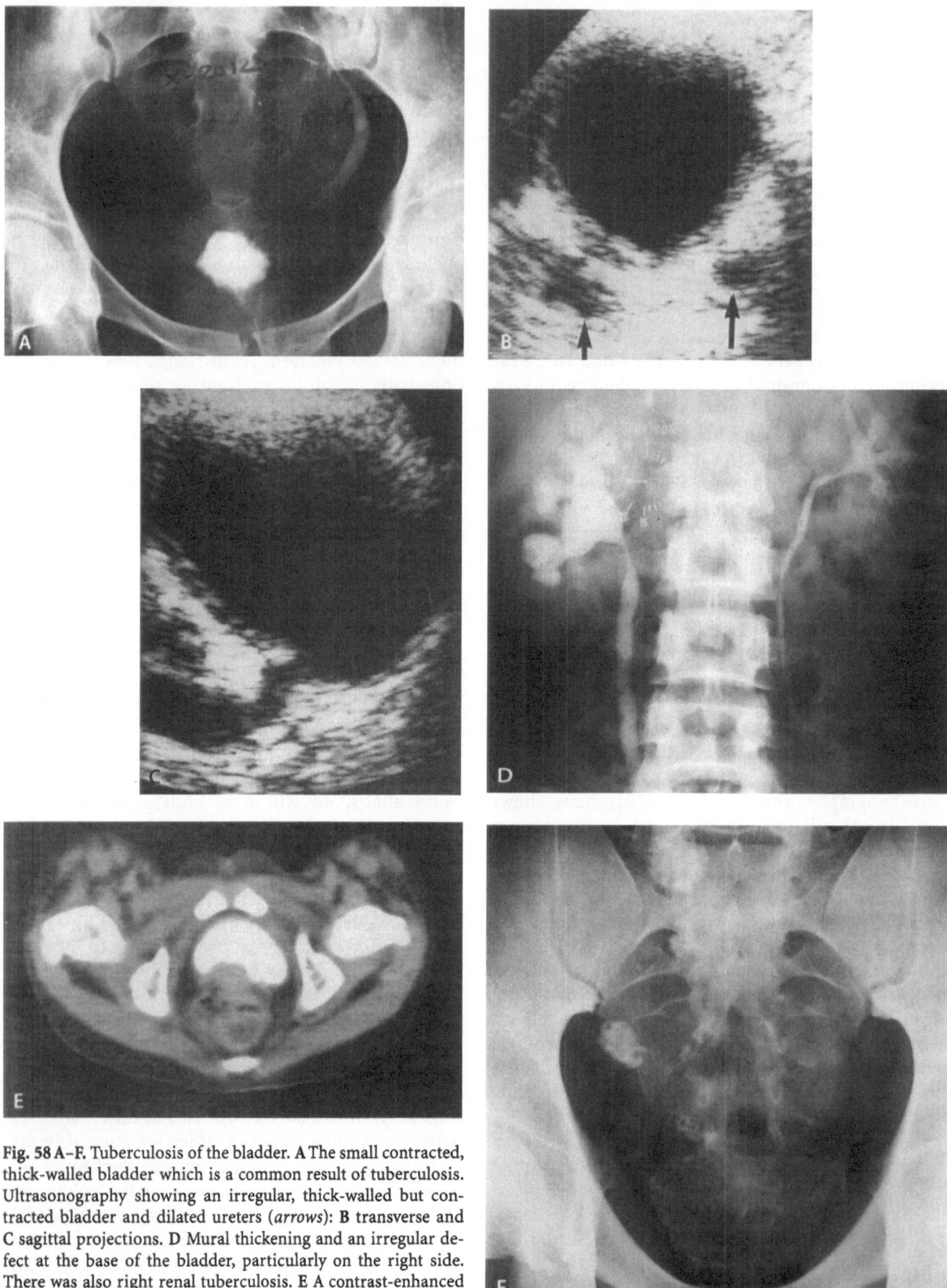

Fig. 58 A–F. Tuberculosis of the bladder. A The small contracted, thick-walled bladder which is a common result of tuberculosis. Ultrasonography showing an irregular, thick-walled but contracted bladder and dilated ureters (*arrows*): B transverse and C sagittal projections. D Mural thickening and an irregular defect at the base of the bladder, particularly on the right side. There was also right renal tuberculosis. E A contrast-enhanced CT scan of a child (not the same patient as in D) with tuberculous thickening at the base of the bladder. F Speckled calcification throughout the bladder and the lower left ureter. There are calcified granulomas in the lymph nodes in front of the sacrum on the right. It is unusual for tuberculous calcification to be so extensive: it is more commonly present in one bladder segment only. This was proven to be tuberculosis, without any evidence of schistosomiasis. (B, C, E from Cremin and Jamieson 1995)

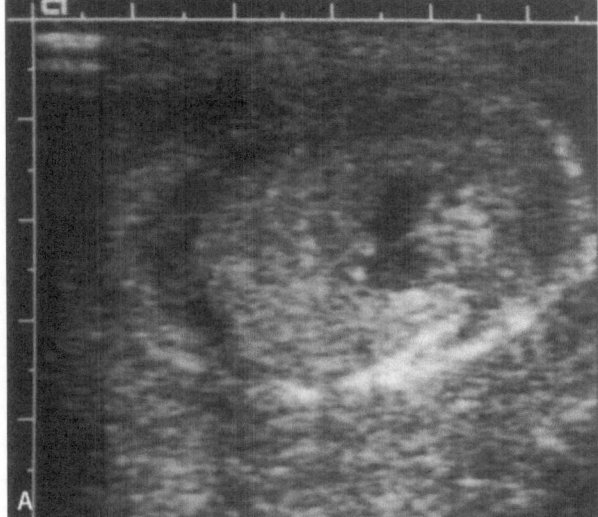

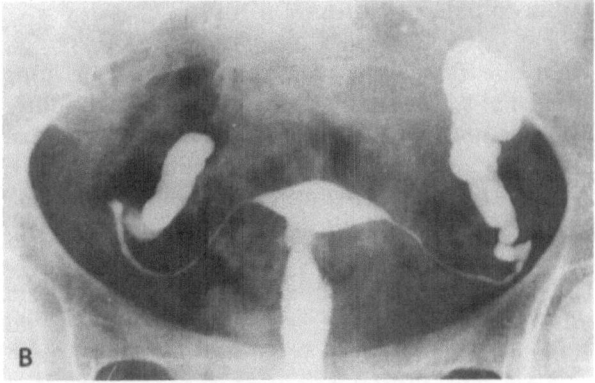

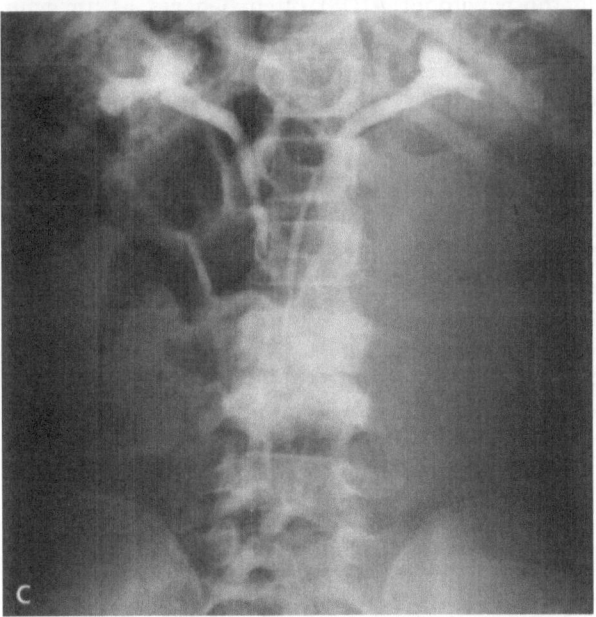

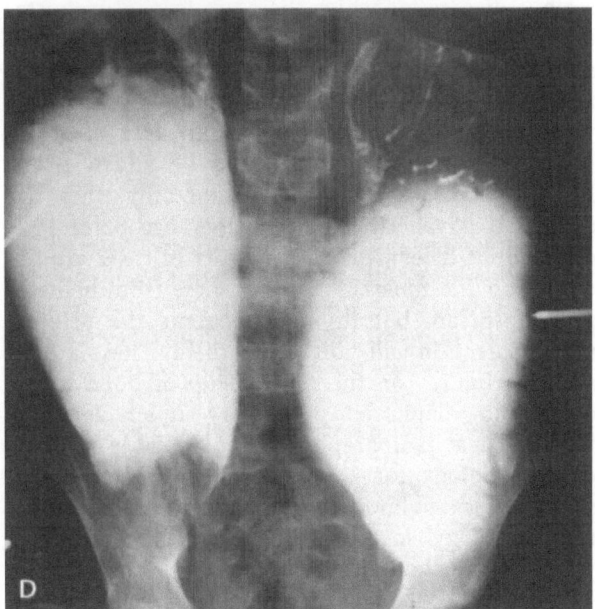

Fig. 59 A–E. Genital tuberculosis. **A** A sonogram of a tuberculous abscess in the testes: there is central necrosis with fluid and debris. Tuberculosis must be considered in the differential diagnosis of almost any testicular mass, even when there is no other evidence of tuberculosis. **B** Fallopian tuberculosis is a common cause of infertility but cannot be easily distinguished from any other infection. This patient from Kenya has bilateral hydrosalpinx and roughening of the contour of the cervix uteri. **C, D** This 13-year-old girl from Zimbabwe presented clinically with bilateral cystic abdominal masses. **C** Urography showed displacement upwards of both kidneys and medially of both ureters. The third and fourth lumbar vertebrae are both sclerotic, with loss of joint space between them. **D** The cystic masses were drained bilaterally and filled with contrast, showing two huge dilated fallopian tubes which were subsequently removed surgically. The infection in the spine responded to antituberculous therapy and a year later the urogram was almost normal. **E** A tuberculous abscess often forms a sinus or fistula. This histerosalpingogram shows a fistula between the left fallopian tube and the sigmoid colon. The uterine cavity is contracted and distorted and the right fallopian tube is dilated and blocked distally

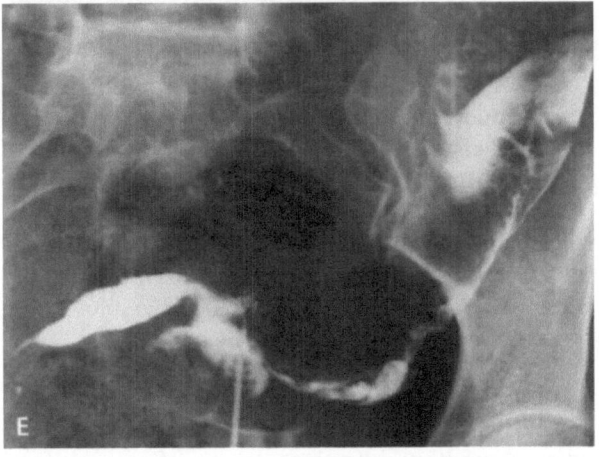

Tuberculosis of Bone

There are three important characteristics of bone and joint tuberculosis in the tropics which differ from the non-AIDS cases in North America and Europe: it is often an acute osteomyelitis or arthritis, it is often a destructive condition which can affect any and every bone in the skeleton, and it is frequently multifocal.

Bone and joint tuberculosis is extremely common in tropical countries; there is a tendency to focus on pulmonary tuberculosis as the major public health problem, but skeletal infection is a neglected and serious disease for which there are few accurate statistics. It is common in India, Asia, much of tropical Africa, and South America. It is fueled by malnutrition, inadequate housing, poor medical services, and, perhaps most importantly, lack of health education; to this list must now be added AIDS. Many patients report to the doctor only when the disease is advanced and when they are paraplegic, crippled, or deformed. The majority of cases are not due to bovine tuberculosis; they arise from hematogenous spread, but in many patients the original site of infection will not be identified. There are no accurate figures on the relationship of bone tuberculosis to diagnosed pulmonary tuberculosis. A normal chest radiograph does not exclude a tuberculous etiology for a bone or joint infection, but neither does pulmonary tuberculosis mandate the same etiology for any osteomyelitis or arthritis; pyogenic infections are as common as tuberculosis. Whatever the state of the lungs, when there is a skeletal infection in the tropics tuberculosis should be considered in the differential diagnosis, whether the patient is HIV negative or positive.

No age is exempt. Skeletal tuberculosis may occur in babies from 6 months upwards and become an advanced disease; the elderly are similarly affected. It is most common in children and young adults, but is not rare in any age group. It occurs most frequently in the spine; the hip is the most common joint affected, followed by the knee, ankle, elbow, wrist, and shoulder in decreasing frequency. Tuberculous osteomyelitis can affect the skull, mandible, pelvis, ribs, and scapula. No bone escapes. Next to trauma it is the commonest cause of crippling and deformity and is such a common disease that it is cared for in general hospitals all over the tropical world. Drug therapy is usually effective, often dramatically so, but the late stage at which the patients appear necessitates skilled orthopedic correction which is not always available. Although the clinical presentation may be acute, the radiological findings often indicate that the infection has been present for a considerable time.

The majority of patients with skeletal tuberculosis (if they do not have AIDS) will be positive tuberuclin reactors, but a small percentage, perhaps less that 2%, will have negative skin tests. This can cause difficulty in diagnosis. Equally, because so many of the population will be positive reactors anyway, the finding of a positive tuberculin test in a patient with bone or joint infection is only really significant in infancy and the first few years of life. The diagnosis must be suggested by a high clinical index of suspicion, and confirmed by biopsy and culture. Unfortunately, not all biopsies will yield positive results; it may be necessary to treat the disease because of the clinical and radiological findings, despite a negative culture. A "sterile" culture is most likely to be tuberculosis and is useful in that it helps to exclude a pyogenic and typhoid etiology.

Because of poor host immunity, tuberculosis (even in those without AIDS) often behaves as an acute disease and must be treated in the same way. Surgical biopsy, even aspiration, is not a benign procedure and may result in widespread hematogenous dissemination and septicemia. A potentially tuberculous bone or joint should be treated with the same caution as a pyogenic dental abscess and surgery should be carried out under a therapeutic "umbrella" of antituberculous drugs; this rule should be violated only in cases of paraplegia due to an abscess or some similar acute emergency. The clinical and radiological diagnosis is surprisingly accurate in the majority of patients; biopsy will be confirmatory and provide a culture for the sensitivity of the organism, rather than being an immediate necessity. In the spine, removing the "pus" may be therapeutic also.

Tuberculous osteomyelitis and arthritis are often associated with marked peripheral lymphadenopathy (also tuberculous). This combination is particularly found in spinal tuberculosis and conversely, when children have gross lymphatic involvement, concomitant bone or joint disease should be carefully excluded. In one reported series from the Philippines, 7 out of 39 children with lymphatic tuberculosis were found, on radiological examination, to have spinal tuberculosis which had been clinically unsuspected. This combination occurs throughout the tropics, particulary in India and Africa; skeletal involvement can be remarkably silent in its early stages and may only become obvious because of a complication. In AIDS, skeletal tuberculosis becomes yet another part of the onslaught of infections.

The best way to image any part of the skeleton will depend very much on the site involved, but because tuberculous bone infections are so often multifocal and may be present in different parts of the skeleton and yet be clinically silent, scintigraphy is a very valuable early investigation. However, when such a scan shows multiple active sites in the patient who is clinically ill and has lost weight, the possibility of malignancy will also have to considered. To make the differential diagnosis more difficult, the tumormarker CA 125 may be positive in destructive skeletal tuberculosis and even in tuberculous peritonitis.

Tuberculosis of the Spine

Synonyms

Caries. Pott's disease. Pott's paraplegia. *Ger*: Pott-paraplegie. *Fr*: paraplégie de Pott.

Clinical Characteristics

The description and statistics in this section refer to HIV-negative patients. For those with AIDS, the clinical and imaging findings are similar, but often more advanced, and the site incidence may be different. Tuberculosis in AIDS is "tropical tuberculosis," but more progressive and often more advanced. In most tropical countries tuberculosis of the spine accounts for more than 50% of all cases of skeletal tuberculosis. Clinically, paraplegia is the commonest presenting symptom. It may be complete and acute, with bladder paralysis and inability to walk; depending on the level of the infection, it may present as quadriplegia. In other patients the symptomatology may be less severe, but neurological involvement occurs in half of those who have tuberculous spines. The remainder complain of the deformity, of pain, or of general weakness and ill health which bring them to the physician. Some will complain of a mass in the groin, which may be a hernia, but the possibility of a psoas abscess should be remembered in every such case. Whatever the presentation, a careful examination will reveal some neurological deficit (often unsuspected) in the majority of patients when first seen; yet they may have very gross spinal deformity but still be ambulatory.

Clinico-pathological-radiological Correlation: Spinal Tuberculosis

If the radiological findings in tuberculosis of the spine are to be understood, it is important to have a clear idea of the underlying pathological changes. These were beautifully described in 1936 by Compere and Garrison, working at the University of Chicago. They made longitudinal whole-body sections of tuberculous spines and correlated these with radiological findings and the histological examination. Their autopsy research has since been confirmed at surgery by Hodgson and others in Hong Kong (1969). Now, in recent years, CT and especially MRI have imaged the pathophysiology, and provided three-dimensional confirmation. There are two basic premises: (a) cartilage resists destruction by tuberculosis and (b) there is no blood supply to the intervertebral disc, although there is a rich blood supply to the vertebrae. The vertebral blood supply does not cross the articular surface; the disc obtains its nutrition from lymphatics.

The intervertebral disc is a fibrocartilaginous ring, the annulus, blended with hyaline cartilage plates above and below and enclosing the nucleus pulposus. The nucleus is an interlacing matrix of fibrous tissue and fibrocartilage, with semigelatinous substance within the mesh; because this is liquid it is incompressible, and therefore narrowing of the disc can only take place when there is extrusion or destruction of the nucleus or, alternatively, dehydration. Tuberculosis commonly starts by hematogenous implant and more rarely by direct extension from infected lymph nodes. It is essentially a bone-destroying infection, with little evidence of repair in the early stages. When it occurs beneath cartilage, the cartilage is eventually destroyed but the annulus and nucleus tend to survive much longer. It is only when there is fissuring of the annulus that the infection can involve the disc proper. Even when the vertebral body has become caseous, wedged, and destroyed, careful histological search will nearly always show the nucleus, either intact or in part, having prolapsed into the softened bone or debris. Primary tuberculous invasion of the annulus has not been demonstrated. Narrowing of the disc space occurs when the annulus eventually becomes infected and fluid escapes, or when the nucleus is extruded. At this stage there will be loss of the disc space when imaged.

Compere and Garrison also showed very clearly the spread of tuberculous abscesses under the anterior or, more rarely, the posterior spinal longitudinal ligaments. When this occurs the infection involves the vertebral bodies and the disc remain intact for a considerable time. They illustrated pathologically what may be demonstrated by imaging, i.e., that multiple vertebrae can be involved above and below the original focus and yet the disc and the dura may remain intact. The cartilaginous plate of the vertebra is a barrier, more effective in children than in adults because it is thicker during childhood.

The studies of Compere and Garrison showed pressure on the spinal cord is likely to be from an epidural or subdural abscess, and sometimes from a prolapsed disc. Direct pressure from a collapsed vertebra is rare. Paraplegia thus results from the abscess, from the prolapsed disc or tuberculous debris, or occasionally from edema of the cord due the neighboring infection.

The same authors also pointed out that tuberculous peritonitis from direct spread of a spinal abscess is uncommon and that most cases of tuberculous spinal meningitis are hematogenous in origin.

Compere and Garrison showed that in pyogenic infections, invasion of the intervertebral disc occurs more rapidly, is seen earlier, and is more common. This is because the cartilaginous plate is destroyed by proteolytic enzymes and the nuclear substance is extruded more rapidly, with dissolution of the annulus. This occurs in all pyogenic infections, but particularly in brucellosis. The ability of the intervertebral disc to survive a tuberculous infection during the early stages is important in the differential diagnosis. It is particularly well illustrated in Figs. 60B,C and E, 61A, and 66. The anterior spread of an abscess is well illustrated in Fig. 71B.

Imaging of Spinal Tuberculosis

There are important general principles which guide the imaging of spinal tuberculosis:

1. Plain skeletal radiography provides a great deal of information, but it does not show the extent of the very important intradural abscess or of its paravertebral extension. It does not always demonstrate every locus of bone infection, nor does it show the nucleus palposus.
2. Because spinal tuberculosis is often multifocal, bone scintigraphy is a very valuable early examination; not only will other vertebral foci be seen, but early infection elsewhere in the skeleton may be discovered.
3. Ultrasonography is useful in demonstrating the paravertebral abscess and any associated lymphadenopathy. Beyond this, it does not provide useful information in most cases.
4. CT is excellent for demonstrating the exact extent of the skeletal disease, the deformity, and the size and position of the abscess (if any). It does not demonstrate intramedullary foci.
5. MRI provides very accurate information about both the soft tissue and bone involvement. It is accurate in the demonstration of bone lesions and is the only way to image focal myelitis in the cord. It is a poor method of imaging calcification as it occurs during healing.

Vertebral Body

Although the destructive process of spinal tuberculosis is best imaged by CT or MRI, plain radiography and standard tomography can provide useful information, particularly if the underlying pathological process is understood.

The original tuberculous focus may be marginal, central, or subperiosteal and is usually in the vertebral body (Fig. 60). Infection starting in a pedicle is not all that uncommon, and can be bilateral. In nontropical countries destruction of the pedicle suggests malignancy, either metastatic or lymphomatous, but in the tropics it may be tuberculous. It is not very uncommon for only the neural arch to be affected, the vertebral body remaining intact.

This can be difficult to detect with plain radiography, but is well seen by CT. The transverse processes are rarely involved except as part of the spread of infection; the spinous process is occasionally affected, but causes little deformity.

Many tuberculous spinal infections are multifocal: when suspected in any vertebra, the whole spine should be radiographed (if there has not been previous scintigraphy). Infection may be centered in one, two, three, or four contiguous vertebrae, or there may be normal vertebrae between other sites of infection, above or below that originally discovered. In some patients there will be more than two involved areas, each separated by normal vertebrae.

Fig. 60 A–I. Tuberculosis can affect any part of the vertebral ▶ body or neural arch. In the early stages, the intervertebral disc space may remain normal. A Destruction in the center of the vertebral without any change in shape. A patient from Tanzania. B This vertebral body has also maintained its shape but there is more advanced central destruction and the upper border will probably collapse. C The upper border of this vertebra is beginning to collapse and there is minimal lose of disc space. The central focus of infection was clearly visible only on tomography. D Similar infection in the lower half of the body. On the routine films, the loss of the cortical margin was visible but not the central destruction. E Central destruction of the lower half of the vertebral body, only seen on the anteroposterior view. The disc space is slightly narrowed. F Destruction of the right side of the vertebral body and the neural arch, with the remainder of the body maintaining its shape. The lower disc space is narrowed on the right side; the upper space is almost normal. A small lytic defect is present on the right side of the body below, and there is a small paravertebral abscess. The extent of the bony destruction is usually underestimated by plain radiography and is better seen by CT or MRI (see also Figs. 68–70). G, H Tuberculosis affecting the spinous process only. This is very unusual. I A CT scan of a child from South Africa, showing destruction of the neural arch on both sides, as well as of the vertebral body. *Arrows*, anterior spinal abscess. (A courtesy of Dr. Harold G. Jacobson; I from Cremin and Jamieson 1995)

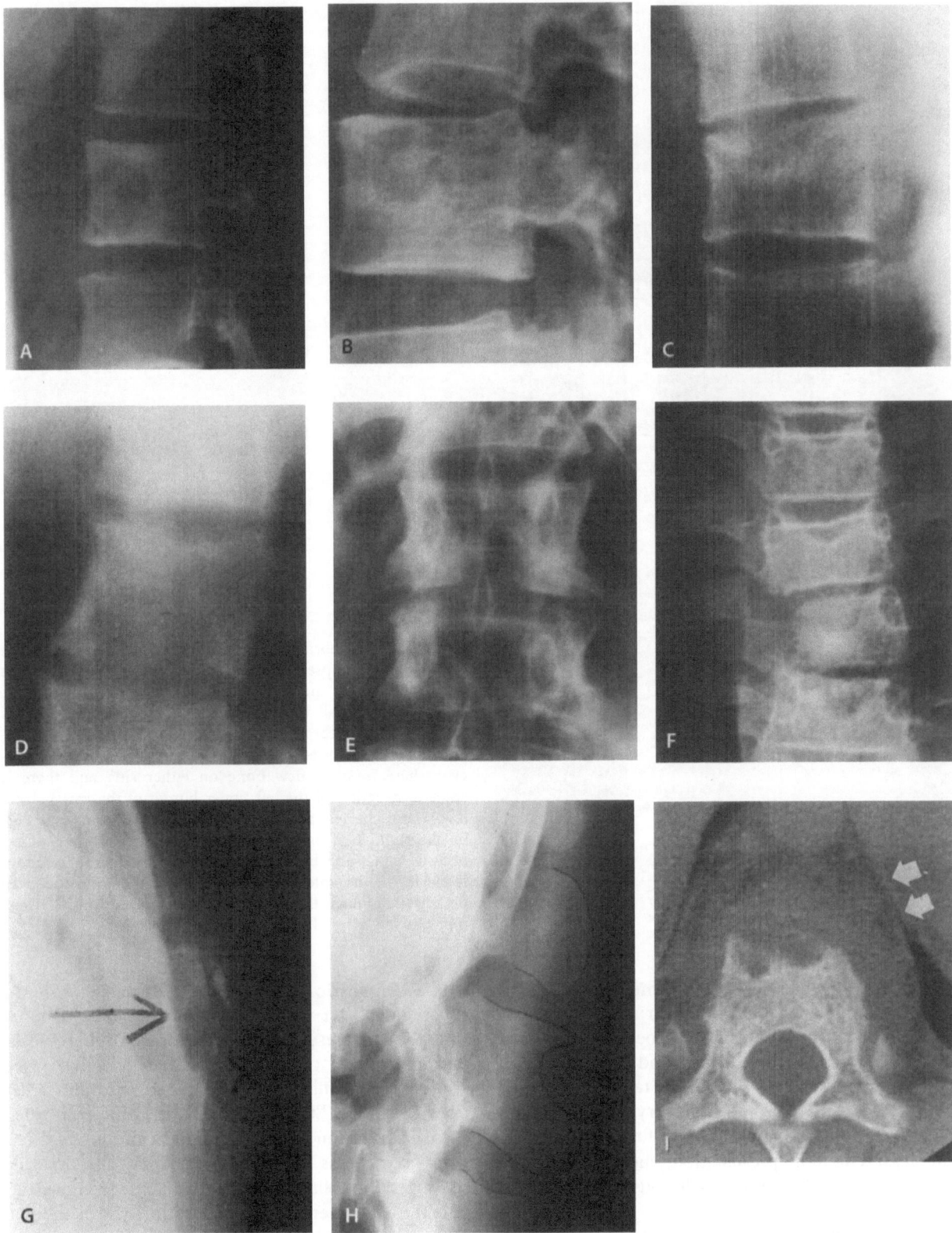

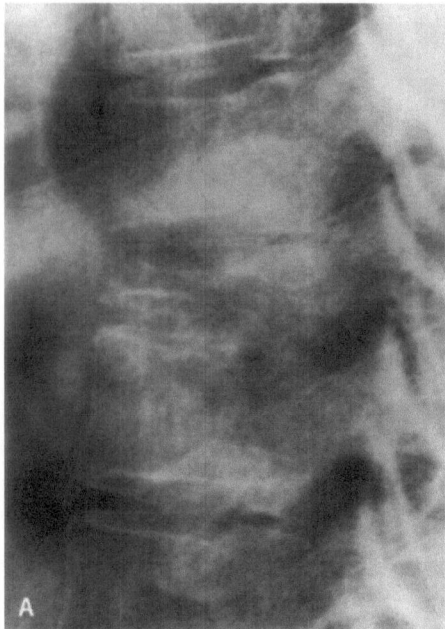
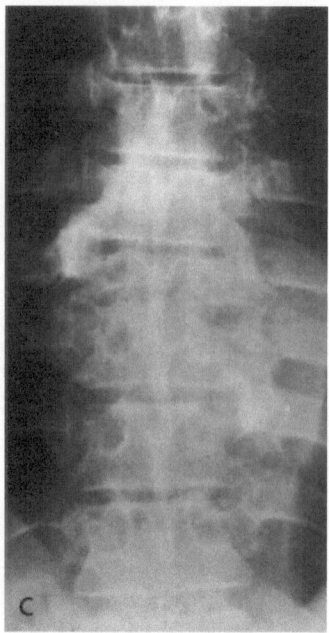
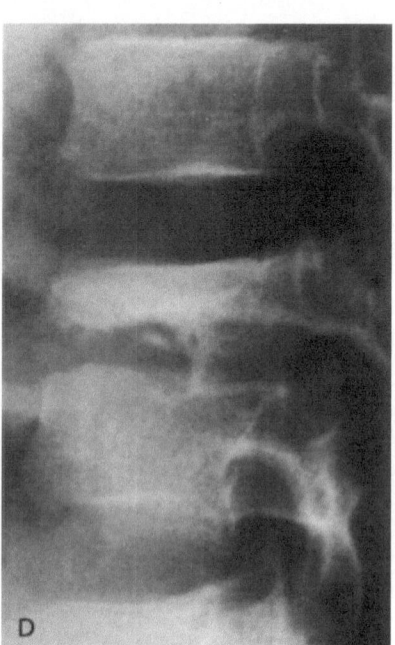

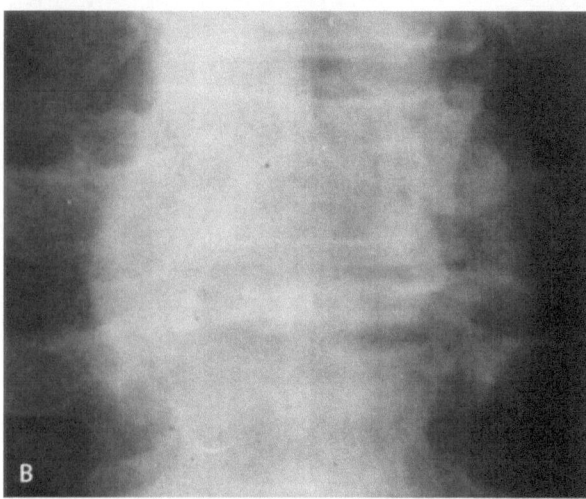

Fig. 61 A–D. Vertebra plana. Four different patients. **A** The vertebra has collapsed and yet the upper and lower margins remain intact and the disc spaces are clear. **B** More extensive collapse with small bilateral paravertebral abscesses. **C** This mid-thoracic vertebra has collapsed in the same way, with intact upper and lower margins, and clear disc spaces. However, there is some new bone on either side and there is some destruction of the left upper border of the vertebra below. There is a left paravertebral abscess, but none is visible on the right. **D** A pseudo-vertebra plana: the top half of the body is intact and becoming sclerotic. The disc space above is clear, but the anterior half of the lower margin has been destroyed although the posterior edge remains normal. These four patients were all Africans from Zimbabwe

The degree of bone destruction when the patient is first radiographed is infinitely variable. There may be an area of increased lucency, occasionally localized by low-grade clinical pain and perhaps identified after radionuclide scanning. Standard tomography, CT, or MRI, may be necessary to find the defect within the vertebral body. Unfortunately, this is the least common presentation; most patients are first seen when the vertebral lesions are easily visible on routine films.

Because of the bone destruction the extent of the change in the vertebral shape is variable. The vertebra may be intact, with no change in its outline or there may be one or more peripheral defects; this can occur in any part of the vertebral body. Alternatively, collapse may be so complete that there is a "vertebra plana" (Fig. 61). In nontropical countries

it is a diagnostic "rule" that vertebra plana is never tuberculous, but we have a personal series of more than 40 such cases, all of them proven tuberculosis, and many other radiologists in the tropics have seen similar cases. Tuberculosis in the tropics and AIDS does not follow European and North American rules. Wedge formation is the commonest finding and is a complex process with many variants. The majority of vertebrae collapse anteriorly; some demonstrate lateral collapse and only in a minority will the collapse first occur posteriorly. (This is said to be more common in pyogenic infections, but it can happen in tuberculosis.) The wedge may be formed of only one vertebral body which has collapsed on itself, either with both the upper and lower borders falling inwards towards the center, or with one or other border remaining intact and the wedging occurring onto it (Figs. 62, 63). However, the radiological appearance of a wedge can be misleading and what at first seems to be one vertebra may be a wedge formed of two or even more. The disc space may seem to be totally obliterated. In such cases, the upper border of the top vertebral body and the lower border of the one beneath remain intact but are no longer parallel, and the central parts of both vertebrae merge to become the wedge. The angulation can vary so that the wedge is around either the upper or lower vertebra, or both. The same destructive process can extend to involve three vertebrae or, rarely, four. When this happens the middle vertebral body disappears. It is thus essential to identify each pedicle when assessing tuberculous wedging, and if possible also to localize the vertebral ends of the ribs to be sure of the number of vertebrae involved. A seemingly severely wedged small vertebral body may be all that is left of three or, at the most, four vertebrae. Fusion may be complete, the cortical margins may be reestablished, and there may be no trace of the original anatomy. The resulting spinal deformity is equally variable. There may be little clinical abnormality, even when there is marked vertebral destruction; alternatively, there may be severe kyphosis (Fig. 64), and if two regions of the spine are infected, the curvature is often bizarre and myelography (see later) becomes a major problem. In the majority of patients the spinal cord manages to adapt to amazing curves, and paraplegia is seldom due to the bony deformity; it usually results from the abscess within the spinal canal. When paraplegia occurs without pus, it is due to vascular or toxic changes, or, as MRI has demonstrated, the intramedullary lesion; very rarely paraplegia is due to vertebral disc displacement. The disc is involved late in tuberculosis, and is more often displaced into the vertebral body than posteriorly into the cord. Even when most of the rest of the disc has disappeared, very late in the infection, MRI can often show the fluid in the nucleus.

Healing of a tuberculous infection in the spine occurs in many different ways. The vertebral body may return to normal and be reconstituted with apparently normal margins and only slight increase in trabecular density to indicate the previous infection. When two vertebral bodies have fused, the resulting bone may also look almost normal, apart from the wedge shape. In other cases vertebrae sclerosis occurs, so that all or part of the vertebral body is increased in density (Fig. 65); this is evenly distributed and is seldom patchy. In temperate climates it is commonly said that tuberculosis does not cause sclerosis, but this is not applicable in the more acute pattern of the disease in the tropics. However, sclerosis does not always follow the expected distribution of the bone necrosis; it may be seen radio-

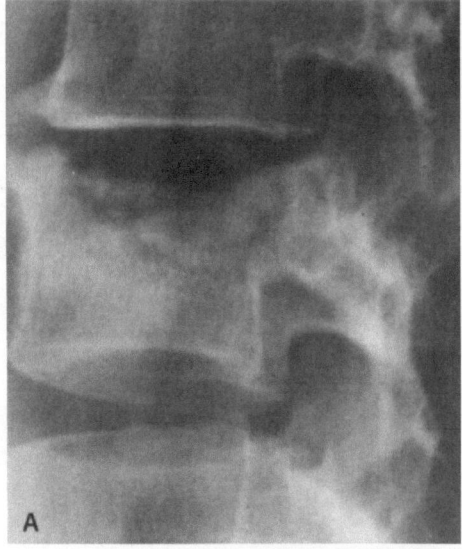

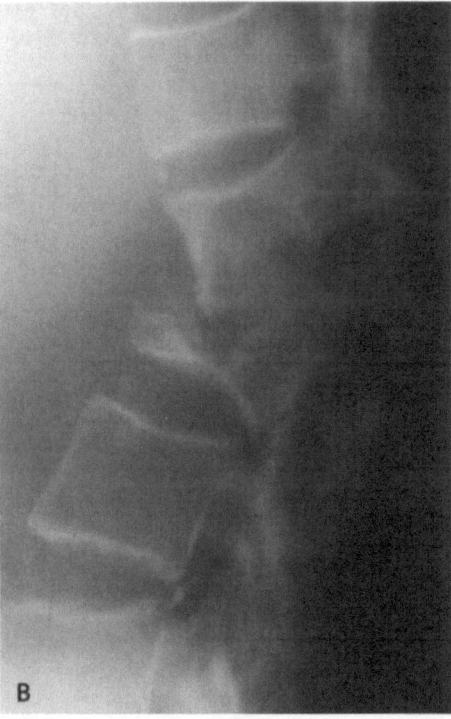

Fig. 62 A, B. In early spinal tuberculosis, the commonest finding is decrease in the disc space and some destruction of the vertebral body. **A** Most of the upper part of the vertebral body has given way and the intervertebral disc has probably sunken into the necrotic bone. The vertebra above is intact, but the disc space is narrowed. **B** This vertebra has given away even more, so that the disc and the vertebra above have caused it to split. Yet the disc spaces above and below are normal and no other infected vertebrae are visible on this radiograph. (Though there is the possibility of other foci of infection above or below and these should be sought and excluded. Increasing kyphosis and, eventually, spinal cord damage are inevitable unless this is treated.)

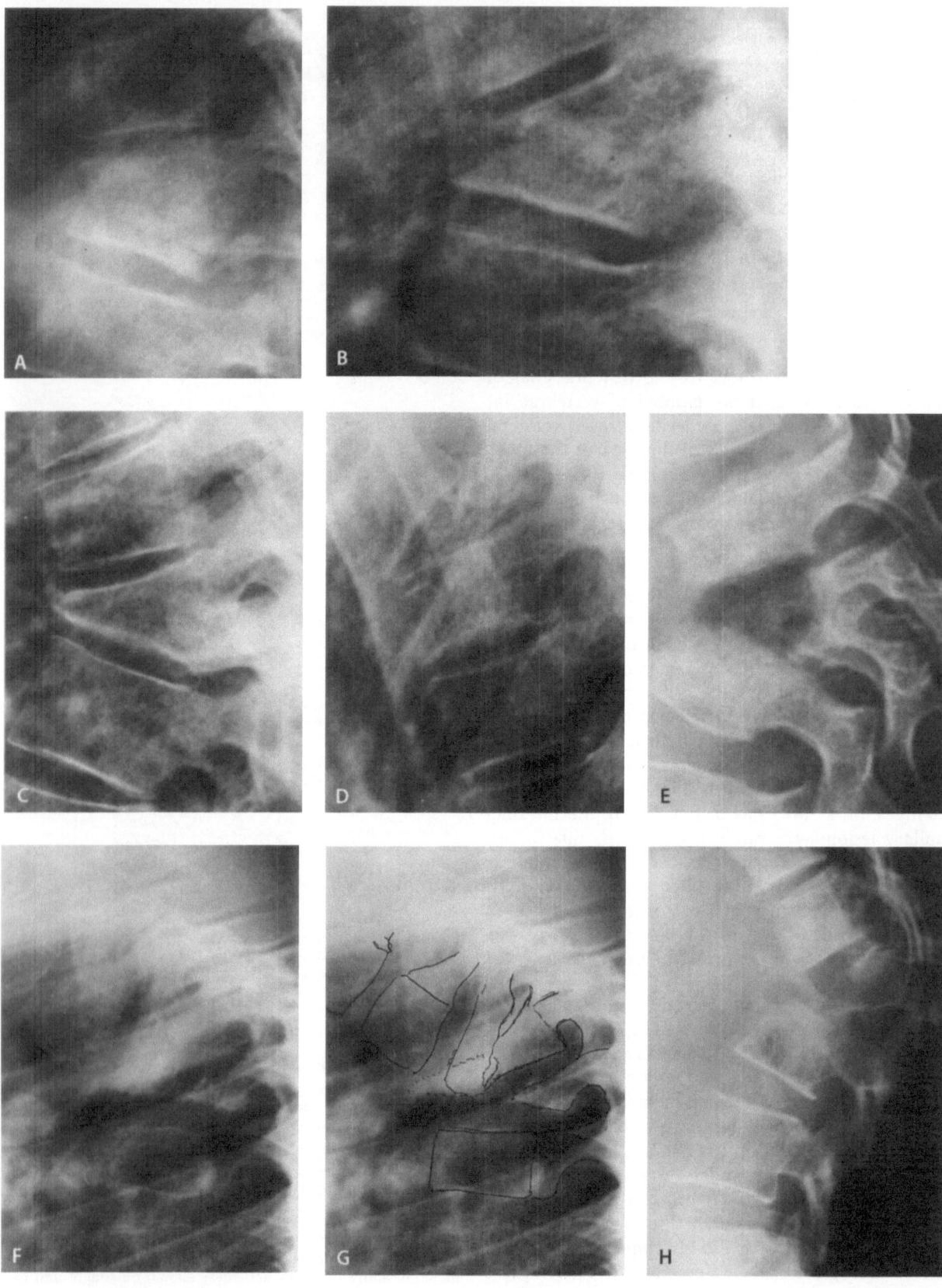

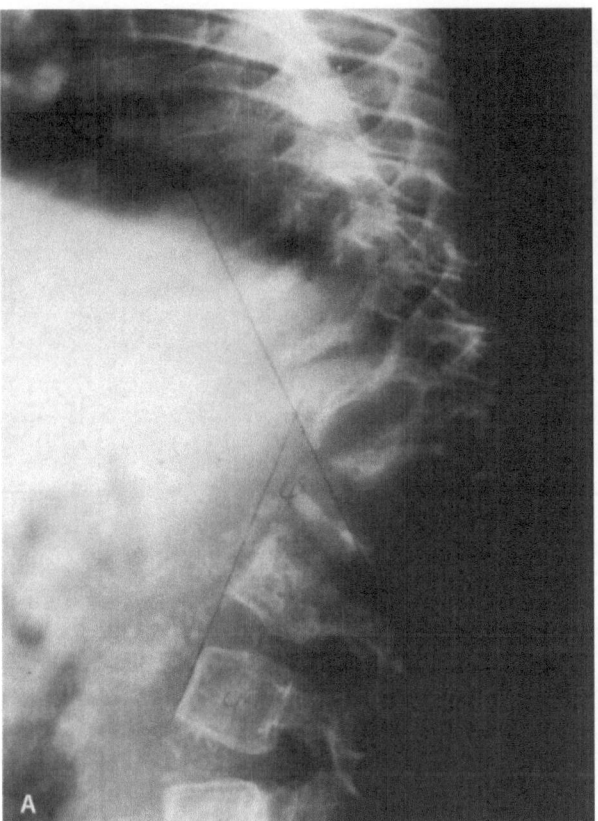 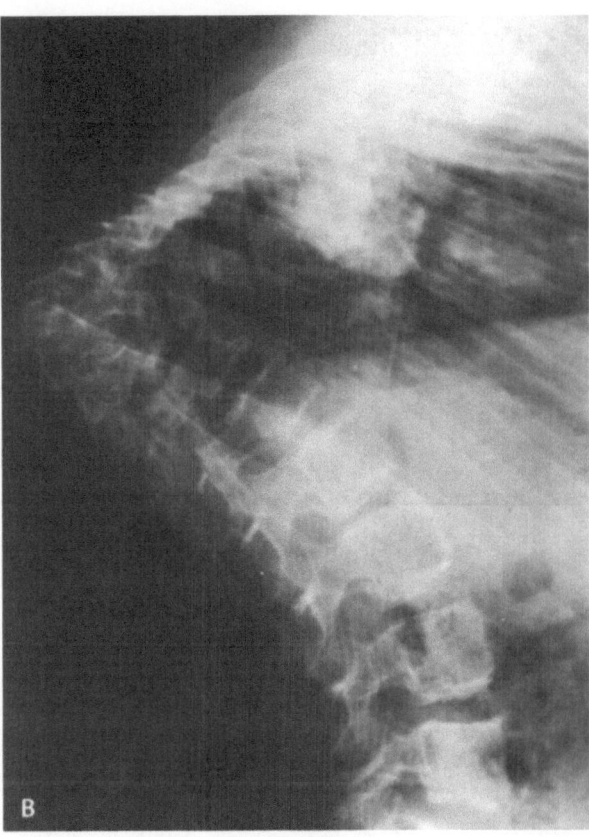

Fig. 64 A, B. Tuberculous kyphosis. Two African patients with severe kyphosis and infection of multiple vertebrae. **A** Eight vertebrae are infected. Three have become individual vertebrae plana. The wedge in the center is formed from the 12th thoracic and the first lumbar vertebrae combined. There is a large calcified psoas abscess extending anteriorly down into the pelvis. **B** Similar mid-thoracic infection involving multiple vertebrae which are now healed. There is a calcified abscess anteriorly. There is no vertebral rotation and the anteroposterior view showed there was very little scoliosis

◄ **Fig. 63 A–H.** The way in which wedge formation in tuberculosis occurs depends on the number of vertebrae involved. **A** Opposing surfaces of two thoracic vertebrale have been infected, and the disc space is no longer visible. **B** Twelve months later (with treatment) the bones are healing and fusing into one. **C** Thirty-one months after film A, there is a solid wedge made from two vertebral bodies, with well-defined edges. The disc spaces above and below the wedge are normal: there will probably be no further change until degenerative arthritis develops. Wedge formation can be more complicated. **D** One complete vertebral body has collapsed into a wedge, but with normal vertebrae above and below. **E** Two vertebrae have become wedged and the opposing surfaces of the vertebrae above and below are also infected. **F, G** There are three vertebrae at the center of this kyphosis and they will probably fuse into one solid wedge. If there is no abscess, this degree of curvature does not immediately cause neurological symptoms in children, but as their growth alters from age 12–13 years onward, there may be pressure on the cord. The end results may resemble **H**. This healed wedge formed from two vertebrae and fused with the vertebral body immediately below

logically in parts of the vertebra which have up to then been considered normal. CT, MRI, or standard tomography of a tuberculous spinal infection usually demonstrates considerably more bone destruction than can be appreciated on routine films, and during healing the sclerosis may follow this distribution. (Except when affecting the upper thoracic vertebrae, the indication to scan or tomograph tuberculous vertebrae, particularly when the diagnosis has been firmly established, should be based more on the neurological examination than the radiographic appearances.) Bone sclerosis does not necessarily indicate secondary infection; it can occur as a result of tuberculosis without any complications. Spinal tuberculosis often heals with marked anterior and lateral bony bridging, to such an extent that it may resemble ankylosing spondylitis (Figs. 66, 67). It may be surprisingly symmetrical, but the normal sacroiliac joints in this pattern of tuberculosis may help to establish the correct diagnosis.

(Text continues on p. 94)

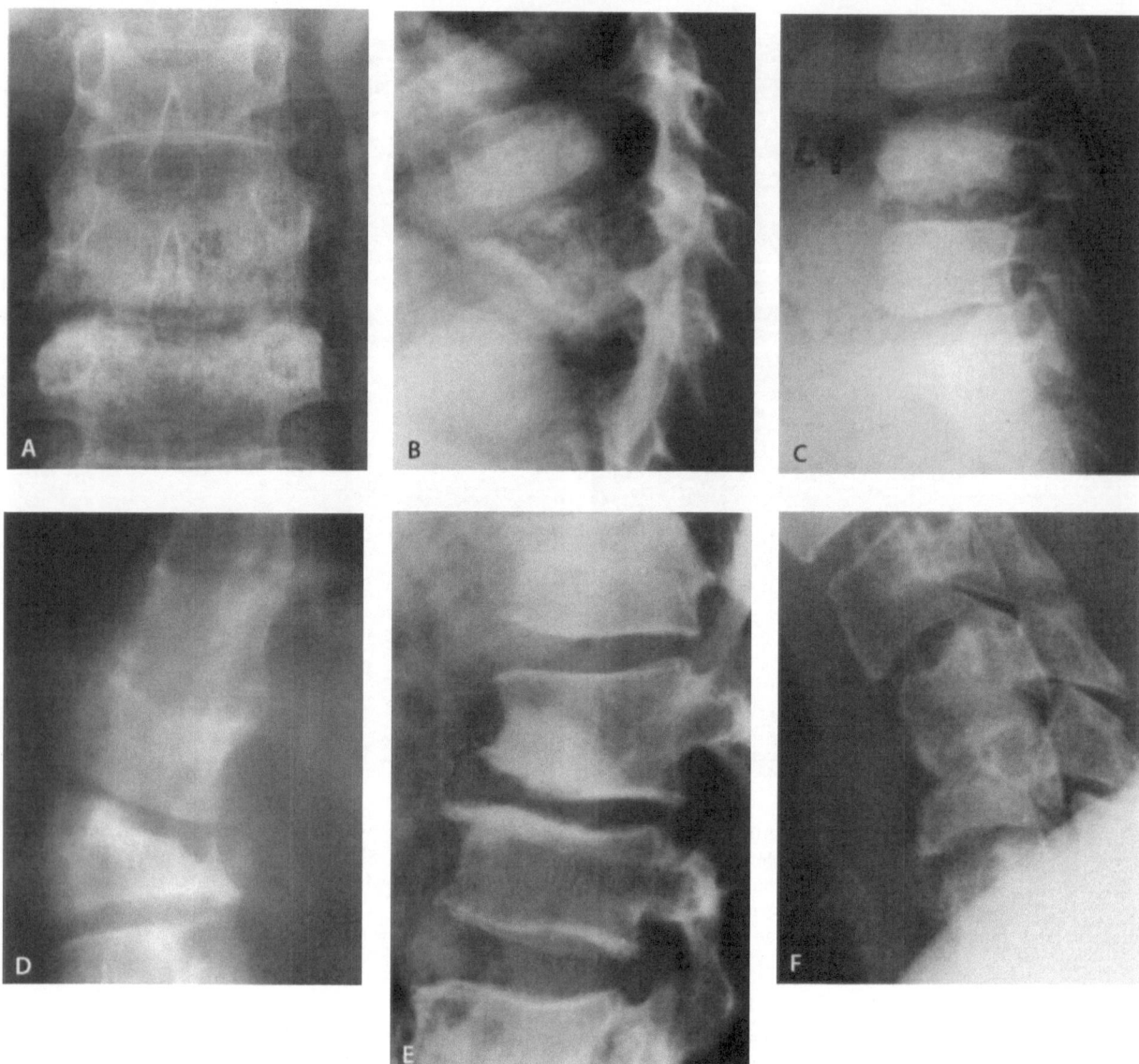

Fig. 65 A–F. The healing of spinal tuberculosis. Sclerosis represents initial evidence of repair but it does not reliably indicate that the infection has healed completely. **A** Increased trabecular density occurring about 3 months after starting antituberculous treatment. **B** Increasing density in the lower part of a wedge of vertebrae. **C** Increased density of a whole vertebral body, which is irregular in outline and smaller than the two normal vertebrae above and below it. The disc spaces are surprisingly clear. **D** The lateral aspect of this vertebra had collapsed and the bone is increasing in density and will probably not wedge any further. The upper margins of the two vertebrae above are also slightly increased in density, suggesting that they were previously infected. **E** A well-healed sclerotic tuberculous focus in the lower anterior portion of a vertebral body, with decrease in disc space and some reactive sclerosis in the vertebra below. This was a young African man who had been treated for more than 2 years to reach this stage. **F** Healing can take place without increased bone density, but will then involve a lot of fibrosis. This could be mistaken for the result of trauma, but was proven tuberculosis. (**C** courtesy of Dr. Harold G. Jacobson)

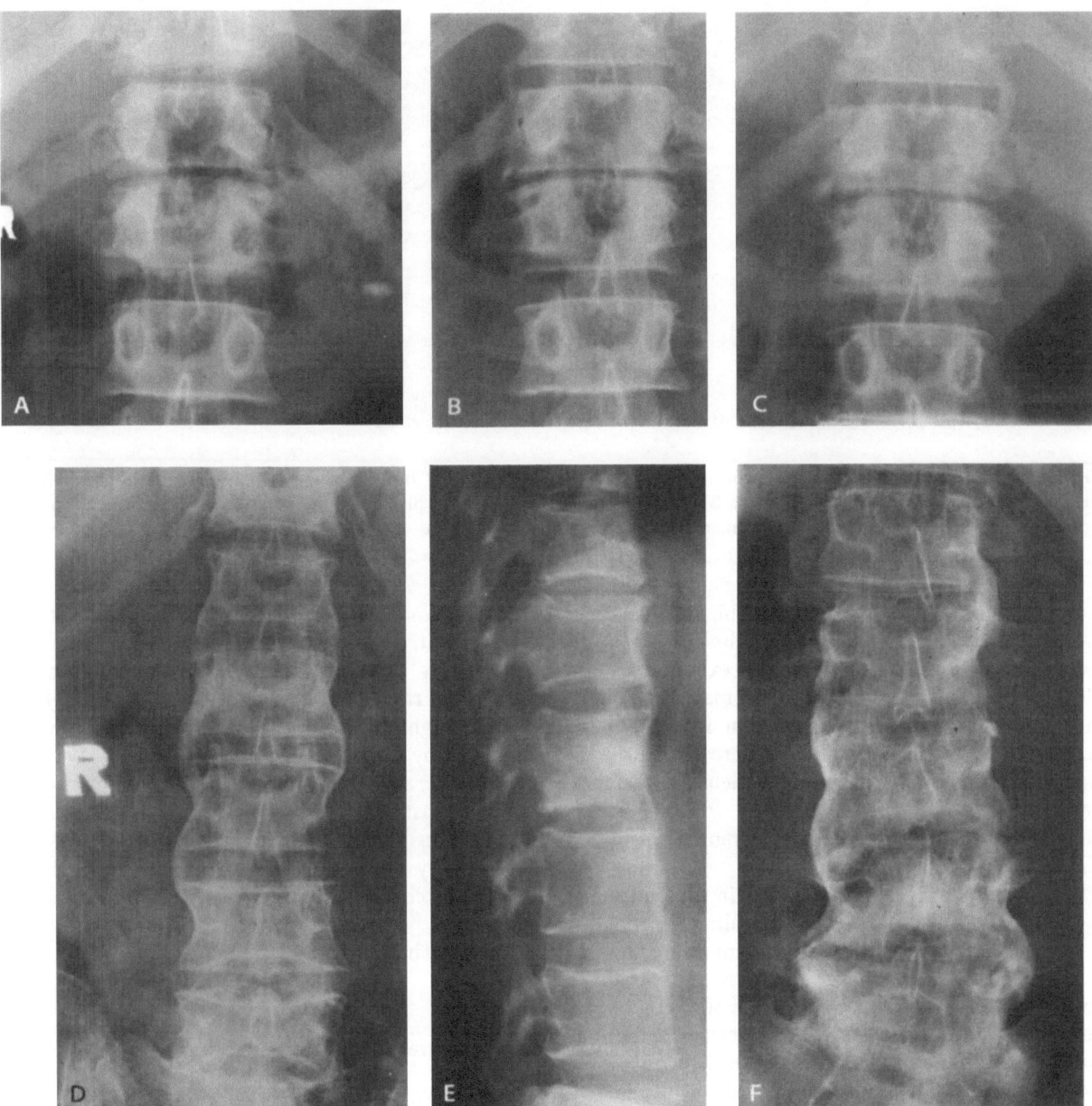

Fig. 66 A–E. Spinal tuberculosis heals with "bridge" formation between vertebrae until, in some patients, it may resemble the bamboo spine of ankylosing spondylitis. The bridges form surprisingly quickly. **A** Loss of joint space between the 12th thoracic and the first lumbar vertebrae. There is little other change except for some slight irregularity of the opposing surfaces. **B** Six weeks later there is new bone forming a bridge between the two vertebrae on the right side and, more surprisingly there are bridges on both sides of the normal space between the 11th and 12th thoracic vertebrae. **C** Three months later there is firm bridging between the 11th and 12th thoracic vertebrae and some increase in thickening of the partial bridges between the 12th thoracic and first lumbar vertebrae. There is now a faint bridge forming on the left side at this level and bridging has started between the first and second lumbar vertebrae, although that disc space is also normal. **D, E** The end result of healing. The sclerosis in the center of the body of the second lumbar vertebra and (in the lateral view) the healed focus of infection on the anterior edge of the 12th thoracic vertebrae show that this is post-tuberculosis. The sacroiliac joints are normal. This African from Zimbabwe has been under treatment for spinal tuberculosis for 3 years. **F** It is difficult to recognize this as the end result of tuberculosis, but this African has also been treated and observed for more than 3 years. Although these patients (**D–F**) show bridging between most of their lumbar vertebrae, the process may be limited to only two or three and occur at any level (see Fig. 67)

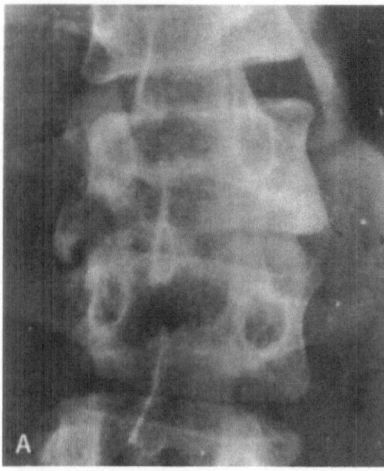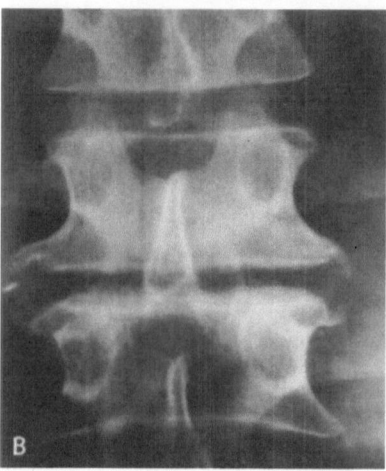

Fig. 67. A Unilateral bridging around a focus of infection in the lower corner of one vertebral body. **B** Small bilateral bridges across a narrow disc space, with marked sclerosis of the vertebral bodies. There is considerable variation in the frequency of bridging and in some countries sclerosis may be less common

Bone formation and bridging can be unexpectedly rapid or may occur slowly; it can be localized between one or two vertebrae, may be extensive at one side of the spine and not the other, may occur anteriorly and not laterally, or may occur in any combination. Many of the bony bridges are arched, leaving a clear space between the vertebrae. In many cases there is no vertebral collapse, which is a distinguishing feature from previous trauma. In others there is marked destruction.

There is some geographic variation in the pattern of tuberculosis of the spine and its healing. In most of Africa, bony bridging is relatively common; in Asia the process is much more destructive and there is much more rarefaction; this heals with either normal bone replacement or some increase in the trabecular density superimposed on the deformity.

One of the most difficult radiological problems is to assess the degree of healing; this can only be done after review of serial radiographs, preferably at 2–3 months intervals, to assess progress. Lack of further change may be the first sign of healing. The difficult surgical decisions are when to operate, to decompress, to use a bone graft, or only to immobilize. The "correct" treatment of spinal tuberculosis is still controversial. It was shown by Konstam in West Africa that a short preliminary period of drug therapy in hospitals could be followed by ambulatory care with a spinal cast. The results were excellent even when there was considerable neurological and urological deficit when the patient entered the hospital. His research was confirmed that the most important factor in the effective treatment of spinal tuberculosis is drug therapy. This must be continued, of course, for an adequate time and patients must be reviewed regularly. If the patient is in a

cast, surgeons must be persuaded to change it at a review visit, so that the progress of the spine may be properly assessed radiologically; attempts at review when radiographs have been taken through the cast are highly unreliable. A more surgical approach was used by Roper in Zimbabwe, Kircaldy-Willis in East Africa, and Hodgson in Hong Kong, combining surgery, cast, and drug therapy. The results are usually excellent, but whether every case needs surgery is very much a matter of opinion. Ambulatory care is successful and thus of great importance in developing countries, where hospital beds and skilled nursing are in short supply.

Paravertebral Abscess

Not every vertebra infected with tuberculosis will have a paraspinal abscess. Tuberculous abscesses present in many different ways:

1. There may be an abscess but no bone lesion (or a bone lesion but no abscess).
2. The abscess may be unilateral or bilateral.
3. The abscess may be lateral, or anterior, or both.
4. The abscess may appear below, or occasionally above, the obvious bone focus.
5. There may be a large abscess with no clinical symptoms or a small abscess with paraplegia.
6. Abscesses may surface almost anywhere on the skin but particularly in the inguinal region, in the thighs, in the loin, near the sternum, or in the neck. A retropharyngeal abscess may cause difficulty in swallowing. A psoas abscess may connect with any part of the intestine and present as a fecal fistula.

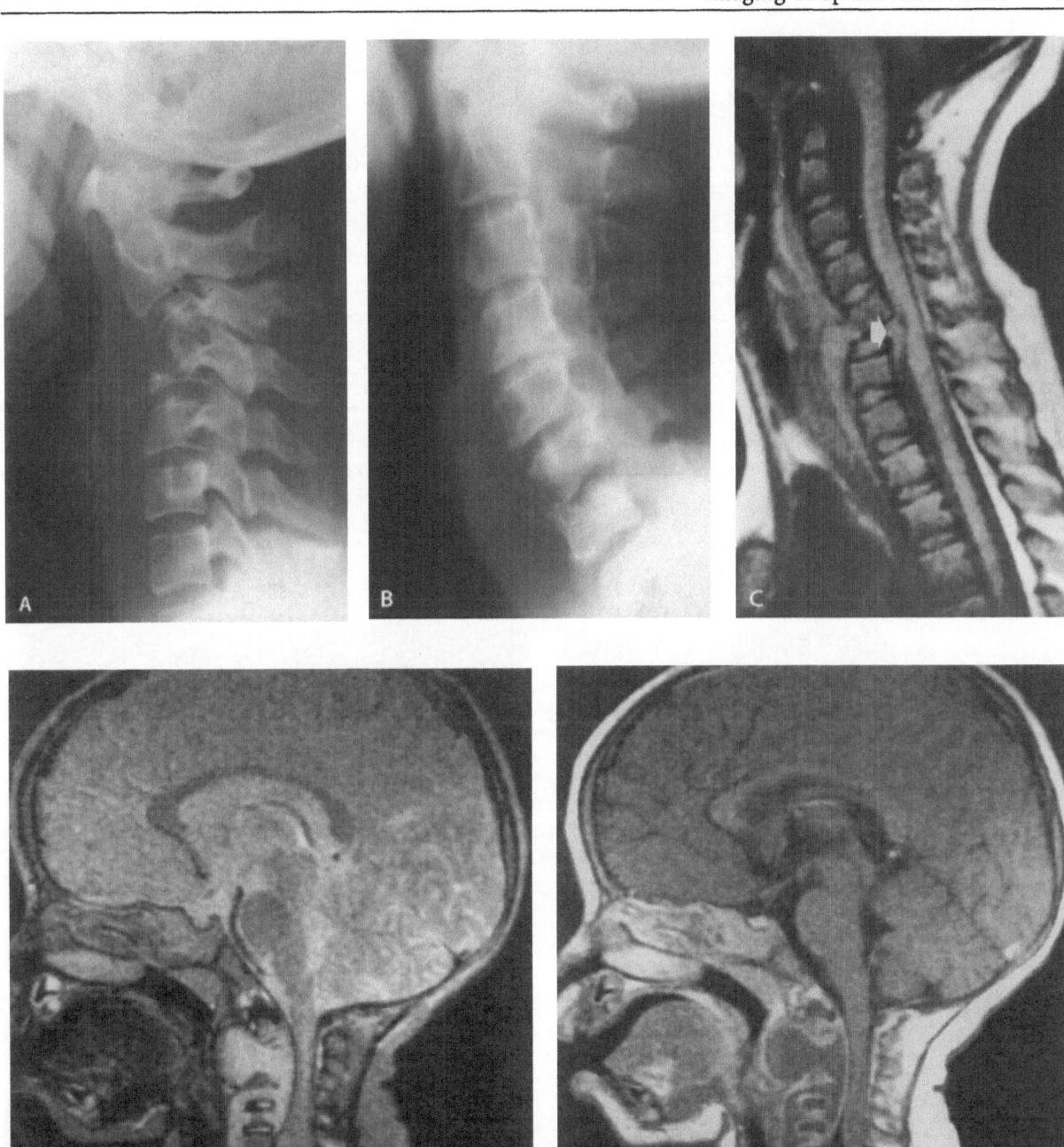

Fig. 68 A–F. Tuberculous abscesses. The best way to image an abscess around a tuberculous spinal infection is by scanning, particularly with CT and MRT. However, when these modalities are not available, abscesses can be seen with plain radiography or, occasionally, with contrast media. They are usually much larger than the plain radiographs suggest. **A** There is a large abscess in front of the cervical spine, displacing the trachea and esophagus anteriorly. **B** Another patient with a similar abscess a little lower in the cervical spine. Both these abscesses could have been demonstrated by ultrasonography, but radiography gives a better image of the spine as well. **C** This 5-year-old child was quadriparetic because of the anterior cord compression from a tuberculous abscess at the level of C5 (*arrow*) shown by T1-weighted MRI. **D** T2-weighted MR-scan in another child shows destruction of C2–4 with anterior and posterior abscesses. **E** with intravenous gadolinium there is peripheral enhancement but the necrotic content of the abscess does not enhance. (**D, E** from Cremin and Jamieson 1995)

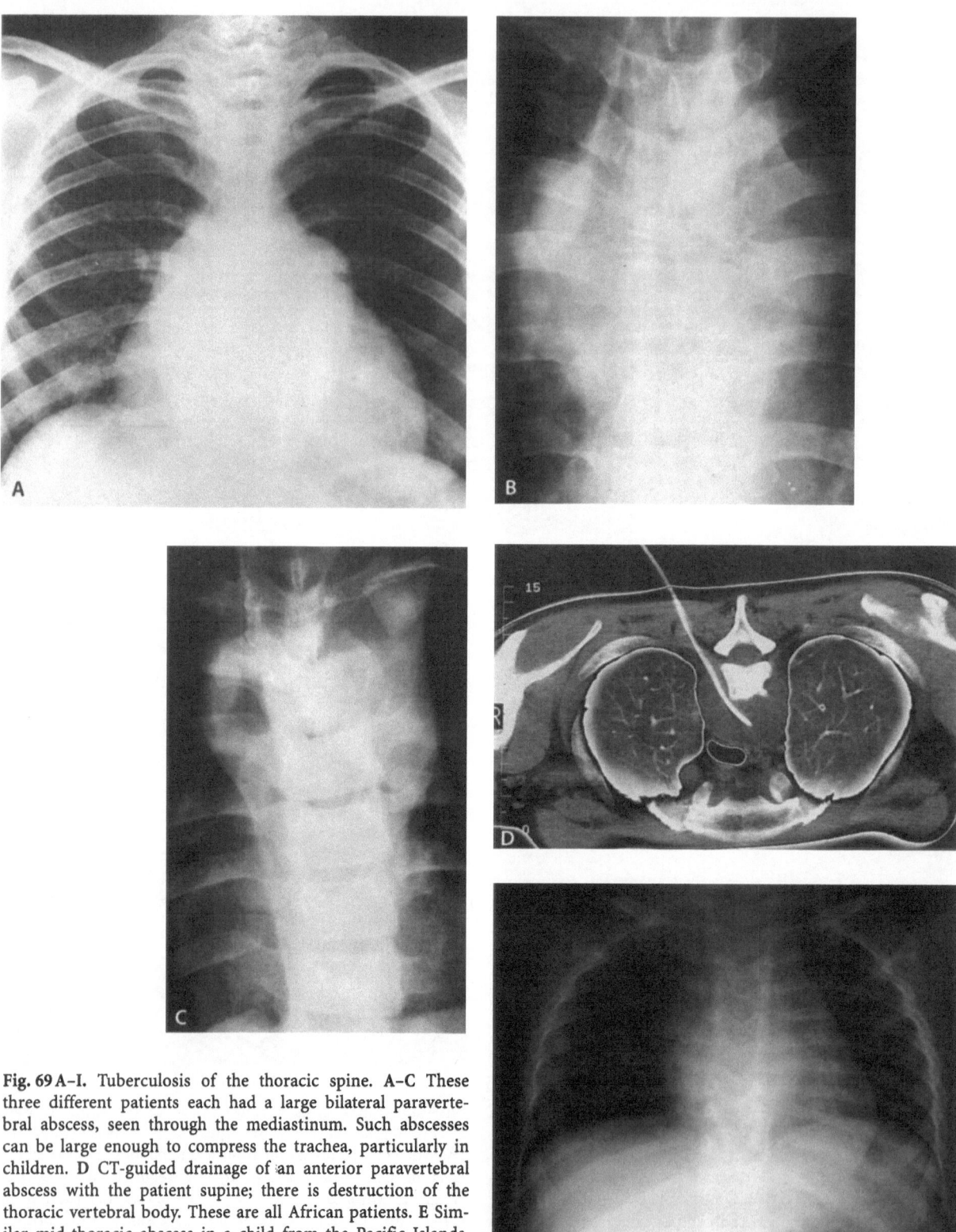

Fig. 69 A–I. Tuberculosis of the thoracic spine. **A–C** These three different patients each had a large bilateral paravertebral abscess, seen through the mediastinum. Such abscesses can be large enough to compress the trachea, particularly in children. **D** CT-guided drainage of an anterior paravertebral abscess with the patient supine; there is destruction of the thoracic vertebral body. These are all African patients. **E** Similar mid-thoracic abscess in a child from the Pacific Islands. **F, G** The CT scan shows left upper cavitating pneumonia (F) and, at a lower level, destruction of the vertebrae with fluid on both sides (G). **H** The MR scan confirms the pneumonia and shows the size of the abscess and the destruction of the vertebrae. **I** The lateral MR scan shows the cord compression. F–I see p. 97

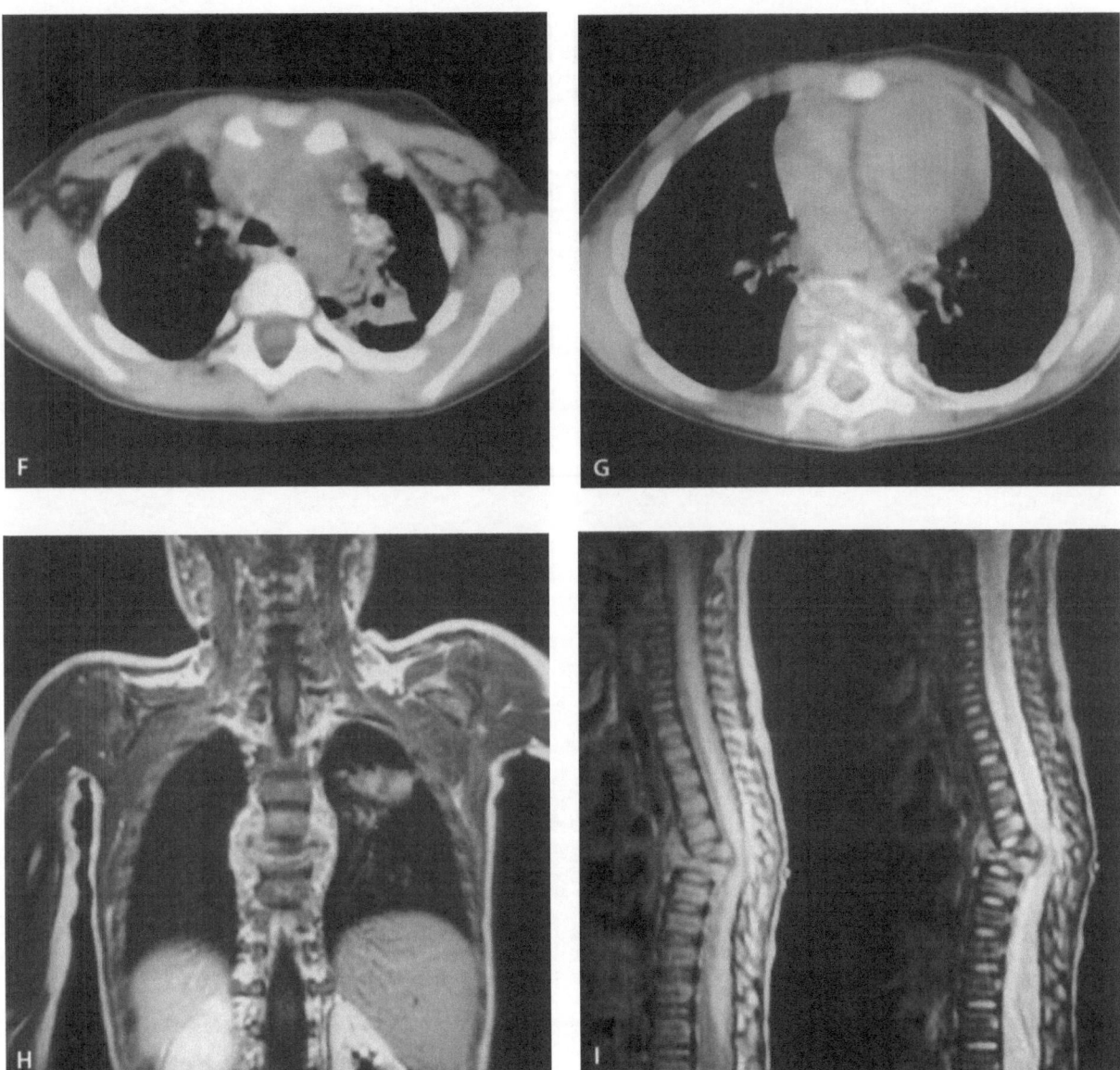

Fig. 69 F–I. Legend see p. 96

CT and MRI have radically altered our knowledge of these tuberculous abscesses by showing them, and bone foci, where they cannot be imaged on plain radiographs (Figs. 68–70). In particular, the full extent and position of any fluid can be demonstrated by CT or MRI.

Wherever the abscess presents, there should be careful imaging of the whole spine, starting preferably with isotope scintigraphy. Paravertebral abscesses occur at all levels and may be present on one side only, on both sides symmetrically or asymmetrically, or only in front of the spine without any lateral extension. Some may be dumb-bell in shape, protruding wherever space permits. They vary considerably in size and shape. When there is an ante-

rior spinal abscess, underneath the spinal ligaments, the anterior edge of the vertebral bodies may become concave because of the transmitted aortic pulsation. The vertebral cartilages are not affected and the pressure absorption of the body may be minimal or considerable, so that it resembles the concavity caused by an aortic aneurysm. MRI is the best way to demonstrate what is actually happening, and CT is also suitable for this purpose.

A psoas abscess may connect with any part of the bowel, may be extensive and surround the kidney or the liver, or may burst through the skin onto the surface almost anywhere (see Fig. 71). In many cases sinography is required, but careful technique is essential. The examination should be carried out

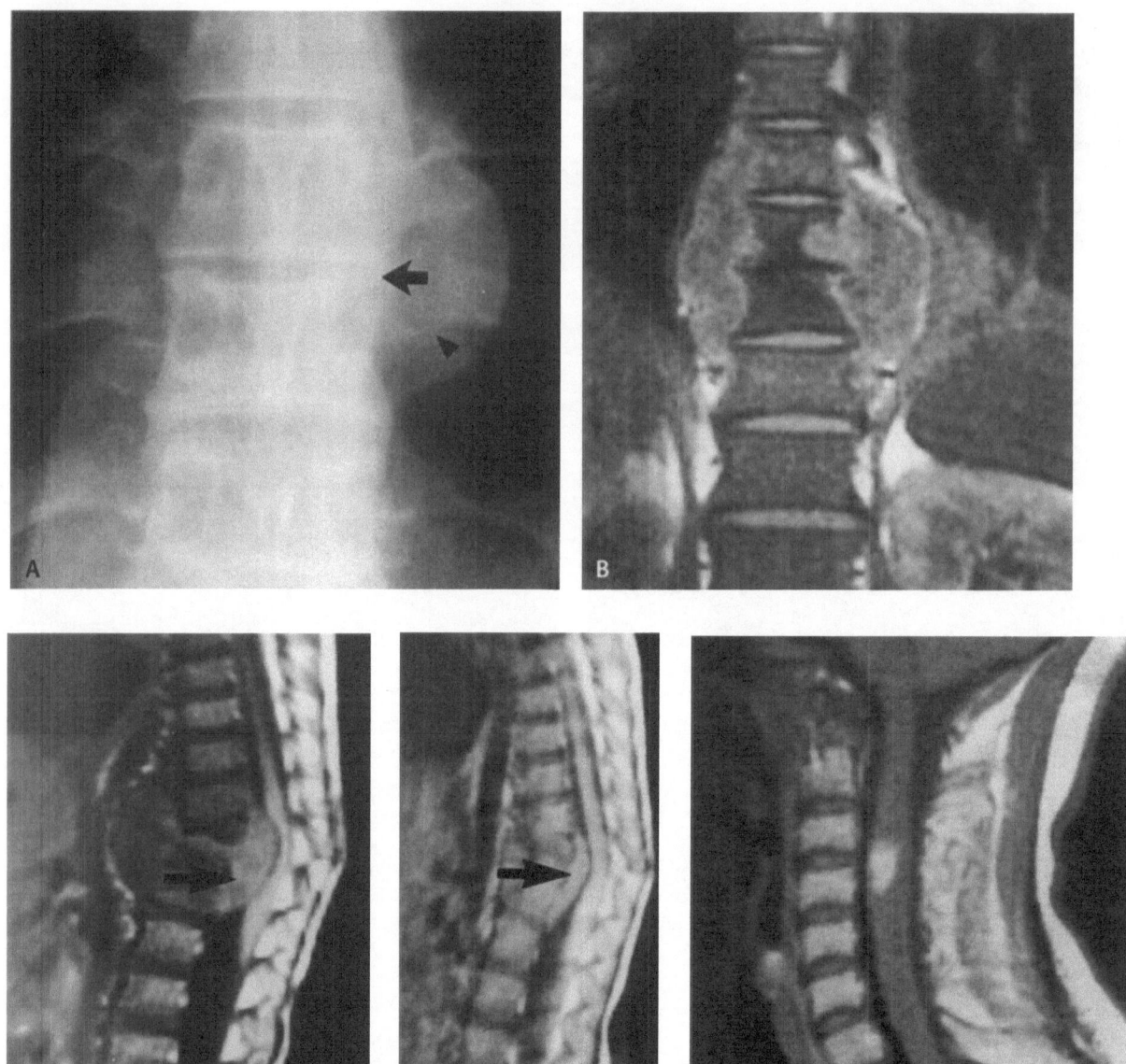

Fig. 70. A A 25-year-old African complained of backache ans was found to have a left paravertebral abscess and erosion of the medial end of left tenth rib (*arrowhead*). **B** A T1-weighted coronal MR scan showed bilateral abscesses and low-intensity signals from the marrow of T8 and T9. **C** The T1-weighted sagittal MR scan showed the large abscess anteriorly and posteriorly, with destruction of the vertebral bodies and compression of the cord (*arrow*). **D** After 6 weeks of antituberculous treatment the abscess is much smaller and the cord is less compressed (*arrow*). It was realized more than 30 years ago (Konstam) that many of the neurological signs are caused by the abscesses rather than the spinal deformity. CT and MRI have confirmed this dramatically. (Courtesy of Dr. Peter Corr, Durban) **E** Apart from pressure on the cord from an abscess, there may be a tuberculoma, as shown at the level of C4 in a different patient, a child. This is an intravenous gadolinium-enhanced T1-weighted MR scan. (From Cremin and Jamieson 1995)

under aseptic conditions and under fluoroscopic control. The sinus should be blocked by using the largest catheter that can be inserted; a balloon catheter will often help in achieving a tight connection. Pus should be aspirated, if possible, and pressure on the surrounding tissue or the suspected site of the abscess may help this drainage. Contrast material should then be injected and considerable pressure may be required if the full extent of the abscess is to be visualized. The case illustrated in Fig. 71 A and B was a patient who presented with a sinus in the left side of the lower abdomen, without any obvious etiology. The only organism cultured from it was *E. coli*. A barium enema and an upper gastrointestinal series with a follow-through examination did not show any connection to any part of the bowel.

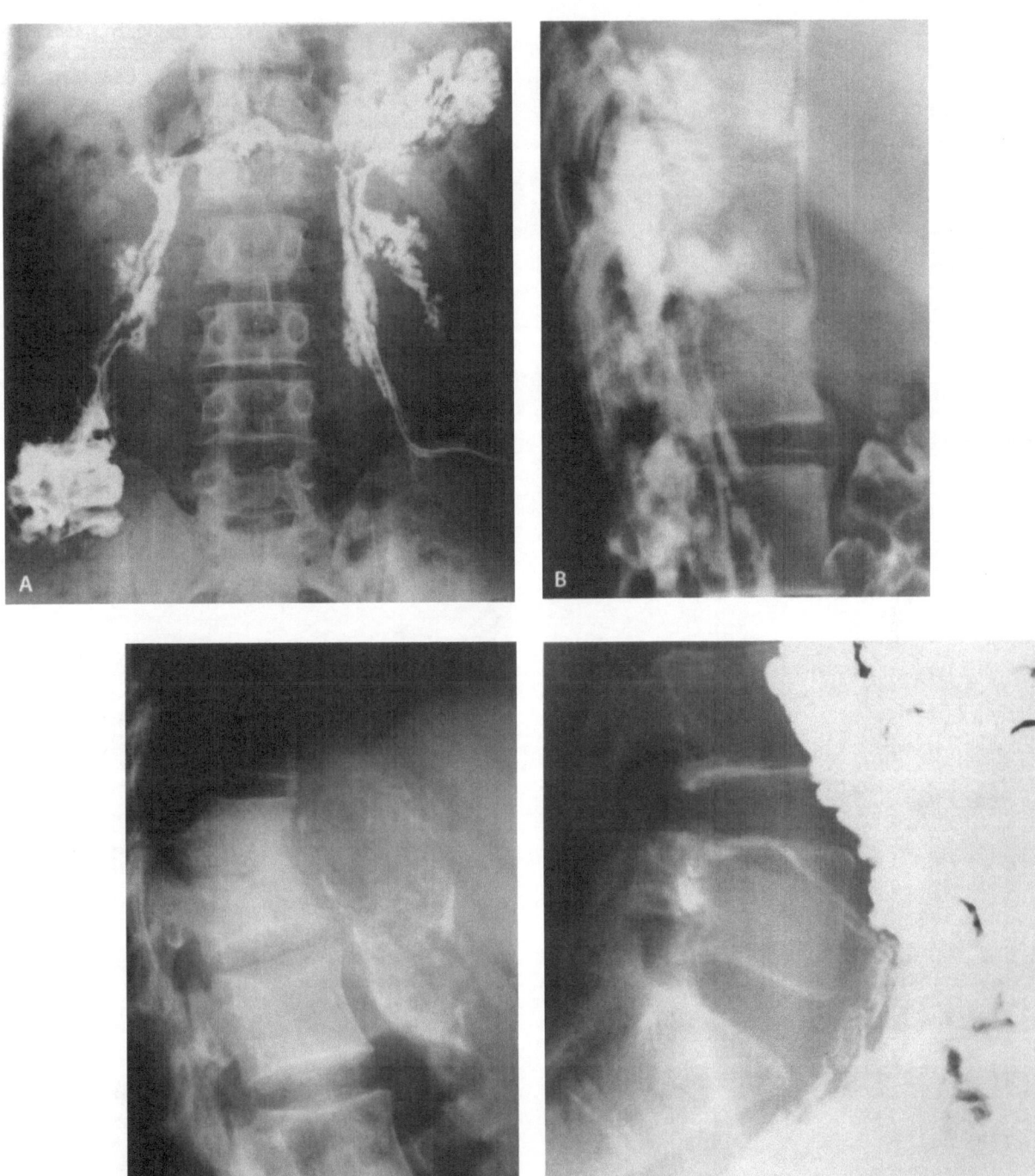

Fig. 71 A–D. Psoas abscesses can be very extensive and have unexpected connections. This young African has a persistent left flank sinus contaminated with *E. coli*. **A** The sinogram from the left lower flank, close to the iliac crest, showed that the fistula was connected to an enlarged left perirenal tuberculous abscess, which had started in a tuberculous infection between the 12th thoracic and first lumbar vertebrae. The abscess had also drained down in the right psoas and connected with the cecum, which was the source of the *E. coli* infection. **B** The lateral view of the same patient shows contrast running in front of the vertebrae, above and below the infection. This demonstrates how anterior abscesses can transmit aortic pulsation and also the etiology of the anterior bone bridging which occurs when the periosteum and anterior spinal ligaments are lifted. The aortic pulsation can bow the front of the vertebrae, as is shown in **C** in a much older patient with advanced aortic athromatous calcification. A previous tuberculous infection has resulted in fusion and increased density of the two lower thoracic vertebrae. **D** Another African patient with a fistula connecting the terminal ileum with a paraspinal abscess, which originated in lower thoracic tuberculosis. It is not always possible to demonstrate these fistulae with a barium gastrointestinal series. *E. coli* contamination is usually sufficient proof, even if the actual fistula cannot be imaged. (**C** courtesy of Dr. Harold G. Jacobson)

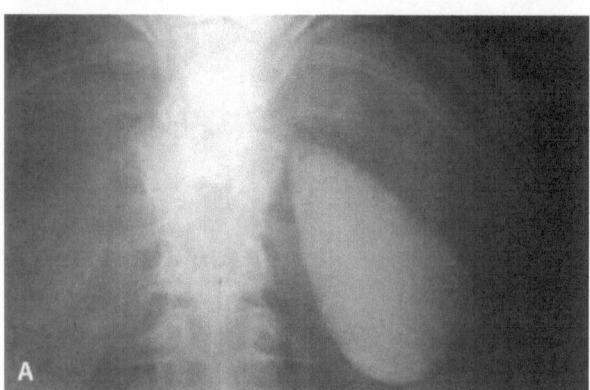

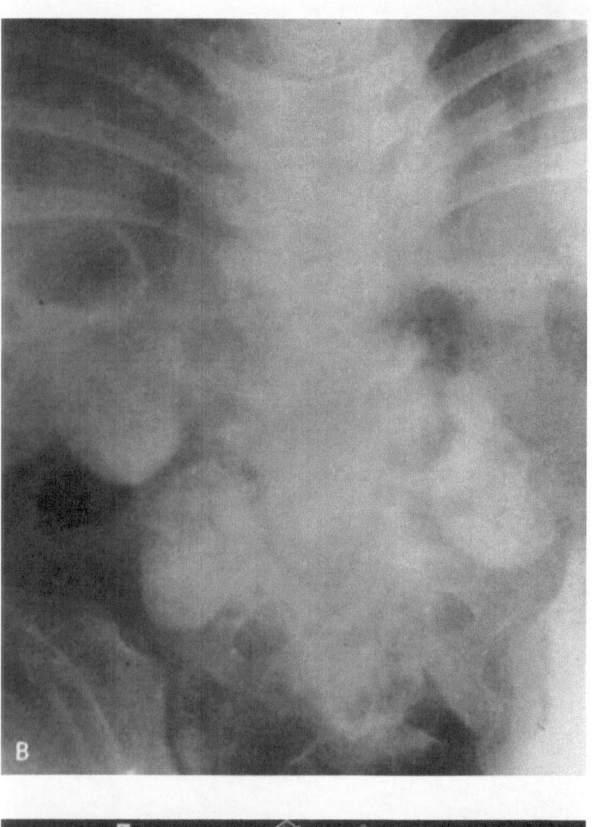

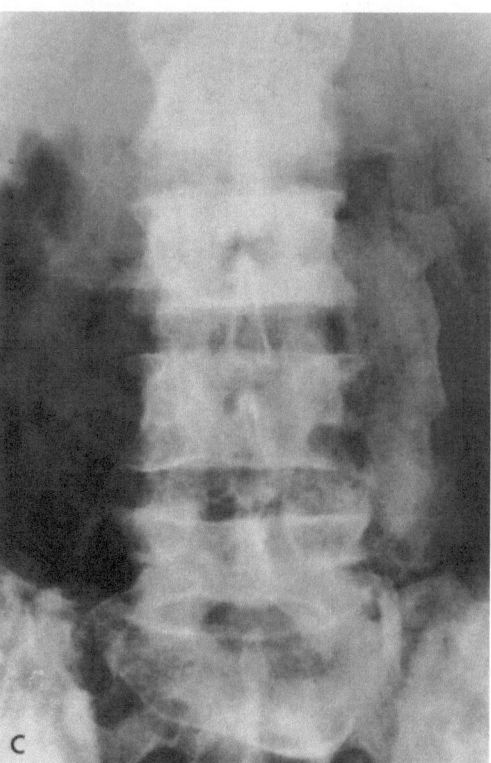

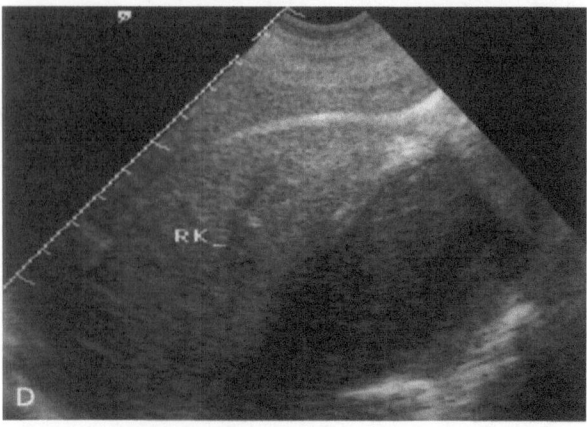

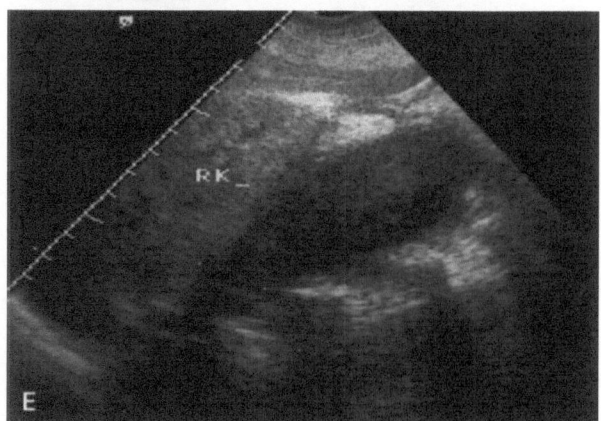

Fig. 72 A–E. Tuberculous abscesses of the lower spine. Thoracic and lumbar abscesses track along the psoas and can present at the inguinal region. When they heal, they often calcify. **A** Well-defined, heavily calcified abscess starting at the lower thoracic region in a patient with severe kyphosis. **B** Bilateral loculated and less densely calcified psoas abscesses. **C** A unilateral calcified psoas abscess in the lumbar region. There is bridging between the upper lumbar vertebrae, well above the visible abscess. **D, E** Sonograms of a right psoas abscess, displacing the kidney, in an African patient from Zimbabwe. *RK* right kidney. (**A** courtesy of Dr. Harold G. Jacobson; **D, E** courtesy of Dr. Sam Mindel)

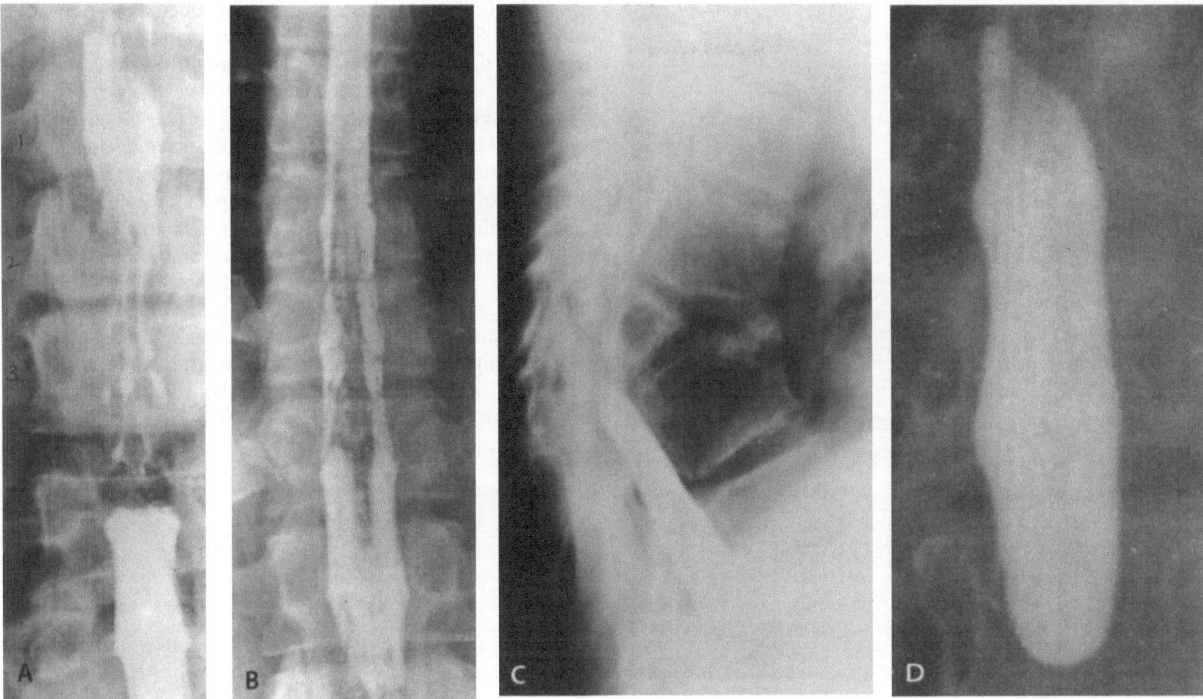

Fig. 73 A–D. Myelography in tuberculous paraplegia. Where scanning is not available, there is no contraindication to myelography under fluoroscopic control. **A** There is narrowing of the disc space between the second and third lumbar vertebrae, but the myelogram shows distortion not only at this level, but also one space above. **B** A patient with thoracic paraplegia. Myelography shows long irregular defects, be-tween the tenth and 12th thoracic vertebrae. **C** Myelography in a kyphotic patient can be difficult. In spite of the angle, in this patient there was no obstruction on myelography: it can be difficult to retain the contrast at the level of the curve when it flows freely in either direction. **D** The myelogram may show complete block. (**D** courtesy of Dr. Harold G. Jacobson)

Two limited sinograms were performed before the patient was referred to the author, who blocked the sinus and, using considerable pressure, showed the abscess extending up the left psoas and surrounding the left kidney, across the midline between T12 and L1, and then coursing down the right psoas to connect posteriorly with the cecum. This was, therefore, a sinus draining from the left lower abdomen and yet connected with the cecum on the opposite side through a tuberculous abscess of the spine. (A very similar case has been seen in an American who has never been near the tropics!)

Paravertebral tuberculous abscesses shrink and often show calcification as they heal and fibrose (Fig. 72). Unfortunately, calcification does not always indicate sterility. There may be active caseous material contained within the fibrotic bag.

Scanning, Paraplegia, and Myelography

As already mentioned, MRI is the ideal way to examine a patient with a tuberculous spine. A T1-weighted scan shows tuberculous spondylitis as an area of low signal intensity where there is normally a higher intensity from the bone marrow, adjacent to the end plate of the vertebra. On a T2 weighted scan these lesions have a high intensity. Liquid pus also yields a high-intensity signal on a T2-weighted scan. MRI gives an accurate image of the spinal cord compression associated with intraspinal abscesses. At surgery, almost all patients with paraplegia will have at least a small abscess; even if it has not been visualized on plain radiography, it will probably be clearly seen on MRI (Fig. 73). Paraplegia indicates that there is about 50%–60% extradural compression above the conus, provided there is no vertebral instability. Peripheral nerves below the cauda or extraspinal nerves are not likely to be affected by compression. In some cases of paraplegia, the cord will appear to be intact, in which case there is likely to be a vascular deficit, a toxic reaction, or, as MRI can demonstrate (by showing a

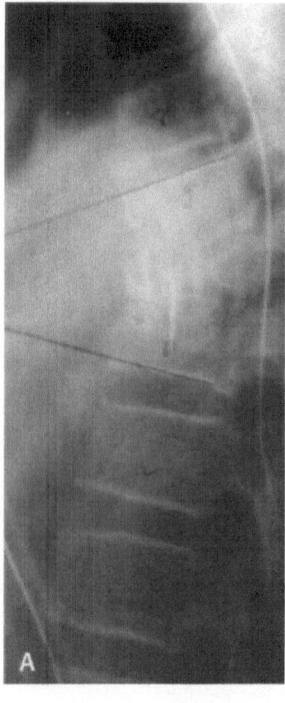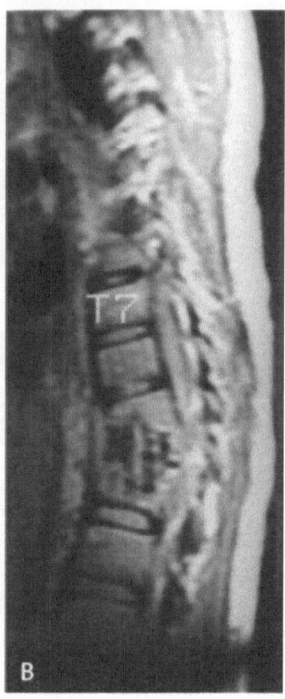

Fig. 74 A, B. Postoperative spinal tuberculosis. It is very difficult to assess healing, which may take years. These are the postoperative images of a 2-year-old girl from the Pacific Islands. **A** The lateral radiograph shows the graft across the infected and wedged thoracic vertebrae (T 9–11). **B** T 2-weighted MRI shows the residual abscess and granulation tissue and the exact extent of the bone destruction. (Courtesy of Dr. Cheryl Sisler, Hawaii)

Fig. 75 A–H. The case history of spinal tuberculosis. A young ▶ man from Saudi Arabia had been complaining of pain between his shoulder blades for some months before he came to hospital one weekend, with rapidly developing paraplegia. Plain radiography (**A**) showed wedging of two upper thoracic vertebrae. Myelography (**B**) showed almost complete block at this level and CT scan (**C**) showed destruction of the vertebra, particularly on the right side, and of the medial end of the rib at the same level. Fluid was present in the right side of the chest. **D–G** Pre- (**D**) and postgadolinium (**E–G**) MR scans show the full extent of the paraspinal abscess, and of the epidural granulomatous tissue. The findings on the CT scan are confirmed but shown more clearly. Subsequent surgical decompression revealed both caseation and granulomatous tissue with *M. tuberculosis* on direct smear microscopy. The patient made a good recovery. (Courtesy of Dr. Frank McGuiness, Ryadh)

high-intensity focus on a T2-weighted scan), damage to the cord, probably due to prolonged compression. This focal myelitis has a poor prognosis and the paraplegia is likely to be irreversible. However, as Konstam demonstrated, rest and drug therapy may relieve paraplegia which is of short duration, and simple decompression, either by needle aspiration or surgery, may be all that is needed in the early stages. Whenever any intervention is planned, it is a wise precaution to have the patient already on antituberculous therapy to obviate the risk of tuberculous septicemia.

Whatever method of imaging is used, it is essential that the whole cord or spinal canal is examined because the obstruction may not occur at the site of the obvious bone lesion, but rather be above or below it, depending on the extent of the fluid, necrotic tissue displacement, or granulation tissue.

If MRI is not available, myelography, with or without CT, is a very useful investigation and should be performed as early as possible and always before surgery to localize the exact site of any obstructions (Fig. 74). There are no contraindications to myelography although in some patients the kyphosis may make it extremely difficult. Where there is a double curve, careful manipulation of the patient (especially in the lateral decubitus position) may permit the column of contrast material to travel throughout the length of the spinal canal. Sometimes this is technically impossible, even when there is no blockage, because of severe spinal angulation. Fortunately, paraplegia is less often due to kyphosis or bony deformity than to an abscess or soft tissue pressure. The decision as to whether myelography needs to be repeated during treatment should be based on the clinical examination.

Following surgery in patients with spinal tuberculosis it is very difficult to assess healing, which may take years. Again, MRI is the best method for this purpose (Fig. 75).

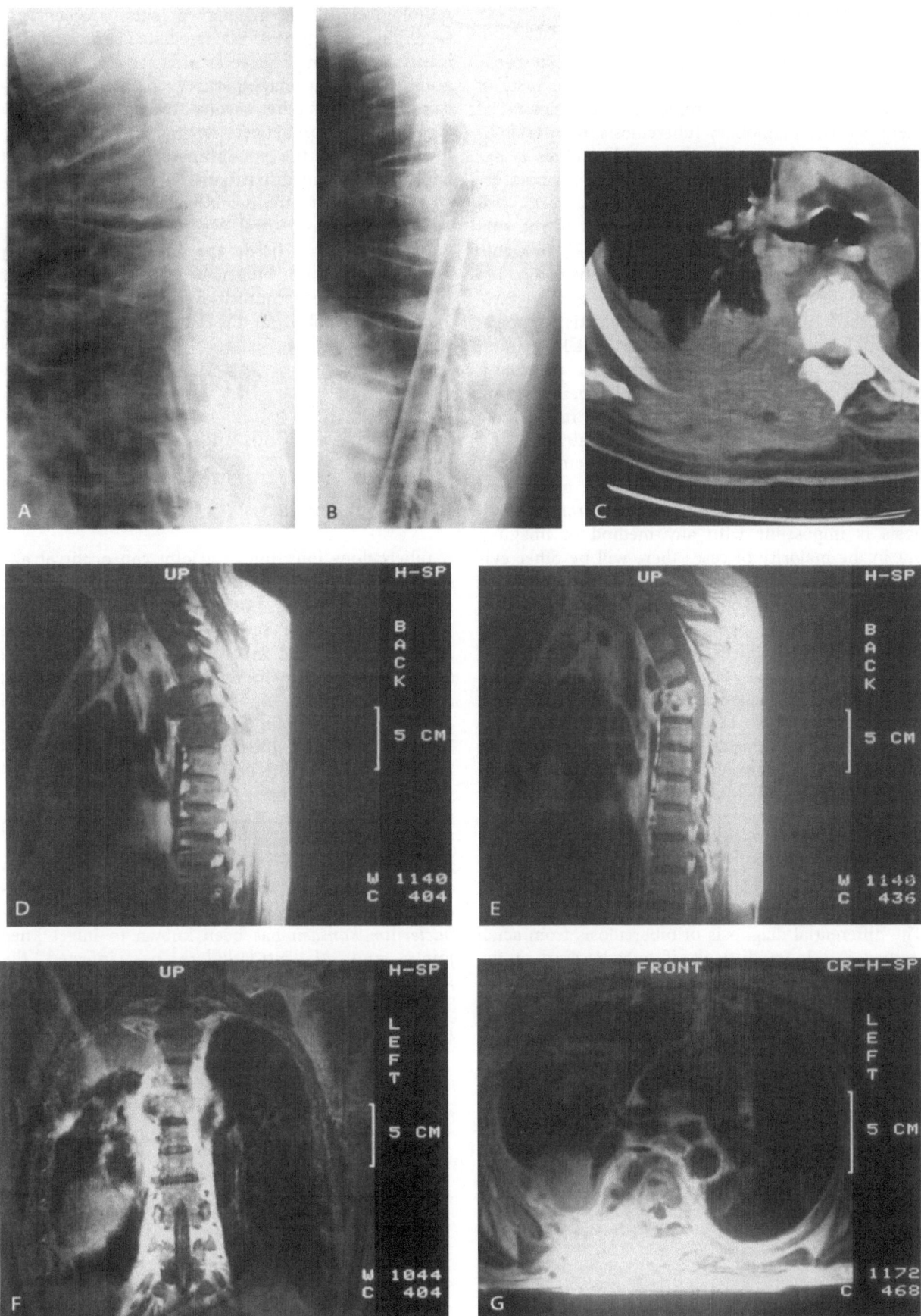

Differential Diagnosis

Theoretically any type of bone infection may cause a paravertebral abscess. In practice, the majority of spinal abscesses in the tropics are tuberculous; if there is active pulmonary tuberculosis, the possibility of spinal infection also being tuberculous is enhanced. It is unusual for a pyogenic or typhoid infection to cause a large paravertebral abscess, and even less often do such abscesses point to the surface as occurs with tuberculosis. On MRI, pyogenic disc infection usually has a high intensity on T2-weighted scans, but in tuberculosis less than one-third show this. However, pyomyositis (tropical myositis) can present in this way, usually without any osseous lesion. Brucellosis may also produce a paravertebral abscess. There may be paravertebral thickening in lymphoma (particularly in Burkitt's lymphoma) which may resemble an abscess, and the bony destruction may also be very confusing because Burkitt's lymphoma may involve a vertebral body or a pedicle. Sometimes the differential diagnosis is impossible with any method of imaging, but in the majority of cases there will be other evidence of Burkitt's or other lymphoma to establish the correct diagnosis. Wedging, particularly wedging and fusion of two or three vertebrae, is very uncommon in any infection other than tuberculosis. Because tuberculous osteomyelitis in the tropics (even in patients without AIDS) is often an acute disease, it is difficult to differentiate from other infections and the clinical condition of the patient may be very important when giving a radiological opinion. Nontuberculous spinal infections usually cause more severe clinical symptomatology than tuberculosis, with more systemic reaction, a higher pyrexia, and a raised white cell count, but none of these criteria are absolute and biopsy or aspiration and culture are essential.

The differential diagnosis of tuberculosis from acute trauma is seldom difficult, if only because of the clinical history. Trauma does not cause as much bone loss and destruction as tuberculosis. Long-standing cases need serial review, both clinical and radiological. In the majority of such problem patients the etiology is of no immediate clinical significance, permitting a more cautious approach. Scintigraphy or other imaging survey of the rest of the spine may show other involved vertebrae and allow the differentiation between trauma and infection.

In all patients, the possibility of malignancy will have to be excluded, particularly when the bone destruction involves the pedicles. Many malignant tumors cause paravertebral swelling resembling an abscess; obtaining a tissue specimen, either through needle aspiration or biopsy, becomes essential. Most errors are made because the possibility of tuberculosis is overlooked in the elderly or in the very young.

Tuberculosis of Bones and Joints (Nonspinal)

Tuberculous Arthritis

A tuberculous infection of a joint can occur at any stage, but is more common in the younger age groups. Even babies from 6 months onwards may present with quite severe skeletal infection. The joints most affected in order of frequency are the hips, knees, ankles elbows, wrists, sacroiliac joints, and shoulders (Fig. 76, 77). In 98% of HIV-negative patients the tuberculin skin test will be positive.

The way in which tuberculosis affects a joint depends on whether the original focus is in the synovium or in the adjacent bone. Spread of the infection from bone into a joint is more common and has a poorer prognosis.

Nontuberculous mycobacteria may cause arthritis and bone infection. This is uncommon, but may become more frequent in patients with AIDS. *Mycobacterium kansasii* has been known to infect knee joints, and *M. intracellulare* has infected the shoulder, the spine, and, in one patient, both wrists. The differential diagnosis will only be made by joint aspiration or synovial tissue culture.

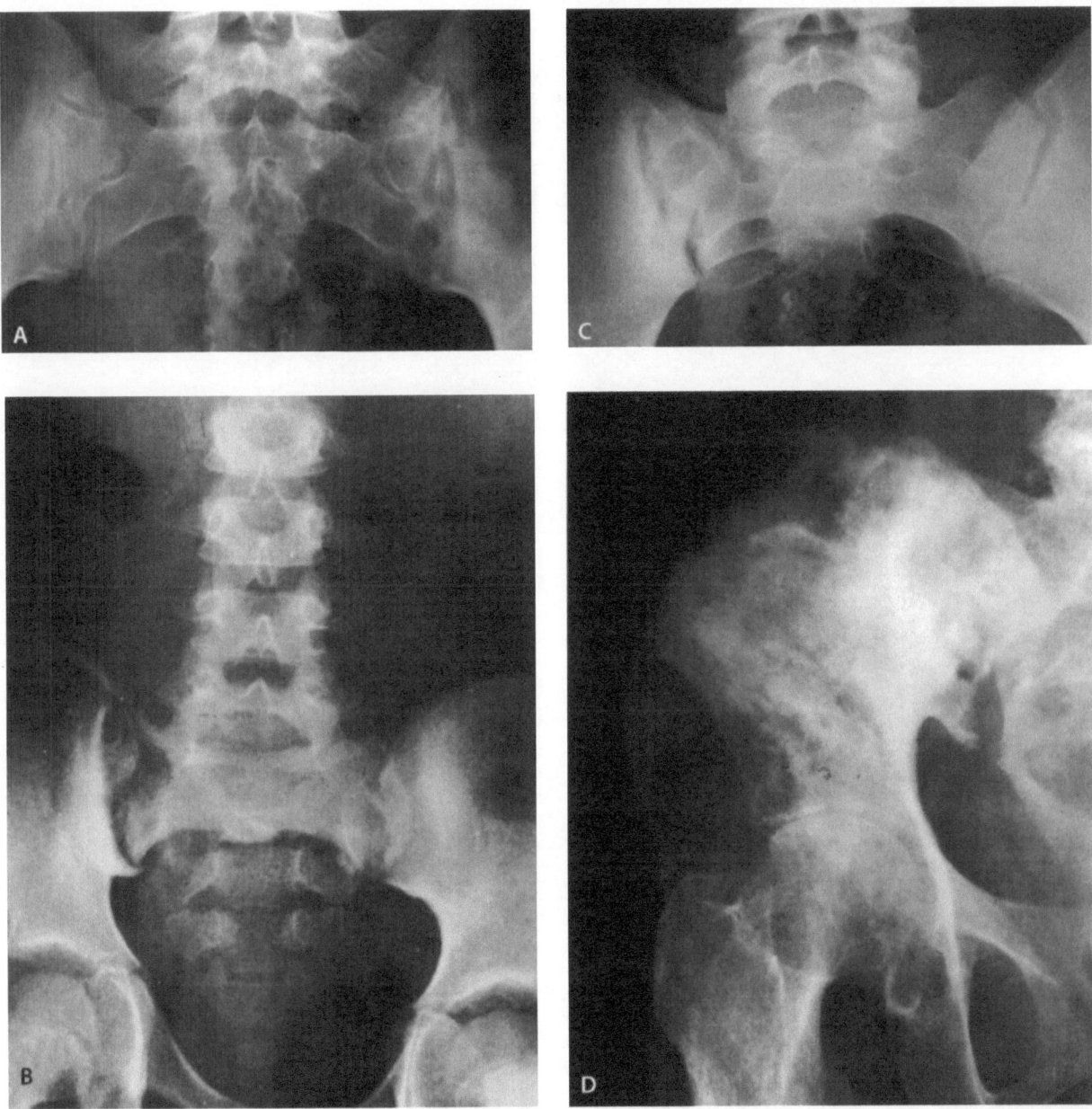

Fig. 76 A–D. Skeletal tuberculosis. The pelvis. Infection of the sacroiliac joints can be destructive and then, as healing occurs, become sclerotic and fused. **A** Widening and irregularity of the left sacroiliac joint. **B** Sclerosis on the iliac side of the right sacroiliac joints. **C** In another patient there is infection on both sides of the sacroiliac joint, particularly the sacrum. **D** Extensive destruction of a right sacroiliac joint in a different African patient. The infection must have spread through the right ilium, which is becoming sclerotic. The joint is widened throughout its length. The right hip joint is narrowed and probably also infected. (**A, C** courtesy of Dr. Harold G. Jacobson)

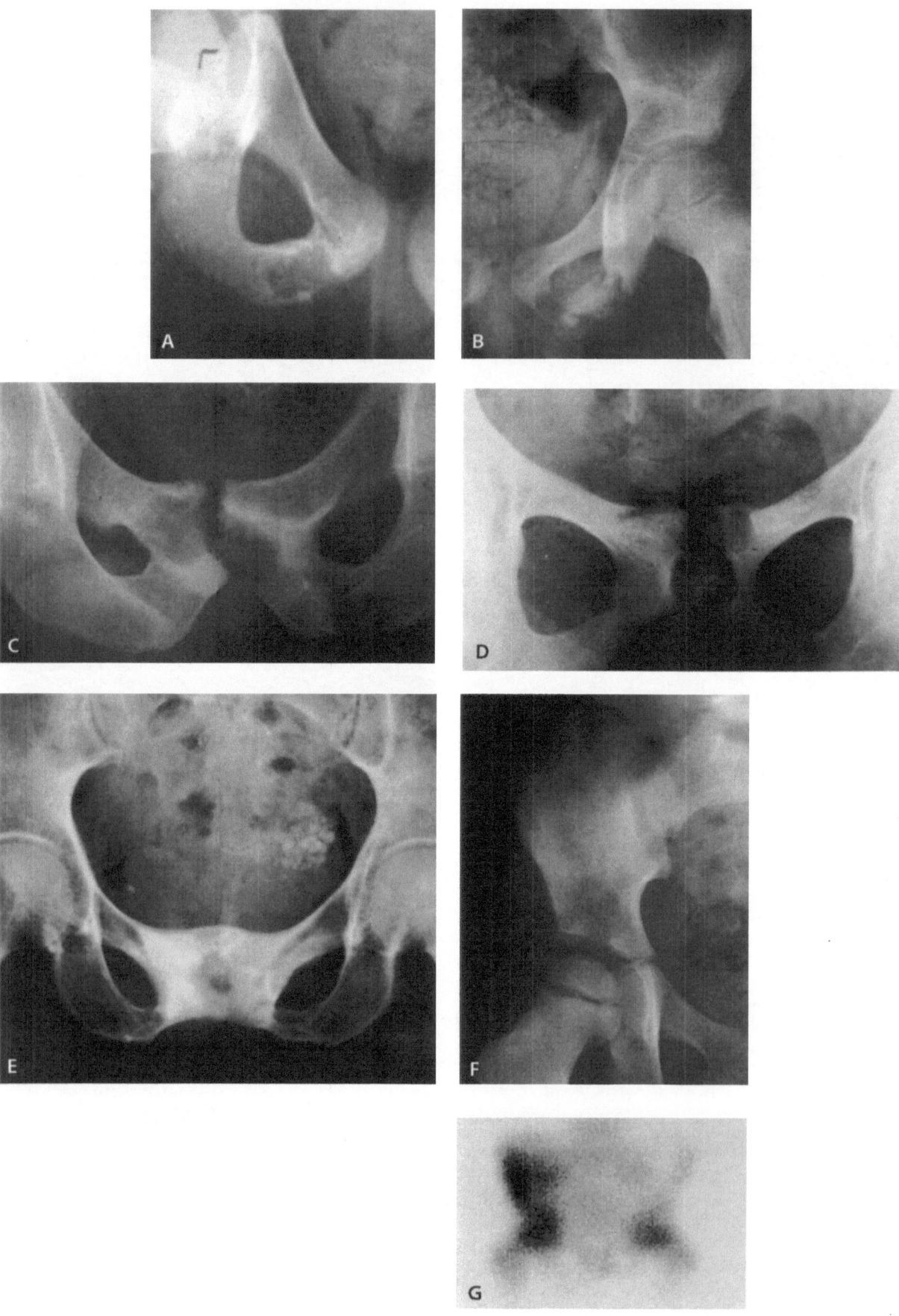

◄ **Fig. 77 A–G.** Tuberculosis of the ischium and symphysis pubis. In the early stages most tuberculous lesion in the ischium are ill-defined lucent areas; they later become sclerotic. **A** A lucent focus of infection in the left ischium, with some early surrounding sclerosis. **B** A more destructive lesion closer to the hip in another patient, with some loose bone fragments. **C** There are foci on both sides on the symphysis and in the left ischium. **D** Both sides of this symphysis have been infected, the joint is widened, and there is a central sequestrum. An African from Zimbabwe. **E** A different patient from the same country in which a similar infection has healed causing sclerotic fusion across the symphysis. **F** There is a lytic area of infection above the acetabulum in this child from South Africa, but the hip joint remains normal. **G** An isotope scan of the same patient shows that the iliac infection is more extensive than seen on the radiograph. Tuberculosis is the proven cause in all these cases. (A–C courtesy of Dr. Harold G. Jacobson; F, G from Cremin and Jamieson 1995)

Synovial Tuberculosis

Patients will first complain of either slight pain or stiffness in the affected joint, and of limitation of movement. In the early stages pain is minimal and it is only as the joint swells that it becomes significant. In many cases there will be a dull, ill-defined ache, often worse at night, but in developing countries such discomfort is often ignored, and it is not until the disease has progressed further that the patient seeks help.

In the beginning, it will be difficult to detect any clinical abnormality to account for the symptoms. Later there will be edema and swelling around the joint, with some limitation of passive movement and increase in local temperature; an effusion often develops early and aspiration of the fluid usually shows it to be clear or straw-colored with few cells; it is sterile on culture. The diagnosis may have to be presumptive, because it may not be possible to find any positive proof of tuberculosis. At this stage, scintigraphy is a reliable way of excluding bone infection; ultrasonography and MRI, and to a lesser extent CT, can be useful in demonstrating the soft tissue changes. Radiographically the joints may be entirely normal in the early stages, and it may take 2–6 weeks before there are positive radiological findings (Fig. 78).

The earliest change is loss of clarity of the bones forming the joint; their margins become hazy and difficult to define. There is a mild generalized osteoporosis but not localized bone destruction. The joint space may be widened because of the effusion. Progress of the infection is slow, but a little more rapid once the radiological changes have developed. The osteoporosis becomes more pronounced and there will be soft tissue swelling around the joint. It may be difficult to define any bone erosion because

of the osteoporosis and the variable change in bone density. Eventually foci of bone destruction become more definite as the cartilage is destroyed. At this stage the radiological appearance is "ghost-like," particularly when compared with the opposite normal side. If only the synovium is involved, and if treatment is adequate, the disease shows no further progress but slowly heals with a return of the joint towards normality. In the majority of cases, however, there is permanent damage to the cartilage and occasionally to the underlying bone which may lead to degenerative arthritis. When the synovial involvement has been severe, the joint may eventually ankylose.

Osteoarticular Tuberculosis

Scintigraphy is a very useful way of locating tuberculous bone infection and ensuring that it is not multifocal. In the majority of patients, tuberculous infection of a joint is due to spread from a focus within an adjacent bone. An effusion is evidence of this spread but may also occur before actual cartilage destruction has taken place. The bony focus may be close to the articular surface, a few milimeters away from it, or in the metaphysis. The very early bone focus (as well as the soft tissue involvement) can be shown by MRI, and the extent of marrow replacement can be assessed. Gadoliniumenhanced T1-weighted images show differential rim enhancement. Such foci can be single or multiple, and although they are usually opposite one another on opposing sides of the joint, they need to be so closely related. On plain radiographs or CT, the lesion within the bone appears as an area of decalcification, without any definite edge and of no particular shape. This slowly spreads up the shaft, but also across the epiphysis into the joint. At this stage clinical symptoms will be minimal. Some patients may complain of a dull ache or "spasm" in the joint or occasionally some swelling. Careful clinical examination may show some restriction of movement, but this may be so slight that there is little suspicion of an underlying osteomyelitis.

As the infection bursts into the joint, there will be a considerable reaction, with fluid and tuberculous debris in the joint space. The sacroiliac joints can be the most difficult to assess on plain radiography. Scintigraphy can be helpful but the uptake ratio between the sacroiliac joint and the os sacrum can vary with age. In any joint, CT and MRI provide more definite evidence of bone destruction on either side of the joint, and of subsequent joint involvement. Clinically, at this stage, there is no doubt that a limb joint is abnormal. It will be painful, swollen, and immobile.

(Text continues on p. 110)

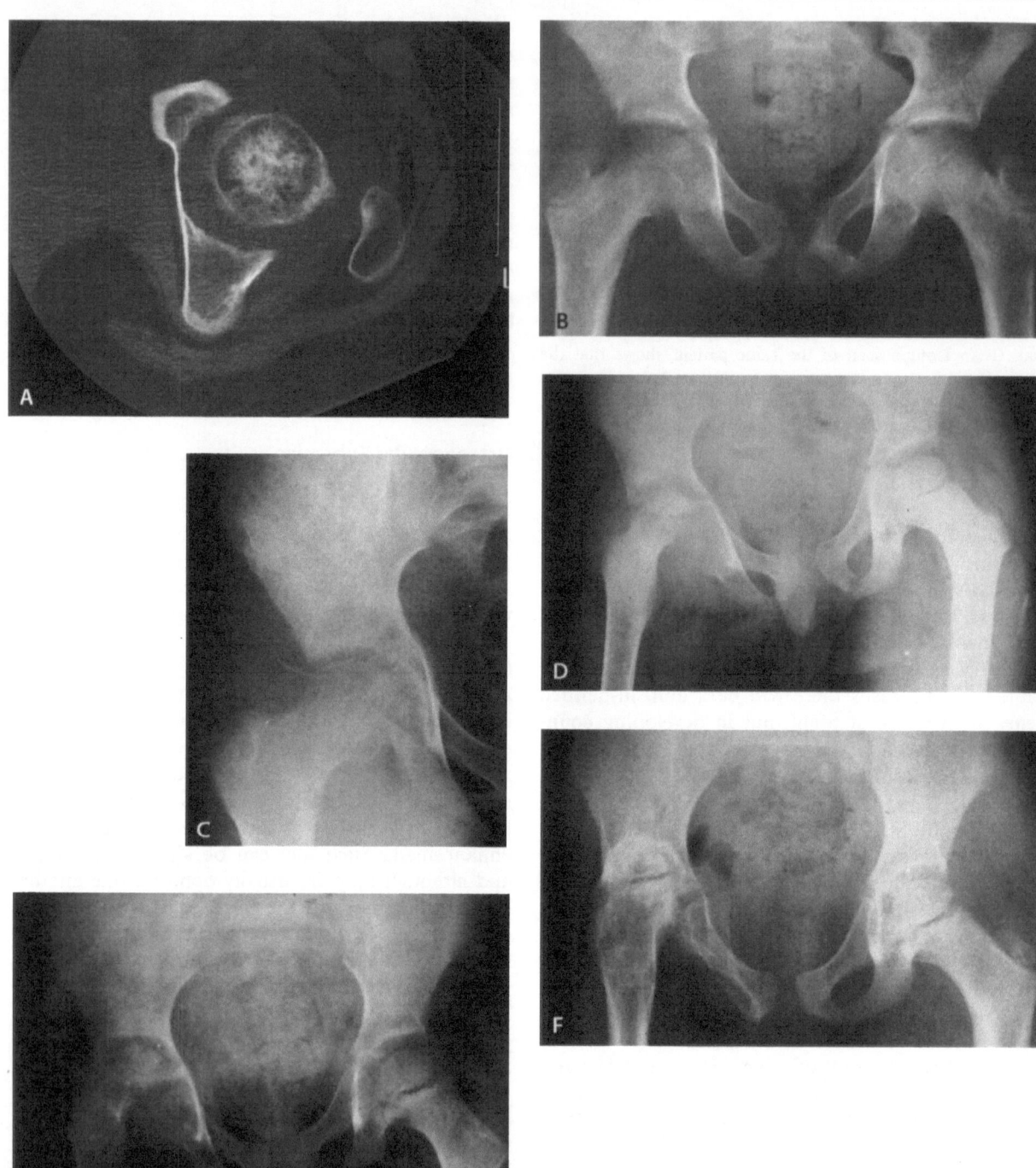

Fig. 78 A–F. Tuberculosis of the hip. The earliest evidence of infection may be an effusion into the joint. **A** The CT scan did not show any bone lesion but there were *M. tuberculosis* in the effusion. **B** In other patients there may be soft tissue swelling, osteoporosis, and loss of bone definition, as can be seen by comparison with the normal left hip. **C** A different but similar infected hip in which the ill-defined, hazy appearance of the joint and the soft tissue swelling around the acetabulum, particularly medially, are clearly shown. **D–F** The same patient (an African from Zimbabwe) over a period of 3 years. **D** There is soft tissue swelling and an effusion in the hip joint with destruction of the acetabulum and femoral head. **E** Six months later there is less fluid and the bone density is improving: in spite of the bone destruction, there are no sequestra. **F** Two years later the joint has fused: there will be shortening of the limb and there is significant deformity of the right side of the pelvis. (**A** courtesy of Dr. Cheryl Sisler, Hawaii)

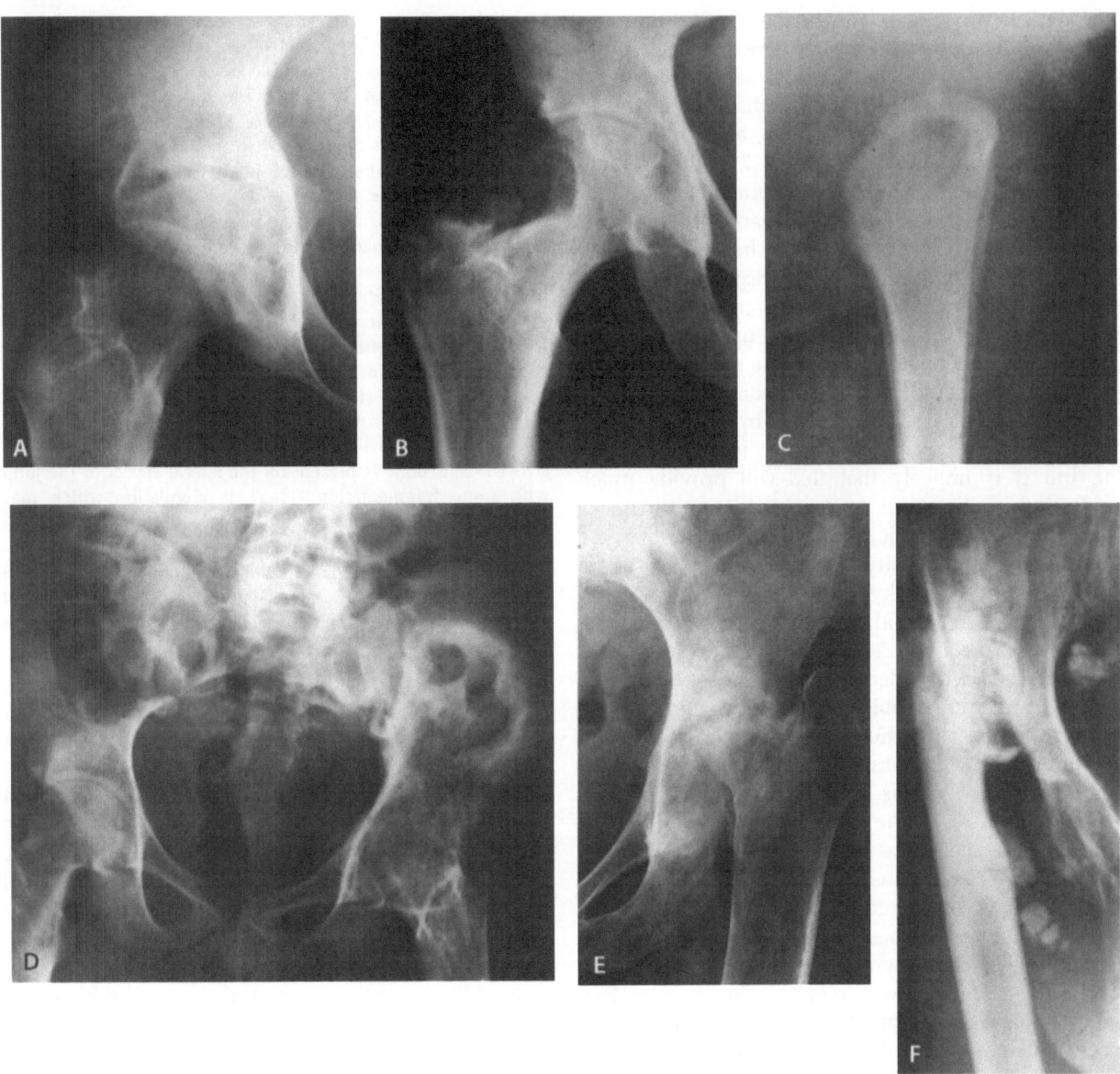

Fig. 79 A–F. Tuberculosis of the hip. In some patients the femoral head remains surprisingly intact, while in others it is eventually destroyed. **A** A cystic lesion in the acetabulum of a young African male: the hip joint is narrowed and irregular. There was also infection in the inferior pubic ramus. **B** There is destruction of the greater trochanter in this adult African and a cystic lesion in the neck of the femur, spreading into the femoral head. The joint space is not affected. **C** In this child there is central destruction of the upper end of the left femur, with a periosteal reaction along the shaft. The joint is infected and the femoral head has been dislocated and will never develop properly. **D** Advanced tuberculosis of the left hip and the left part of the pelvis. The hip joint is fused and there is dense sclerotic bone around the three lytic foci in the left iliac bone. This could be mistaken for a pyogenic infection. **E** The femoral head has disappeared, the acetabulum is irregular and sclerotic, and there are cystic changes both in the femoral neck and around the joint. The end result will be fusion, and there may be fixed fibrosis already. **F** The joint has been completely destroyed and dislocated and is surrounded by calcified debris. There will be marked shortening of the right leg and the hip is fixed. (Courtesy of the University of Capetown Radiology Library)

The disease progresses with further destruction and disorganization; the soft tissues, already swollen, become involved with the tuberculous process, best shown by MRI or CT. The joint may subluxate. In weight-bearing joints such as the hip (one of the most commonly affected: Figs. 78, 79), involvement of the acetabulum may lead to rupture of the femoral head into the pelvis, particularly where treatment has been delayed and weight bearing has been continued. CT and MRI, and occasionally ultrasonography, can show a soft tissue mass within the pelvis when this occurs. There will usually be multiple, very small sequestra within the mass. The margin of the infected lesion in the bone will probably show reactive sclerosis. Angiography has been used to show the soft tissue mass, which is usually vascular, but it is unlikely that this will provide much useful information.

As healing occurs, the joint space usually narrows; the bones increase in density but do not remodel and the deformity persists. Ankylosis is almost inevitable and may in fact be beneficial because it often permits limited use of the limb. The entire process may take months, even with adequate drug therapy, and the principal imaging problem is to assess the progress of healing, because the changes occur slowly and are often minimal. Ankylosis may become complete, but because of fibrous tissue rather than bony fusion. In such cases the infection may be quiescent but not fully healed. As with all other types of tuberculosis, experience must be combined with careful clinical judgment and continuous observation of the patient to obtain the best end result.

This process, either synovial or osteoarticular, may occur in any tuberculous joint and the same pattern will be followed. Variations occur depending on the type and location of the joint, whether or not it is weight bearing, and how much it is used. For example, when foci oppose each other on either side of a joint [usually the knee (Fig. 80), but also the elbow, shoulder, and hip], dense areas of bone may be seen ("kissing sequestra"). There is some individual variation depending on the host resistance, but this is more in terms of the time taken for the changes to be complete rather than in the pattern of the disease.

Tuberculosis of Bones

Osteomyelitis due to the tubercle bacillus is a common condition and need not always involve a joint; in fact it can occur in virtually any bone in the skeleton (Figs. 81–85). It is most common in childhood but no age is immune; it is frequently multifocal. Scintigraphy is very helpful in identifying the possible multiple sites.

Fig. 80 A–H. Tuberculosis of the knee. **A** The earliest signs of ▶ infection may be a large joint effusion and mild generalized osteoporosis, without any visible bone defect. **B, C** Another African patient from Zimbabwe. The soft tissue swelling and effusion are marked but there is a lytic bone focus in the lower end of the femur and periosteal elevation around the full diameter of the femoral shaft. There is accentuated growth of both the femoral and tibial epiphyses. The lytic bone focus is important in the differential diagnosis, because nonarticular rheumatoid arthritis and hemophilia can both give similar appearances before there is any bone destruction. **D** Another patient with a lytic bone focus medially in the lower end of the femur, but also in the femoral epiphysis with a slight sclerotic margin. There is a joint effusion. **E** A tuberculous infection of the patella with a suprapatellar effusion. **F, G** The end stages of a tuberculous joint infection. In F there were foci on both sides of the joint; the lateral femoral and tibial condyles had been partially absorbed and destroyed. New bone is present on the lateral aspect of the joint with some fragmentation. There is angulation which may progress. The joint will probably fuse, with either fibrous tissue or bony anyklosis, but in some patients it may remain a lax unstable joint. **H** A healed but badly damaged joint, with partial destruction of the upper end of the tibia, and a lot of new bone and loose fragments. There is calcified debris in the supra-patellar bursa

As with pyogenic organisms, the infection most commonly begins in the metaphysis. Its effect on the adjacent joint has already been described. Away from the joint the early changes, localized areas of osteoporosis without any surrounding bone reaction, are often more easily seen than delineated. They have no particular shape, but follow the contour of the bone; in the skull they may be large and apparently spreading. The initial lesion may be seen in the diaphysis, but this is rare. Multiple bones may be infected or the same bone may have foci in different places, not initially connected.

If there is a lesion close to the cortex, expansion is common and cortical thinning occurs. There will be an associated periosteal reaction with a fine lamellar pattern, spreading proximally and distally from the original focus. Ultrasound can identify the edema and the periosteal reaction and be used to guide aspiration biopsy to establish the diagnosis. Both CT and MRI can image the infection, but are unlikely to add much to the accuracy of the diagnosis. This type of reaction is most common in the smaller bones, such as the phalanges (tuberculous dactylitis) and is most frequently seen in children (Figs. 86, 87). The changes are more marked at this age because the cortex is relatively soft and expansion occurs more easily. The old descriptive term "spina ventosa" (spina = short bone; ventosa = inflated with air) is thus especially appropriate (see Fig. 86 A).

(Text continues on p. 114)

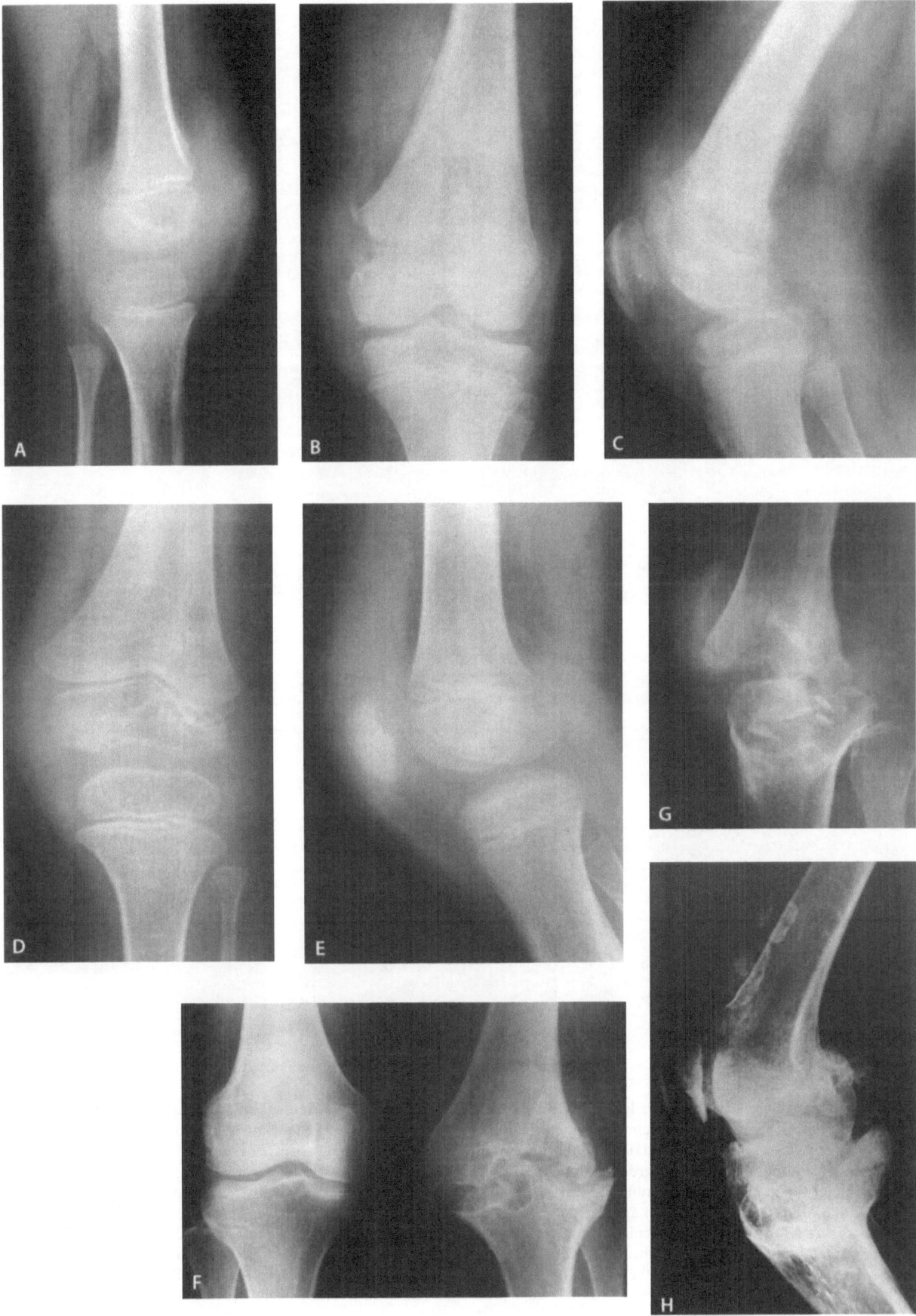

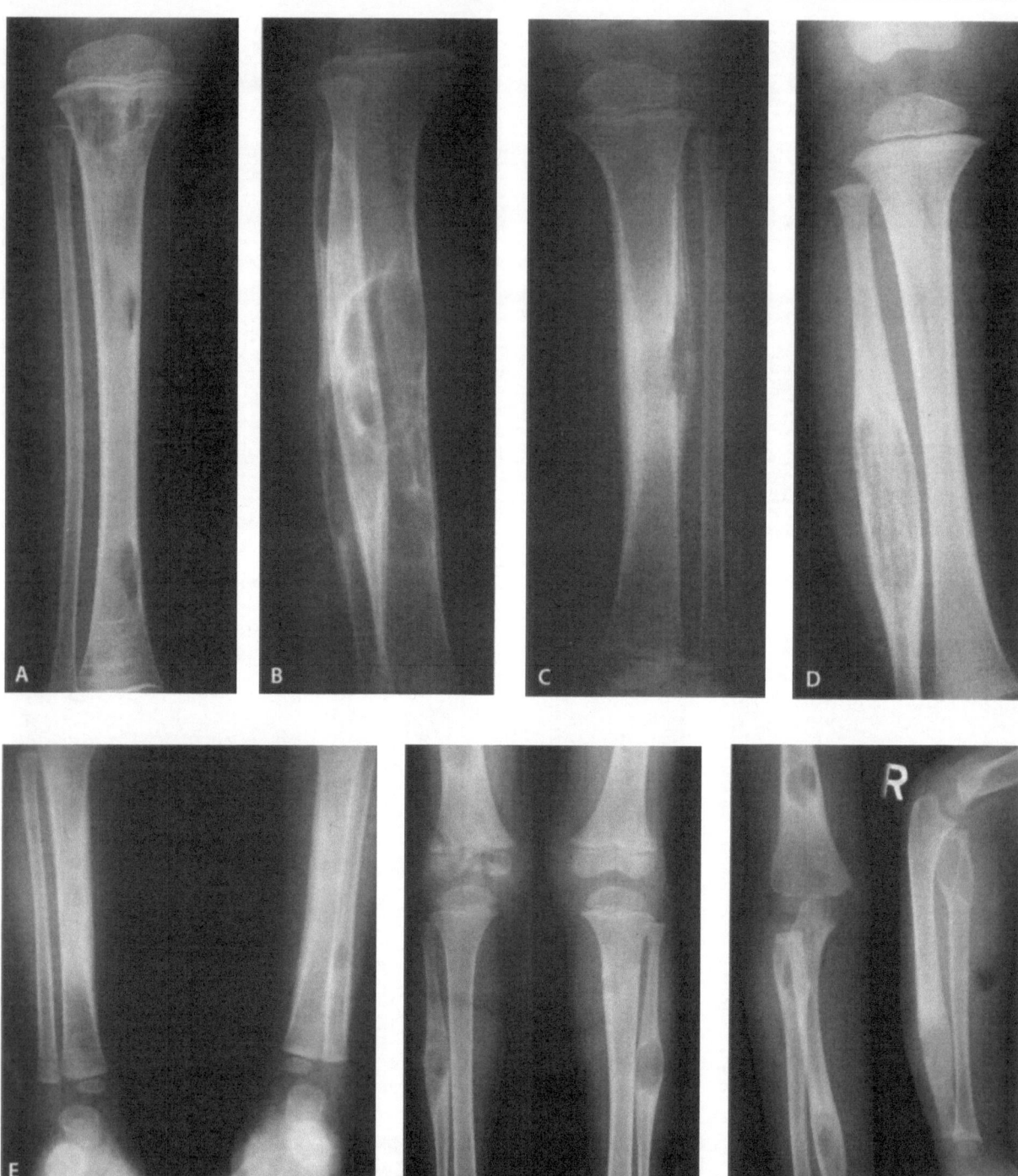

Fig. 81 A–G. Tuberculosis in the lower leg. Many tuberculous foci in the long bones are cystic and are often multiple. There is nearly always periosteal reaction. **A** Multiple cystic foci in the tibia and lower femur. There is minimal periosteal reaction along the upper half of the tibia laterally. **B** Multiple similar cysts in the tibia and fibula of another patient, with a periosteal reaction in the fibula particularly. Some of the cystic foci have minimal sclerotic but incomplete borders. **C** A focus of infection in the center of the tibia with a layered periosteal reaction. **D** An expansile lesion in the fibula, with the ghost of the original shaft running through it. **E** Tuberculous foci may appear in more than one limb. In this African there is a periosteal reaction around the shafts of tibiae and the left fibula, with several faint foci of infection. **G** An African child from Kenya with tuberculous foci in both fibulae, the left tibia, and the lower ends of both femora. He also had similar foci in the right humerus, radius, and ulna. This type of tuberculous bone infection usually responds quite well to appropriate treatment (see Fig. 88). (**B** courtesy of Dr. Harold G. Jacobson ; **A, C, D** from Cremin and Jamieson 1995)

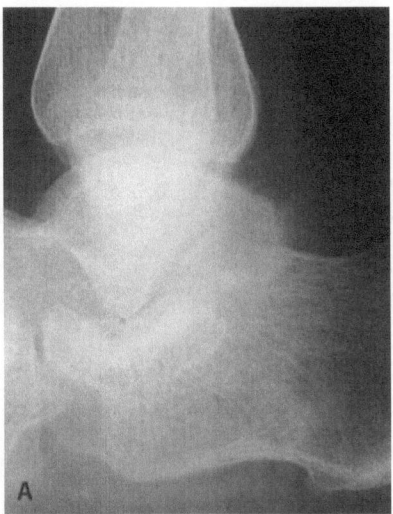

Fig. 82 A–F. Tuberculosis around the ankle joint. **A** Very early infection. There is synovial thickening and a small effusion, as well as a minimal bone defect in the cortex of the upper surface of the talus posteriorly. **B** Infection of the lower end of the tibia, with an effusion into the ankle joint. There is marked soft tissue swelling. **C** A large defect medially and posteriorly in the lower end of the tibia with generalized osteoporosis and a small focus in the fibula. The ankle joint is decreased and a focus of infection is seen in the talus posteriorly. There is marked soft tissue swelling, and a joint effusion. **D** Collapse and destruction of the os calcis with marked soft tissue swelling and generalized osteoporosis. The infection is in the os calcis, but the whole joint will eventually be destroyed. **E** Cystic and sclerotic tuberculous infection in the posterior two-thirds of the os calcis, only affecting the cortex in one small area of the upper margin, near the tendo Achillis. **F** A more localized infection of the os calcis, with a sequestrum and extensive disruption of the upper surface. In **D–F** the differential diagnosis would include pyogenic and mycotic infections. (Courtesy of the University of Capetown Radiology Library)

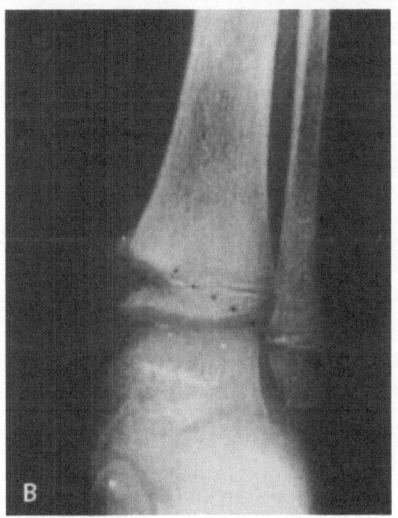

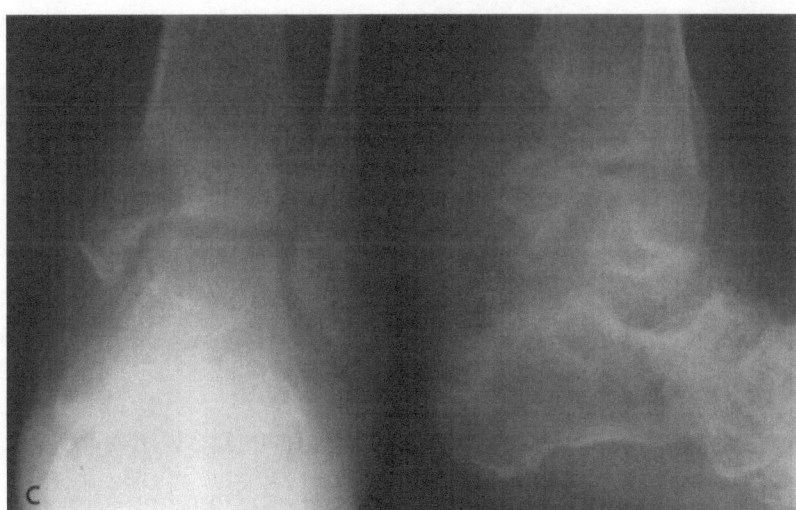

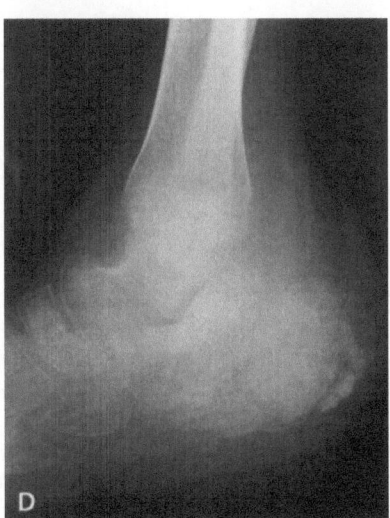

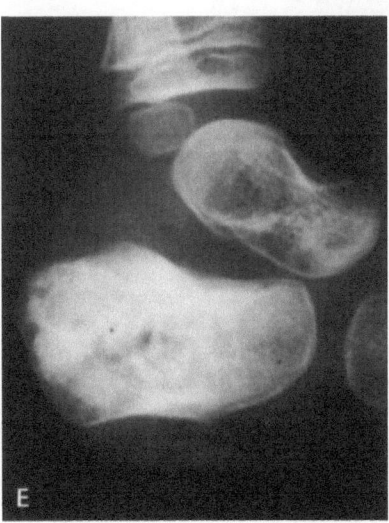

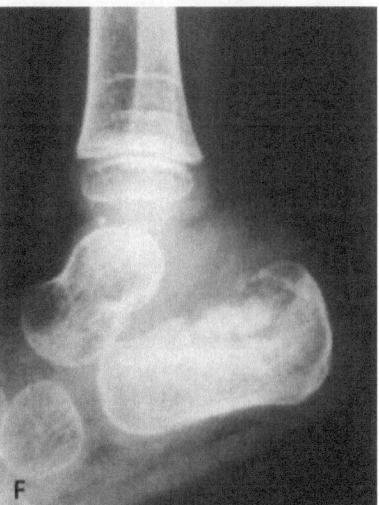

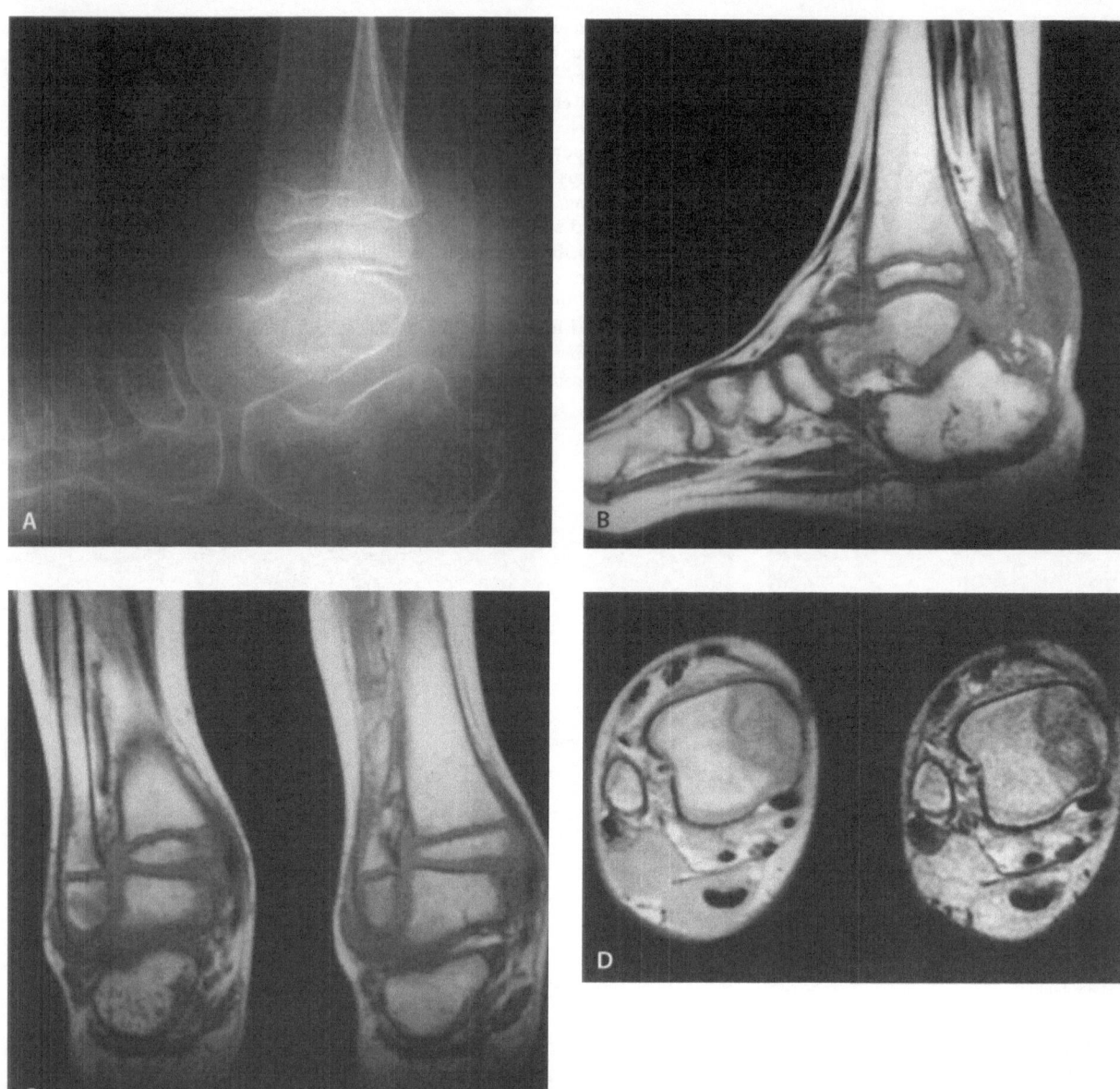

Fig. 83 A–D. The swollen ankle of a 7-year-old boy from the Pacific Islands. **A** There is generalized osteoporosis, but it is difficult to be sure that there is any bone focus. **B–D** MR scans show the joint effusion, and destruction of both the ta-lus and the posterior part of the tibial epiphysis. The unfor-tunate child had tuberculous meningitis and miliary tubercu-losis

The radiolucent areas are filled with tuberculous granulation tissue, which causes the expansion. As the cortex expands and thins, subperiosteal new bone is laid down on the outer aspect; this new bone may be layered, quite thick and dense, and sclerotic. Because there is the appearance of an in-volucrum, the possibility of a pyogenic infection should be considered. The differential diagnosis then depends largely on the patient's clinical condi-tion. In tuberculosis the pain is not marked, pyrexia is minimal, and the whole condition is relatively be-nign. A pyogenic osteomyelitis causing such a reac-tion would be acutely painful, swollen, and in-flamed, with a generalized systemic reaction which is often absent in the tuberculous variety.

Radiographically detectable sequestration does not often occur in tuberculosis because the blood supply has not been affected, but there are exceptions and tuberculous sequestra do occur without any other as-sociated infection (cf. Fig. 82 F). CT may detect very small sequestra when a bone lesion has spread rapid-ly into the soft tissues, e.g., in the pelvis.

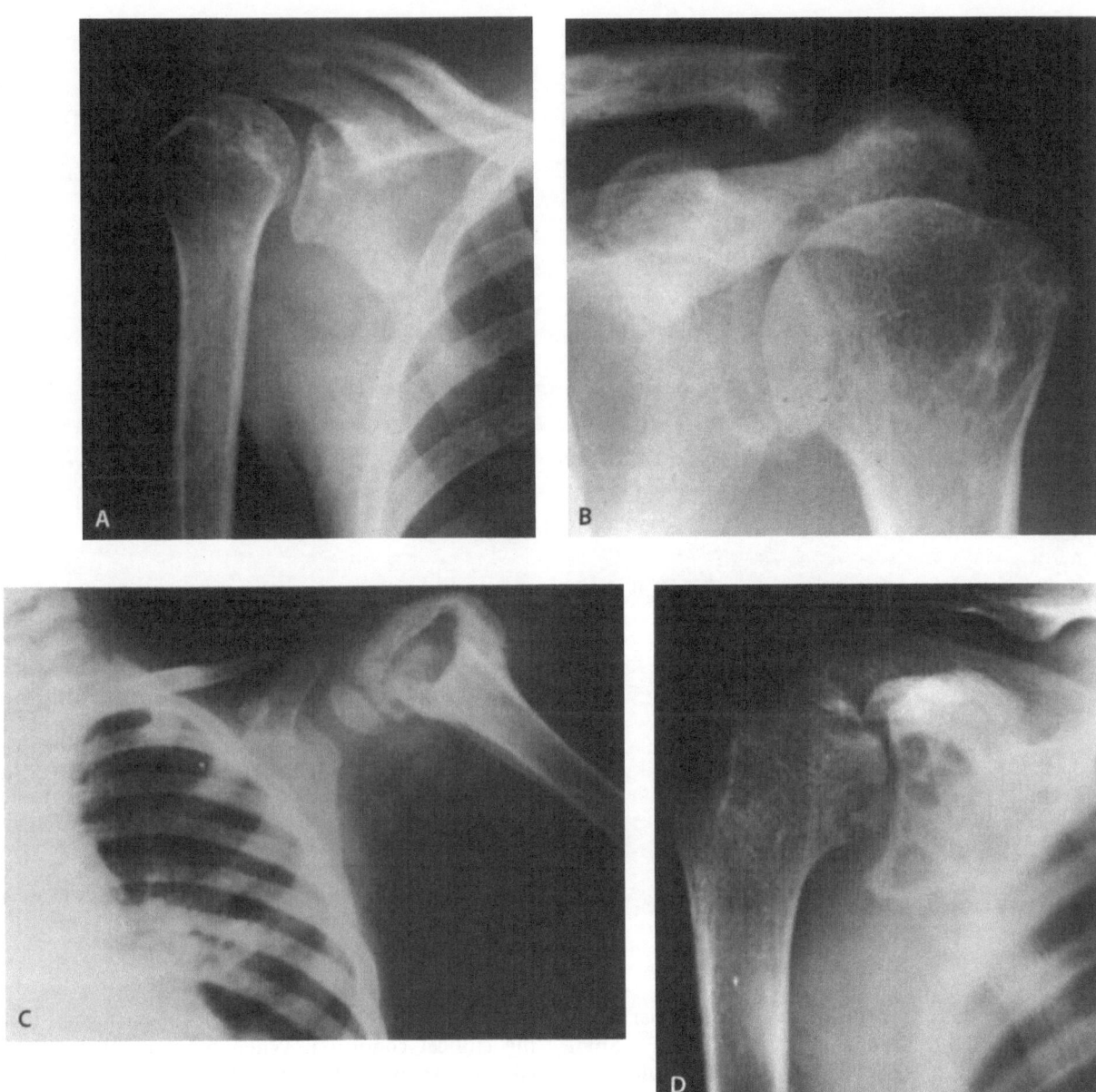

Fig. 84 A–H. Tuberculosis of the shoulder. **A** Acute tuberculosis with marked osteoporosis of all the bones around the shoulder. There is minimal periosteal elevation along the clavicle, but there is no bony defect. **B** A more chronic infection with multiple lytic areas in the head of the humerus, the clavicle, and the acromion process. The acromioclavicular joint is disrupted, but the shoulder joint is still relatively intact. **C** Considerable destruction of the head of the humerus in a child. There is left hilar lymphadenopathy. **D** Cystic tuberculosis in the right shoulder of a middle-aged African. The glenoid is almost destroyed, and filled with multiple cystic lesions. Next to it, the head of the humerus shows similar defects and there is destruction of the articular surface. The joint space is narrowed. The acromion was also affected. **E–H** see p. 116

In Europe and North America, the same multifocal lytic lesions can occur, but they are better demarcated and cyst-like. They are most common in the long bones of children and young adults; sometimes they are symmetrical. Probably the difference between the tropical and nontropical pattern is related to immunity or more prompt treatment. The "tropical" pattern will be seen in AIDS anywhere and will often be more destructive and spread more rapidly.

In the skull, scapula, and ribs, as well as the long bones, there may be similar lytic foci without any bone response. In the skull this results in quite large, destructive lesions. Solitary skull foci also occur, sometimes with a reactive edge and a central "button" sequestrum. Because these solitary tuberculous foci are associated with a fluctant "cold" abscess they were named "a puffy tumor" by Sir Percival Pott in the eighteenth century.

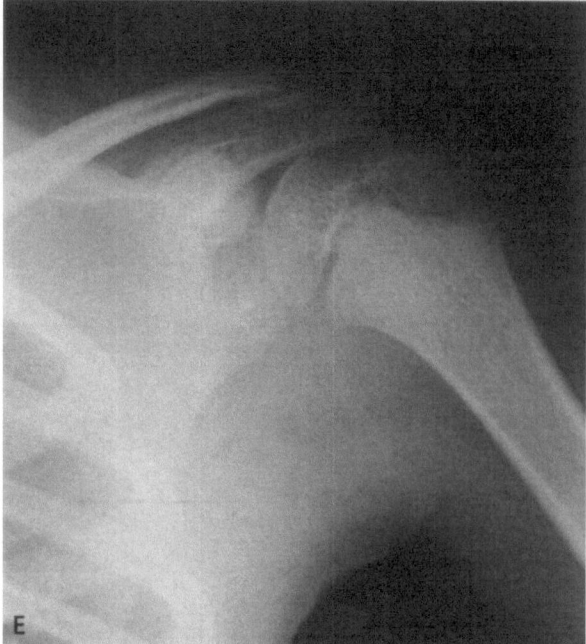

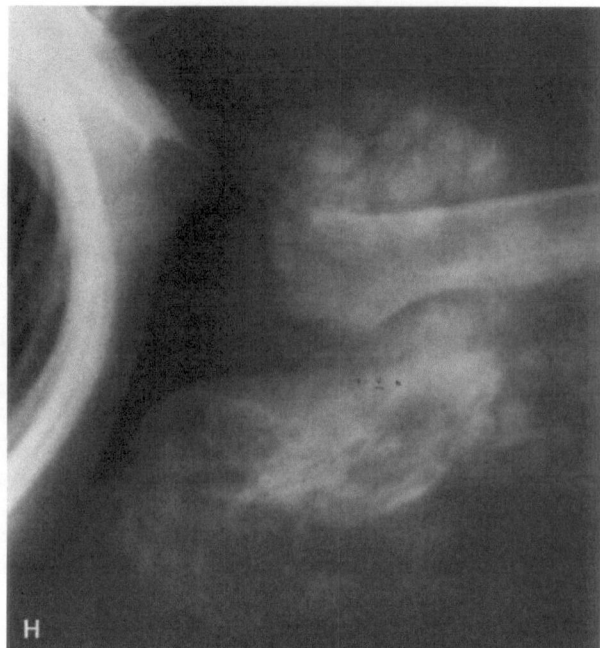

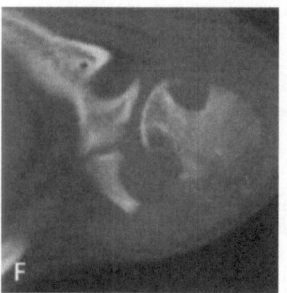

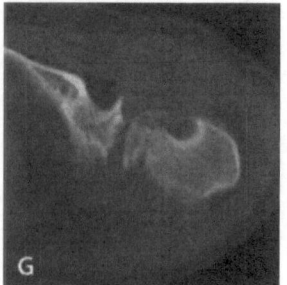

Fig. 84 (continued). E Erosion of the glenoid and humeral head in a child. F, G two CT scans in the same patient showing the defect in the humeral head and the partial destruction of the glenoid (G). H The late stage of tuberculosis; there is calcified debris all around the joint, with partial destruction of the end of the clavicle. The acromion was also involved. The humeral epiphysis is very decalcified. The calcified masses could be mistaken for tumoral calcinosis, but the destruction of the clavicle and acromion provide evidence of infection. (E–G from Cremin and Jamieson 1995)

In the ribs there may be quite a marked periosteal reaction and localized swelling (Fig. 85 E, F). Some care has to be taken to differentiate tuberculous osteomyelitis of a rib from a tuberculous abscess tracking around the intercostal space from its origin in the thoracic spine. Such as abscess will also cause swelling, and there may be some local periosteal reaction on adjacent ribs. Ultrasonography (or CT/MRI) may help to track the abscess more accurately and, like scintigraphy, will locate any intrinsic bone focus. Radiographically, in slender bones such as ribs it may not be easy to see the destructive areas within the bone itself, particularly when the periosteal reaction is marked; an exactly similar process can occur in the long bones. Periosteal new bone is not the "hallmark" of pyogenic infection; it can also be seen in tuberculosis.

In the tropics this type of "cystic" and periosteal tuberculous bone infection (cf. Fig. 81) may be relatively slow and benign, or can be progressive, depending on the host resistance and the rapidity with which treatment is commenced. In young children the imaging changes may appear worse than the clinical condition. When tuberculous osteomyelitis occurs in adolescence and adults, it takes longer to heal and may be more significant. In the skull vault the infection usually spreads outwards; only rarely does it involve the meninges, leading to an intracerebral tuberculous abscess (see Fig. 95).

Healing follows the same slow pattern already described (Fig. 88). The bone density returns; the defect may be filled in with new bone, or may persist for some time. The bone may be sclerotic, heavily trabeculated, or relatively normal. Some areas may be dense, resembling sequestra; the dense sclerosis in some bones, especially in the diaphysis, may mimic a chronic pyogenic (Brodie's) abscess. Healing is more rapid in the long bones and digits than in the flat bones and tends to leave less deformity except when the epiphyseal place has been involved, which commonly results in shortening. Almost all diaphyseal lesions respond well to therapy.

(Text continues on p. 120)

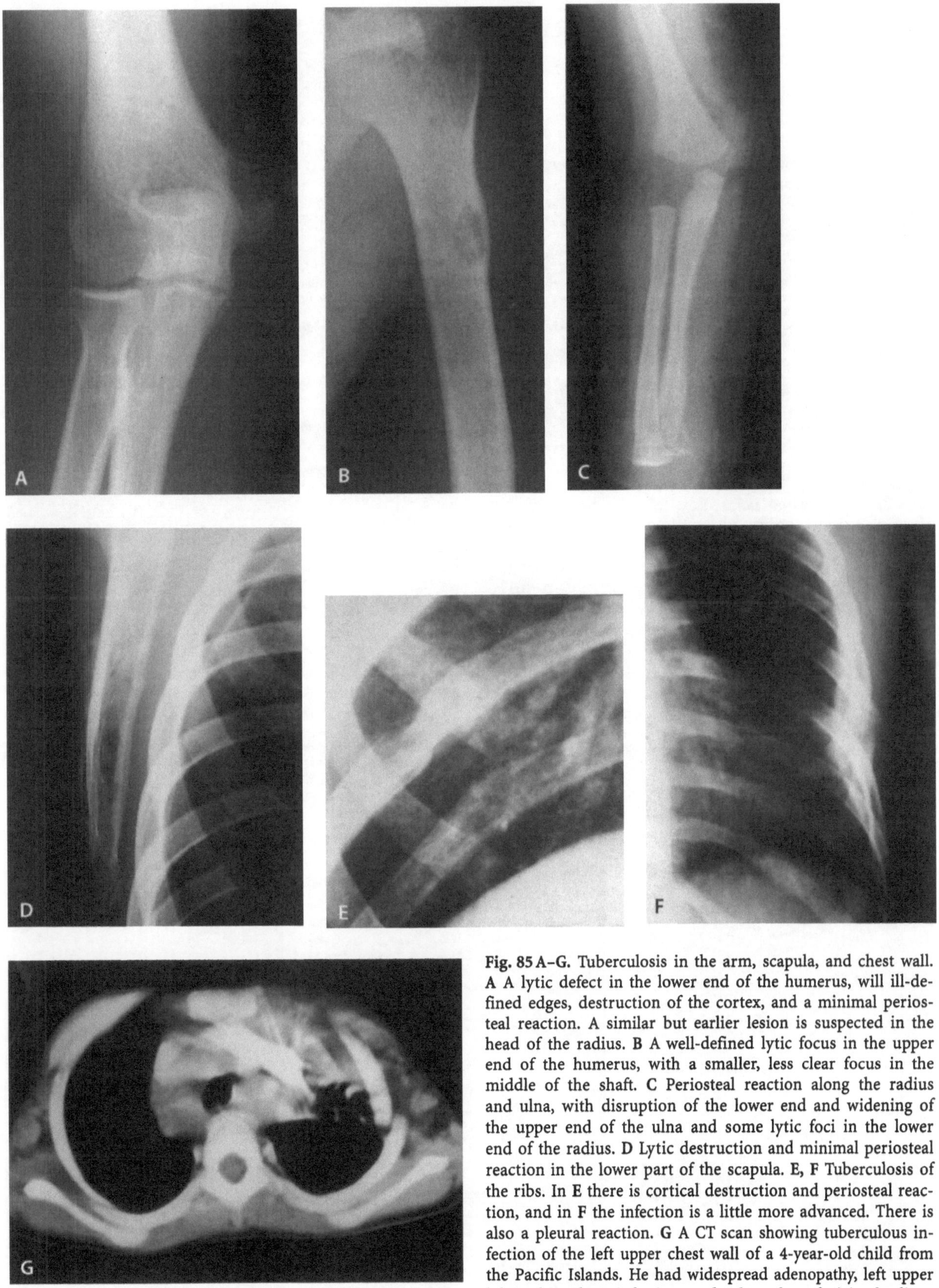

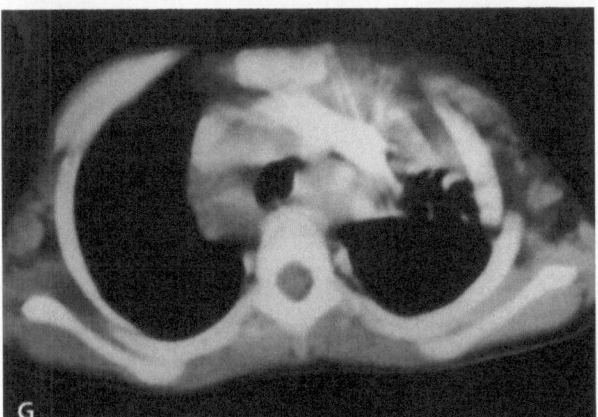

Fig. 85 A–G. Tuberculosis in the arm, scapula, and chest wall. **A** A lytic defect in the lower end of the humerus, will ill-defined edges, destruction of the cortex, and a minimal periosteal reaction. A similar but earlier lesion is suspected in the head of the radius. **B** A well-defined lytic focus in the upper end of the humerus, with a smaller, less clear focus in the middle of the shaft. **C** Periosteal reaction along the radius and ulna, with disruption of the lower end and widening of the upper end of the ulna and some lytic foci in the lower end of the radius. **D** Lytic destruction and minimal periosteal reaction in the lower part of the scapula. **E, F** Tuberculosis of the ribs. In **E** there is cortical destruction and periosteal reaction, and in **F** the infection is a little more advanced. There is also a pleural reaction. **G** A CT scan showing tuberculous infection of the left upper chest wall of a 4-year-old child from the Pacific Islands. He had widespread adenopathy, left upper lobe tuberculous infection, and tuberculous foci in the liver. (A progressive primary infection.) (**D–F** courtesy of Dr. Harold G. Jacobson; **G** courtesy of Dr. Cheryl Sisler, Hawaii)

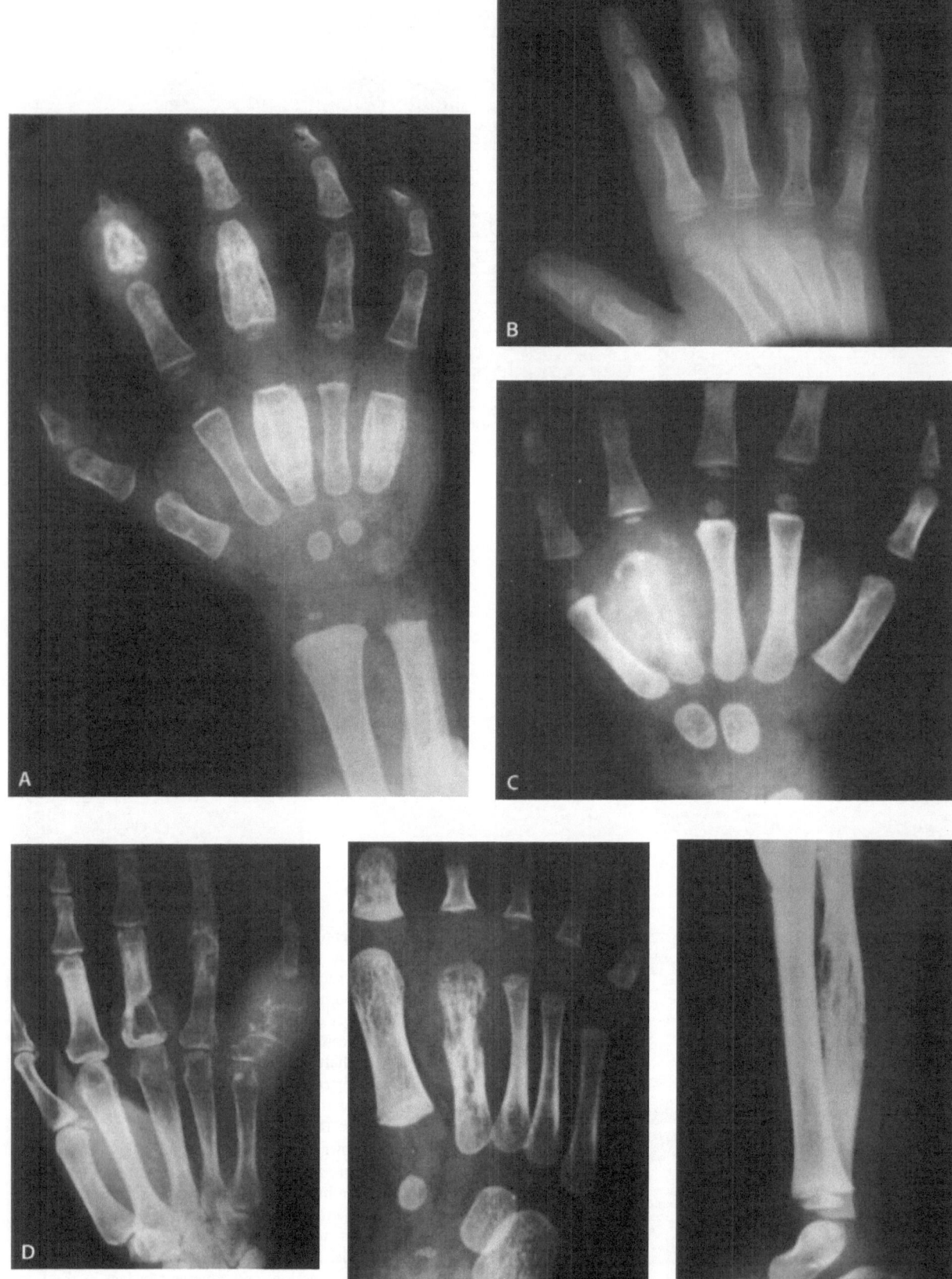

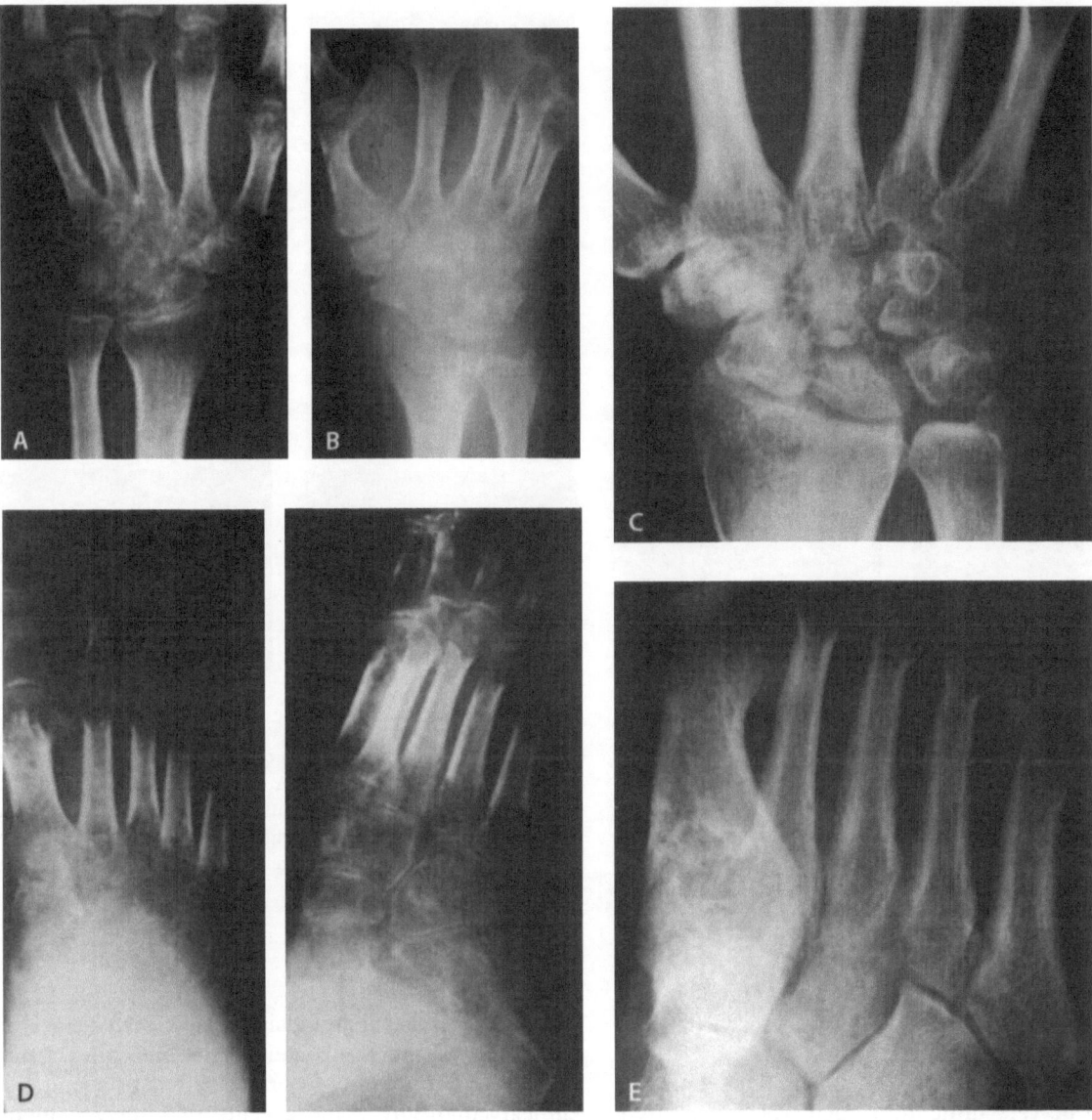

◀ **Fig. 86 A–F.** Tuberculosis of the extremities. Infections of the hands and feet produce similar images. **A** The hand of an infact showing marked periosteal elevation and multiple lytic areas of bone destruction. There is a periosteal reaction along the ulna also. Lytic defects are present without periosteal reaction in some of the digits. **B** In this child there is also some epiphyseal damage and less periosteal reaction. The infection was by atypical mycobacteria, but this could not be recognized by looking at the images. **C** Destructive tuberculous osteomyelitis affecting the fourth left metacarpal. **D** The hand of an older patient with more advanced tuberculosis; the proximal phalanx of the fifth finger has been almost entirely replaced by granulation tissue and there are also multiple lytic foci in the phalanges of the third and fourth fingers. This is an unusual form of tuberculosis and it could be mistaken for multiple enchondromas or even metastases. **E** Osteomyelitis of the second right metatarsal and **F** the fibula of the same patient, a young African girl; her hand is shown in **C**. Her chest x-ray was normal and there were no other bone foci. (**A** courtesy of Dr. Harold G. Jacobson; **C, E, F** courtesy of the University of Capetown Radiology Library)

Fig. 87 A–E. Tuberculosis of the carpal and tarsal bones in different patients. All show generalized osteoporosis, particularly at the ends of the digits and around the carpal and tarsal joints. No bone foci are visible because the disease is mainly synovial. Bone destruction is likely to occur eventually and fusion will be the end result. **A** Osteoporosis is well marked. **B** The wrist of another patient in which bone destruction has started, involving the radius and ulna as well as the metacarpals. The carpus is collapsing. **C** In this patient the cystic and destructive tuberculous infection could be mistaken for rheumatoid arthritis or even severe gout. **D** Similar changes in the tarsus: generalized osteoporosis particularly around the joints. There are a few bone foci in the metatarsals. **E** A more chronic infection at the proximal end of the right first metatarsal and the medial side of the tarsus. The cystic lesions are becoming sclerotic and the joints are fused. There is a periosteal reaction along the shaft of all the metatarsals, but this may be due to the soft tissue reaction rather than local bone infection. This African patient was treated for many months but there was no significant change in his foot. (**B** courtesy of Dr. Harold G. Jacobson)

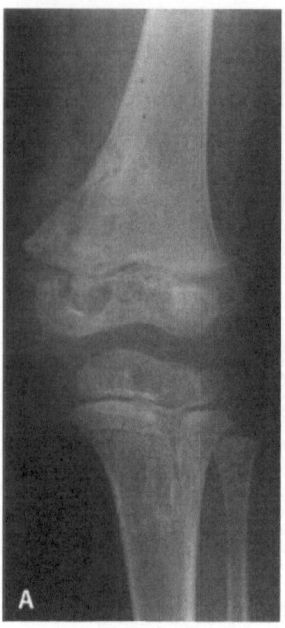
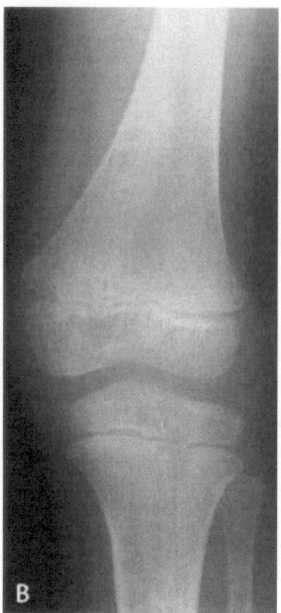
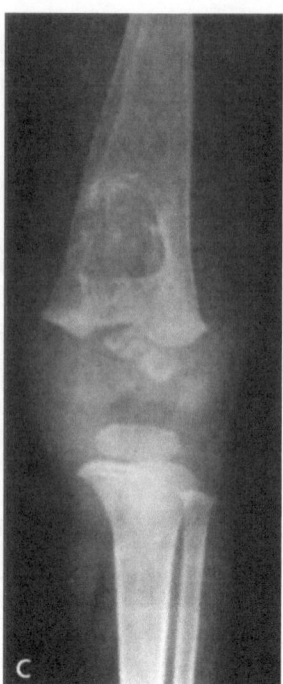
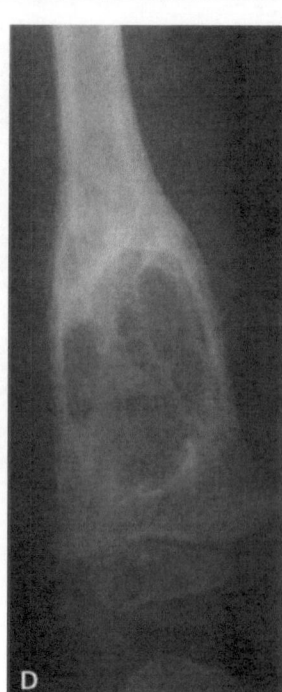

Fig. 88 A–D. The healing of tuberculous bone foci. **A** Cystic changes in the lower end of the femur and the upper end of the tibia, including their epiphyses. Eighteen months following the completion of 6 months of antituberculous treatment, the changes have responded well. They will probably leave only small dense foci. **C** This young African child had been on treatment for 3 months and the large cystic focus was healing with sclerosis. The periosteal reaction along the medial aspect of the femur has also thickened. Unfortunately, there is considerable disruption of the normal growth pattern of the femoral epiphysis. **D** A longstanding lesion in the lower end of the femur of another child with a multicystic appearance, periosteal thickening, and widening of the lower end of the femur. In this case the epiphysis is small but otherwise appears normal

Differential Diagnosis

The differential diagnosis of tuberculous bony lesions can be very difficult, particularly if this possible etiology is not considered early. The patient's clinical condition may help to differentiate tuberculosis from acute pyogenic infections, but even this may not ensure accuracy. The mycotic infections, such as North American blastomycosis, can simulate tuberculosis, particularly in their early stages. The multifocal destructive lesions of histoplasmosis duboisii may resemble the aggressiv form of tuberculosis.

In the early stage osteoporosis is more marked in tuberculosis compared with pyogenic infections, but tropical (or AIDS) tuberculosis (unlike tuberculosis in Europe or North America) may behave quite acutely and change can be rapid. Sequestration is not common in tuberculosis, but can occur. Spread of infection across an epiphyseal plate is uncommon in pyogenic infections, but is frequent in tuberculosis. Eosinophilic granuloma may mimic tuberculosis (or vice versa) and infections associated with sickle cell disease, such as typhoid osteomyelitis, may closely resemble tuberculosis in their early stages in the bones of the skull vault or in the spine. Later there tends to be more periosteal reaction and the clinical condition is more acute in nontuberculous infections. The frontoparietal lytic lesions of tuberculosis are usually ill defined, compared with the clearly beveled edge of histiocytosis. Syphilis can produce very similar osteolytic lesions.

Particularly in children it is important to differentiate early rheumatoid arthritis, especially the monarticular variety. Cystic lesions could be mistaken for hydatid disease of bone, but this is uncommon in the midshaft of long bones.

The pattern of tuberculosis is the same in the bones of the skull, and may be found in the sphenoid, mastoid, zygoma, as well as the calvarium (Figs. 89–95). The maxilla and the mandible can also be involved and the differential diagnosis can be extremely difficult.

Tropical tuberculosis affects the skeleton in many ways, and this possible etiology must be remembered when any bone infection, "acute" or "chronic", is being reviewed.

(Text continues on p. 125)

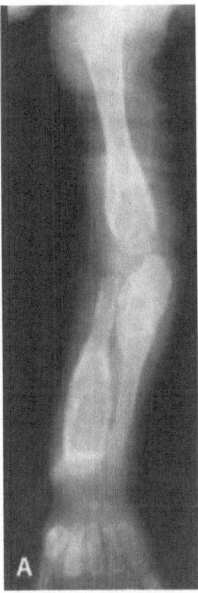
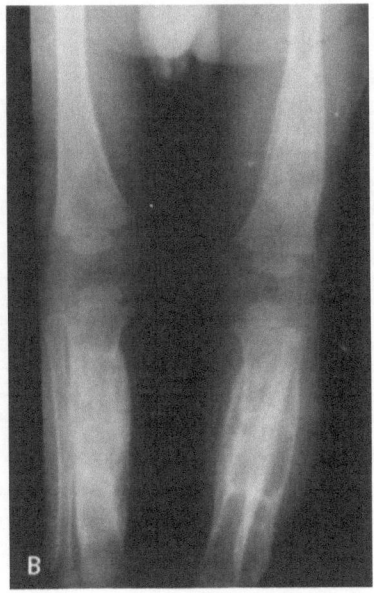

Fig. 89 A, B. Tuberculosis is very frequently multifocal, particularly in patients who have a depressed immune system for any reason. Although multifocal infection is seen most commonly in children as part of the progressive primary tuberculosis, it may also appear in adults, even in old age if the infection is reawakened. A This 10-month-old African baby had widespread skeletal tuberculosis. There are cystic lesions with marked periosteal reaction in almost all the long bones. In the right arm, the lower end of the humerus, the upper end of the radius, and the lower half of the ulna are affected. There is dactylitis of the fifth right metacarpal. There were similar lesions in the left arm. B There cystic lesions in both lower limbs, in each femur; the left is more severely affected and shows a greater periosteal reaction. In the lower legs only the right fibula seems to have escaped. Whenever there is multifocal disease, a radioisotope bone study may reveal unexpected foci. The only reason that this child was brought to a hospital was because of swelling of the left lower leg. (Courtesy of the University of Capetown Radiology Library)

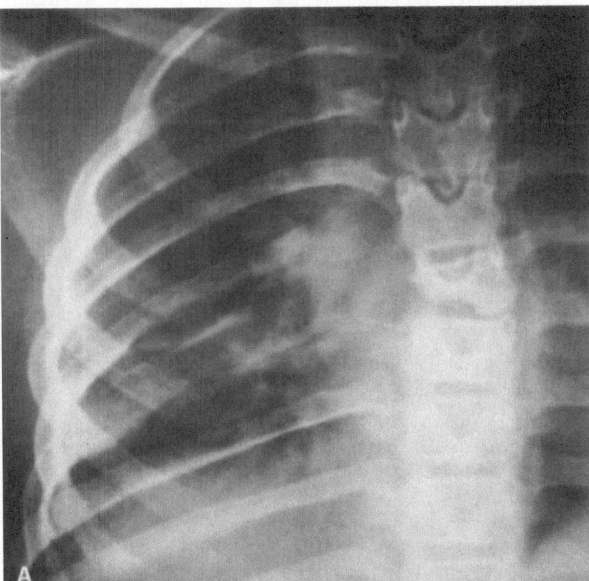

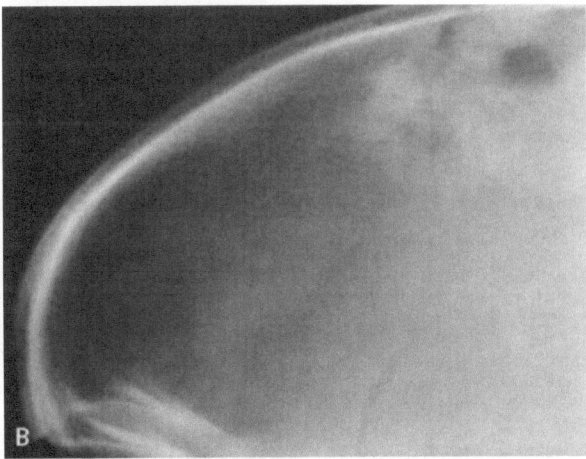

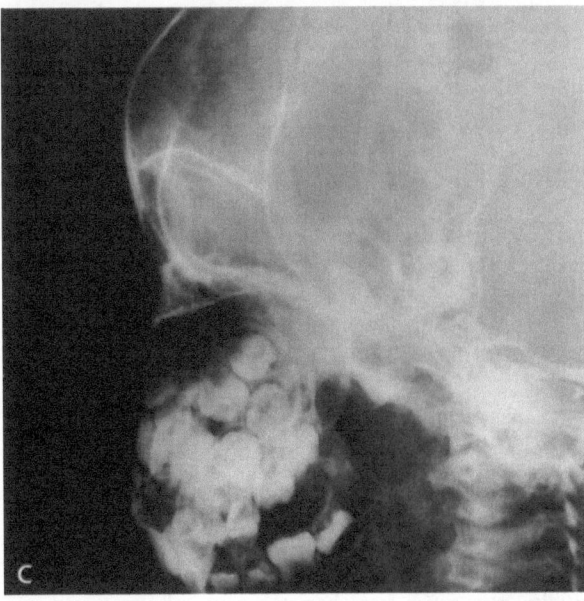

Fig. 90 A–C. Tuberculosis in the spine may be the clinical reason why a patient comes to hospital, but there are often foci of infection elsewhere. A In this small African child there is destruction of the inner ends of the sixth and seventh ribs and the infection has spread along the seventh right rib. The right side of the sixth thoracic vertebra has collapsed, although so far the disc space has remained relatively intact. There is a large right paravertebral abscess and a smaller one on the left. It is probable that the rib became infected from the spinal foci, because there is a periosteal reaction along to the anterior ends of both the sixth and seventh right ribs. The child also had two tuberculous foci in the vault of the skull (B) and destruction of the condyle of the mandible (C)

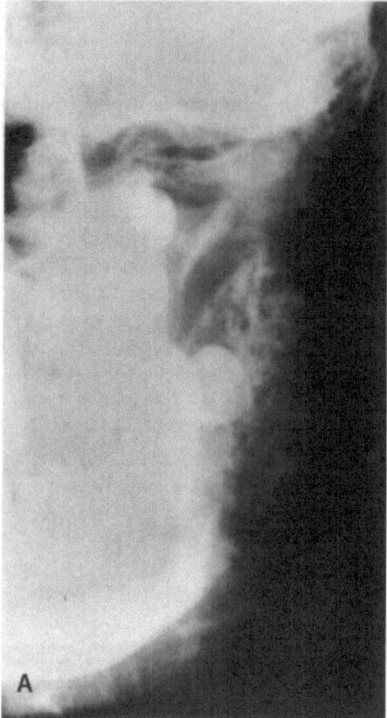

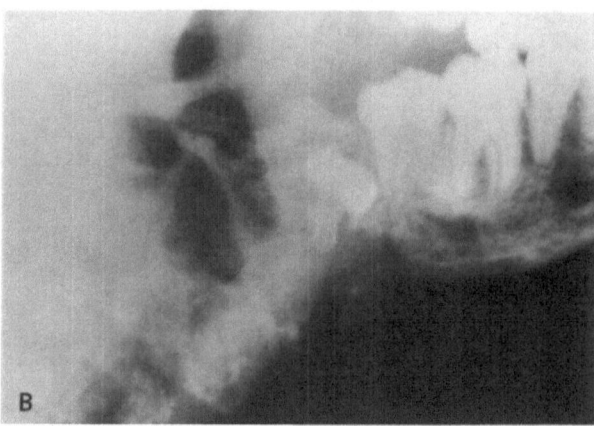

Fig. 91 A, B. A young African from Zimbabwe being treated for pulmonary tuberculosis was found to have a swelling around the jaw. There were no local symptoms. There is destruction of the left side of the mandible with multiple granulomas and very little new bone periosteal reaction

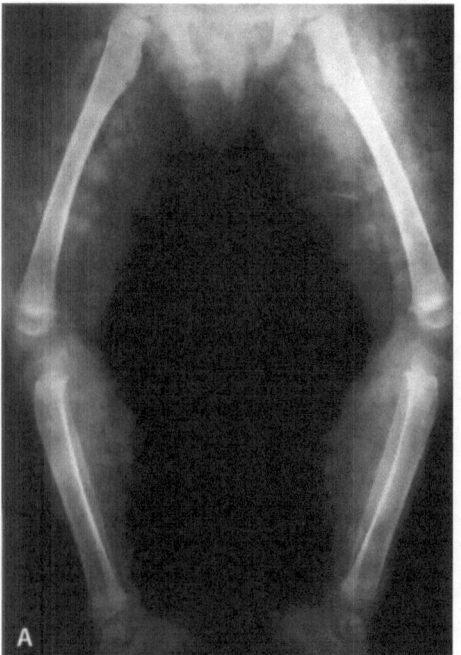

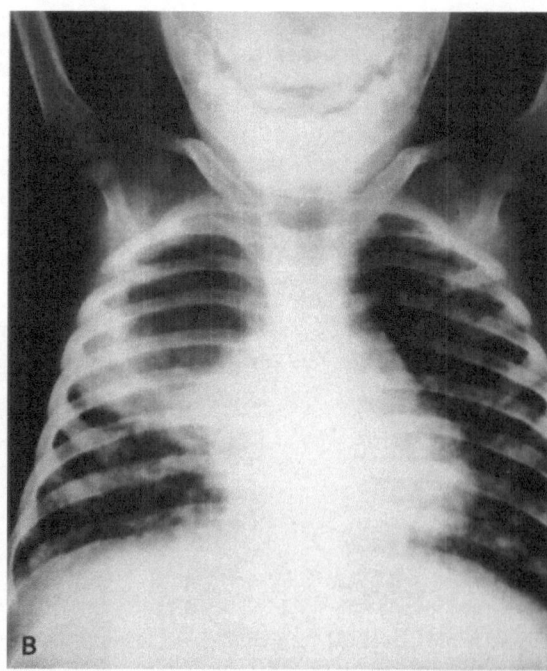

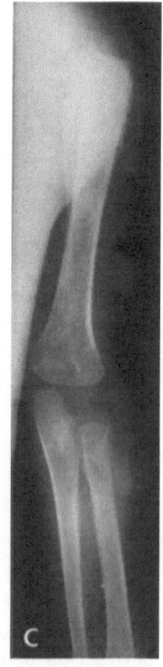

Fig. 92 A–C. This young African child had an overwhelming tuberculous infection and presented (A) with multiple cutaneous lesions covering the whole body. The only bone focus was in the upper end of the left humerus (B), seen on the chest radiograph. There was a large cavity almost replacing the right upper lobe, pneumonia, and multiple tuberculous foci throughout both lungs. Young children with lowered immunity may suffer from hematogenous miliary spread and present with papules, erythematous macules, or purpuric lesions over most of the body. The tuberculin skin test is usually negative and the prognosis is often poor, but this child improved after 2 months of treatment (C). The skin lesions are regressing, and there is a thick periosteal reaction around the upper end of the humerus, seen through the shadow of the shoulder

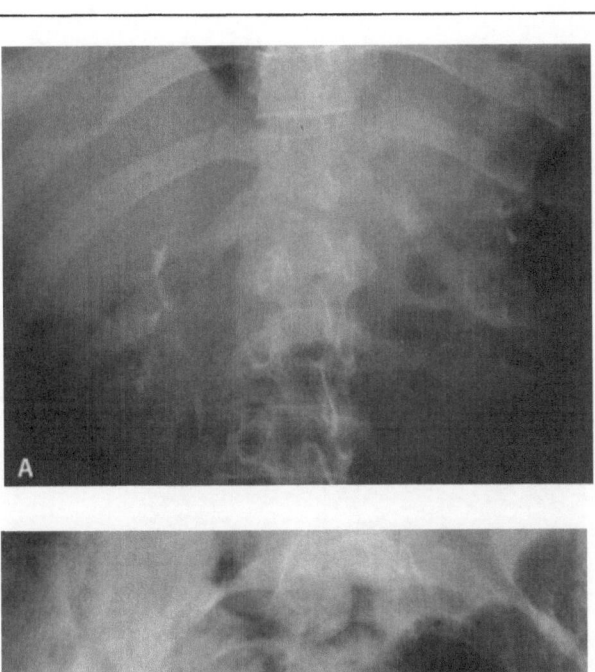

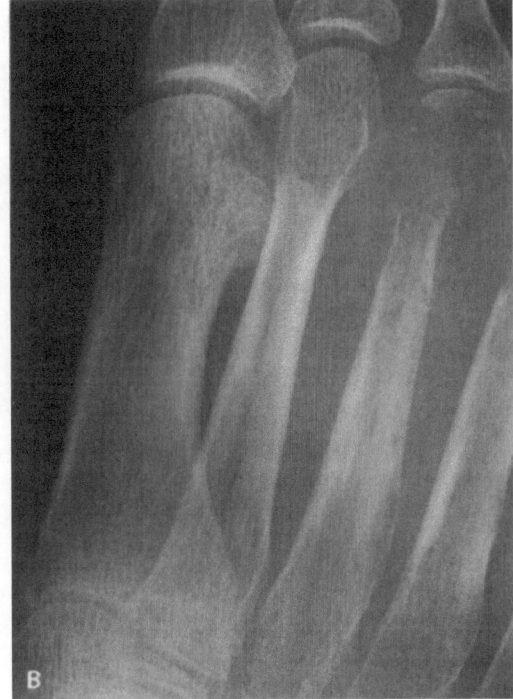

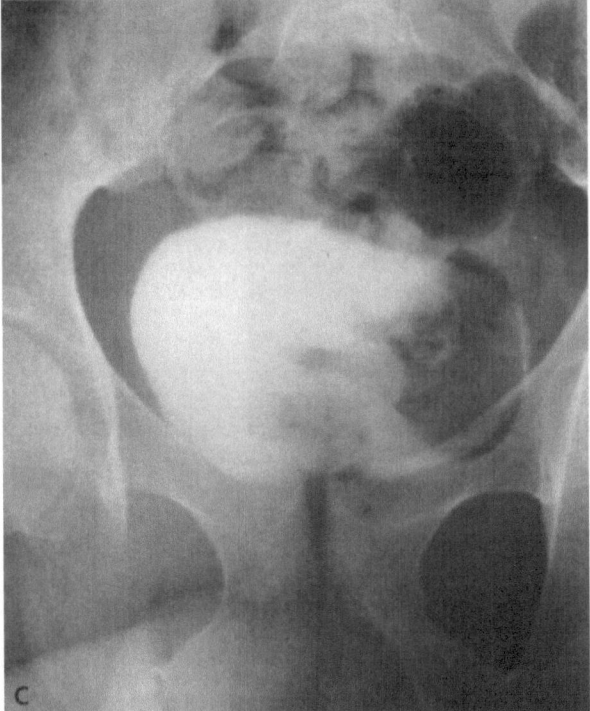

Fig. 93 A–C. This African patient came to hospital with a complaint of frequency of micturition. An intravenous urogram showed that he had a tuberculous infection of the spine (A) with almost complete destruction of the 12th thoracic vertebra. There was a large left paravertebral abscess, and destruction and subluxation of the 12th left rib. A large lytic defect was present in the right 10th rib. The patient was started on antituberculous therapy but later complained of a pain in his foot where there was a similar destructive lesion in the third metatarsal (B). The urogram had also shown tuberculous swelling of the left side of the bladder (C) and another incidental finding was almost complete destruction of the right lower pubic ramus. Despite all these lesions, the patient's main complaint was still urinary!

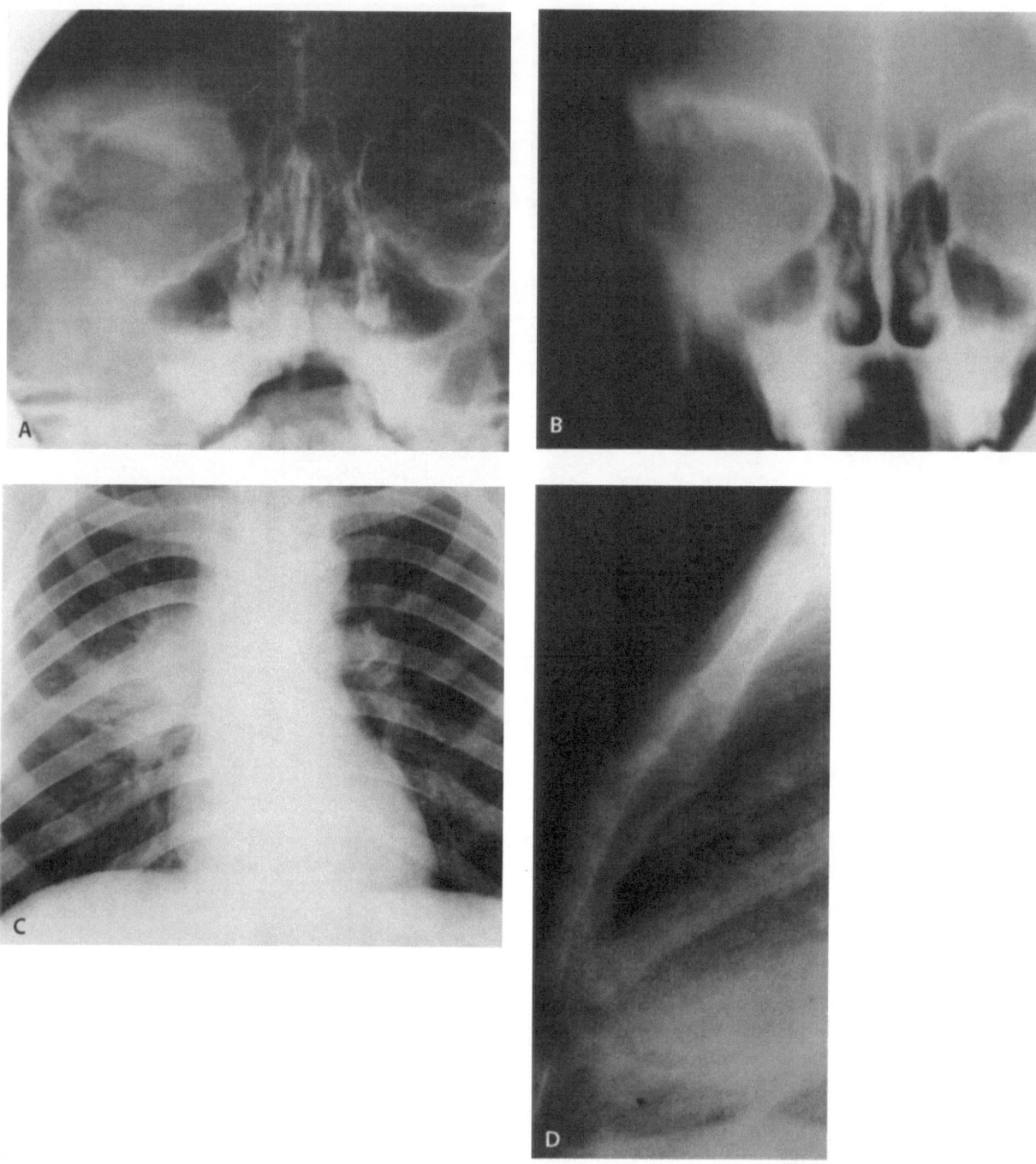

Fig. 94 A–D. When the infection is tuberculosis, patients may only complain of part of the infection. This teenage African came to hospital with swollen right orbit (**A, B**) and was found to have a destructive bone infection of the right orbital margin. Standard tomography (**B**) showed not only the sclerotic thickening, but a sequestrum within a lytic area above the eye. His chest radiograph (**C**) showed a primary tuberculous infection, with lymphadenopathy and the lateral radiograph (**D**) showed that he also had a tuberculous sternum.

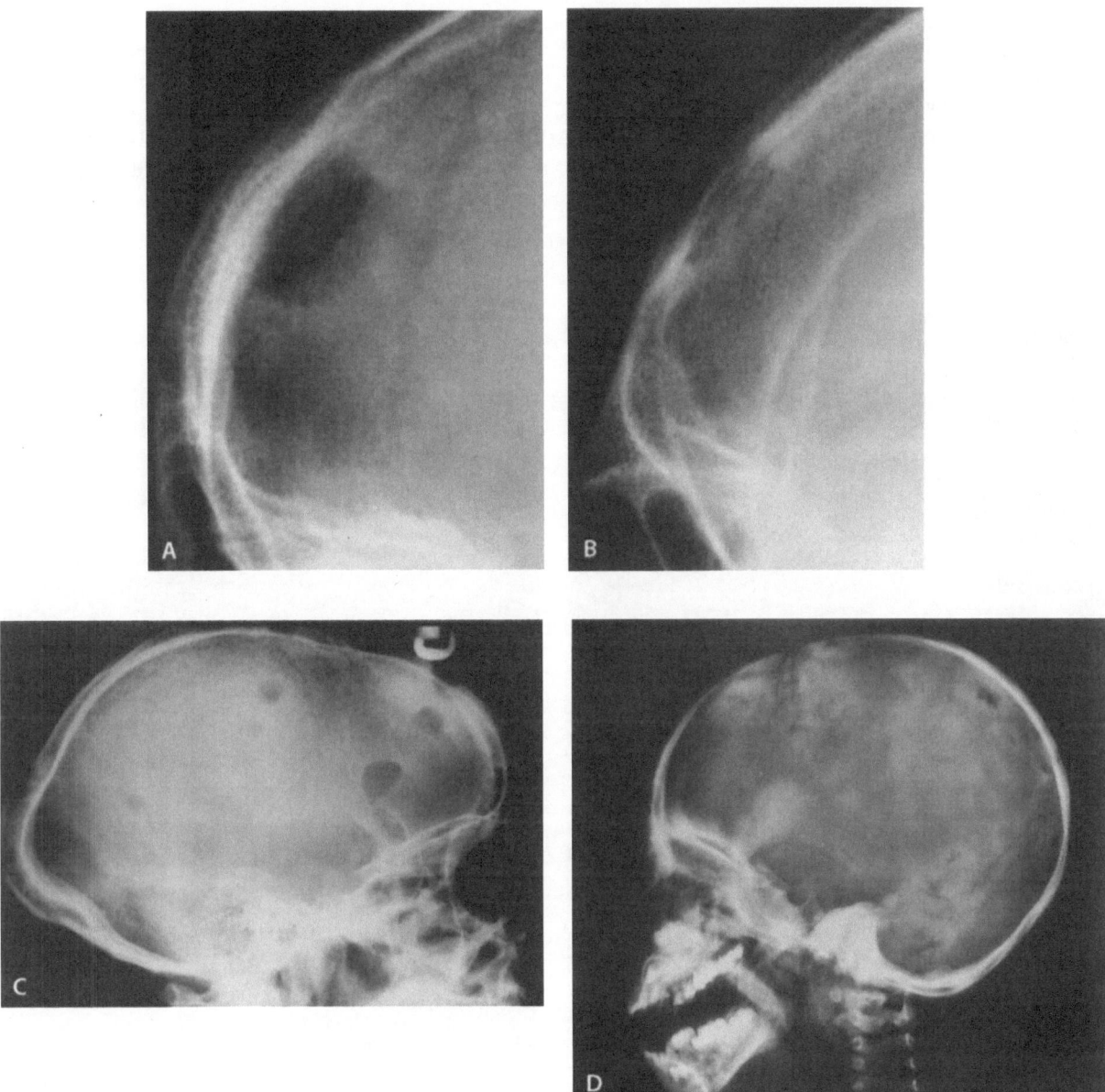

Fig. 95 A–D. Tuberculosis may affect any part of the skull. A, B Two different views of a lytic area in the frontal bone. C Multiple lytic areas across the skull vault. D Similar foci with surrounding bone reaction suggesting healing or a more chronic infection

Tuberculosis of the Central Nervous System

CT and MRI have changed our knowledge of the incidence and significance of tuberculosis of the central nervous system (CNS). It was not possible previously to image satisfactorily the different lesions caused by tuberculosis: plain radiography seldom gives much information or allows an accurate diagnosis (Fig. 96), and many foci that cannot be demonstrated with plain radiography are recognized on CT or MRI.

Tuberculomas of the brain probably have a variable geographic distribution, seemingly being more common in India and southern Africa. But this may also reflect the availability of imaging facilities and, particularly with the AIDS epidemic, CNS tuberculosis is likely to become more common. Tuberculosis involves the CNS by hematogenous spread, except in a few cases which result from rupture of cranial osteitis, e.g., from the mastoid or frontal bones. The tubercle bacilli may be seeded anywhere but are most common superficially over cerebral and cerebellar hemispheres and the ventricular system.

(Text continues on p. 128)

Tuberculosis of the central nervous system affects the meninges (which is probably the most important, because it carries a poor prognosis but if treated early may respond) and causes granulomatous masses (tuberculomas) which vary considerably in size and significance. Tuberculous abscesses are probably the result of central necrosis in a tuberculoma. Most tuberculomas respond to treatment, some may resolve spontaneously and the minority calcify. Plain radiography seldom gives much information or allows an accurate diagnosis. CT and MRI are the important imaging methods

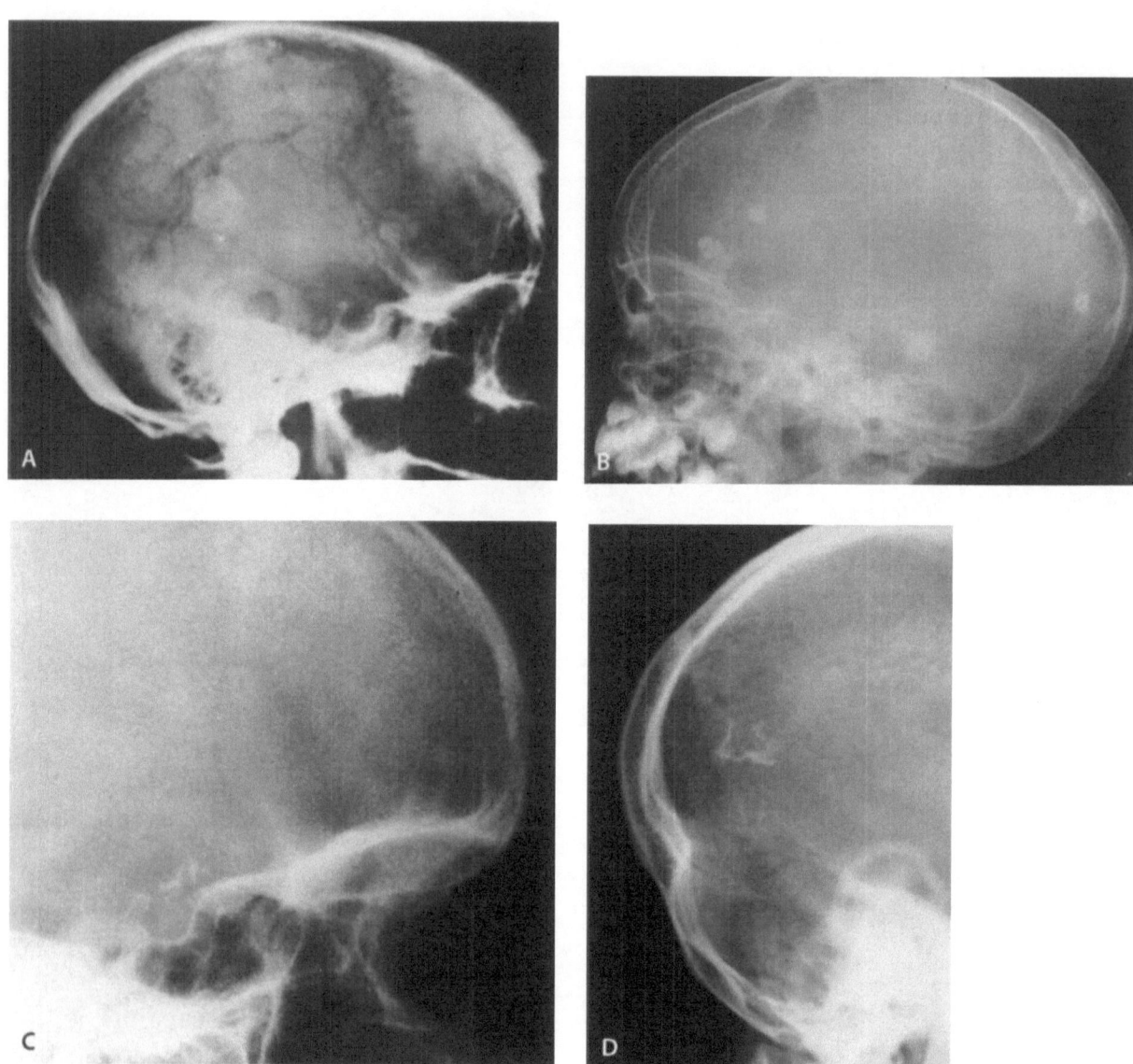

Fig. 96 A-H. Plain radiography of the skull may show calcification in tuberculomas or basal meninges. A normal skull radiograph does not, of course, exclude tuberculosis. **A** Multiple calcified tuberculomas with calcification of the meninges, probably also tuberculous. **B** Multiple calcified tuberculomas in a child. **C** Calcification of the basal meninges. **D** Curvilinear calcification in a tuberculoma in the left occipital region. E-H see p. 127

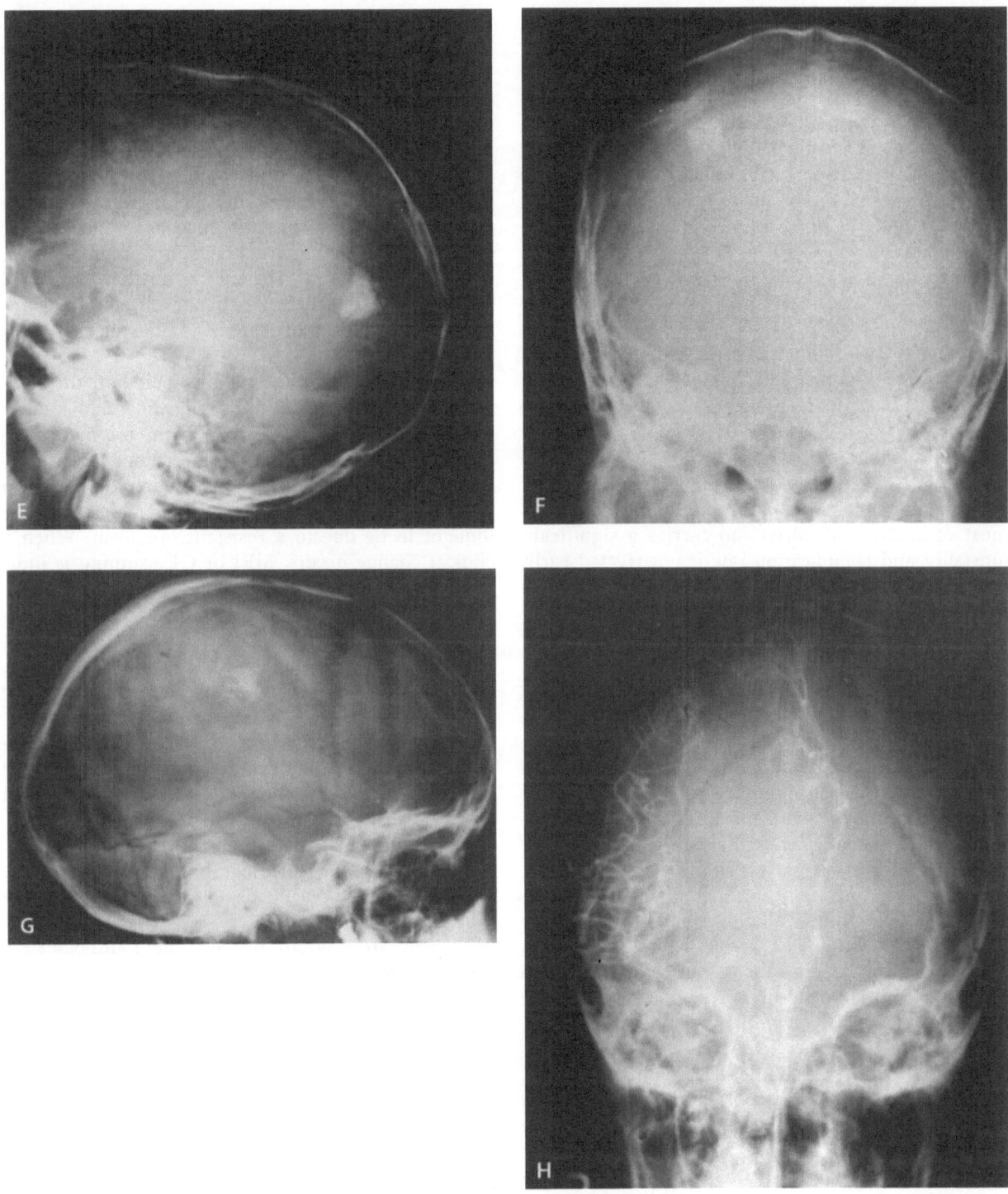

Fig. 96 (*continued*). **E, F** A calcified tuberculoma which was a chance finding in a patient who suffered a head injury. The depressed fracture had to be elevated and the tuberculoma was removed to confirm the histology. **G, H** Angiography is not usually helpful in making the diagnosis. However, it may indicate the size of the mass, if scanning is not available. A lateral radiograph (**G**) shows suture diastasis and a calcified mass. Arteriography (**H**) shows that this is a avascular and indicates that the size of the tuberculoma is much larger than the calcification would suggest. (**G, H** courtesy of Dr. L. Handler, Capetown) None of these radiographs would permit an accurate diagnosis of tuberculoma

As tuberculosis develops and caseates, involving the meninges, the infection can spread to the subarachnoid space. The most extensive reaction is at the base of the brain, where a thick exudate develops and surrounds the cranial nerves, compressing blood vessels and blocking the foramina of the fourth ventricle. The general inflammatory response also causes raised intracranial pressure.

Two main patterns of CNS tuberculosis develop:

1. Tuberculous meningitis
2. Parenchymal or meningeal mass lesions, tuberculoma

In some patients both varieties develop together, becoming a very complex infection.

Tuberculous meningitis

Tuberculous meningitis as a presenting illness is most common in children and carries a significant mortality and complication rate if not treated early. It is often associated with progressive primary tuberculosis and will also be seen in immunosuppressed patients. Cranial nerve damage may occur and ischemic infarction can occur in any part of the brain: these are the two most common causes of long-term complications. The incidence of infarction is high, perhaps 40%–60%. As the blood vessels become involved with the proliferative arachnoiditis, spasm and occlusion may result in infarction.

Plain radiography has little part to play in the imaging of tuberculous meningitis. Small infants may have suture diastasis as intracranial pressure rises, but this is a nonspecific finding. In babies and if the diastasis is sufficient, ultrasonography can be used to demonstrate the basal meningeal changes and the ventricular dilatation. Most children are too old for ultrasonography to be really useful.

Where available, CT and MRI are the most important imaging modalities (Figs. 97, 98). Following infarction, CT will show ill-defined hypodense areas, which may enhance with contrast. MRI is more accurate and a contrast enhanced T2-weighted scan can demonstrate a mass effect and hyperintense areas. These may progress to show cavitation. On T1-weighted scans these areas are well defined and hypointense.

Hydrocephalus develops in most cases of tuberculous meningitis unless treatment is started very promptly (Fig. 99). The obstruction is usually in the basal cisterns or the aqueduct. Immediate surgical shunting may be necessary. The end results, even with treatment, may be fibrosis and continued basal obstruction. Healing may be recognized by lack of contrast enhancement on CT but the end stage is difficult to judge. A small number of cases will resolve completely.

Cerebral arteriography during the active phase of meningitis may show vascular irregularity of the larger vessels basally, probably due to arteritis because of the surrounding inflammation. Sometimes complete obstruction of the vessel can be demonstrated. These findings correlate well with the CT images and it is doubtful whether arteriography is indicated during tuberculous meningitis. Occasionally tuberculous meningitis may form plaques or clusters of thickening around major cerebral vessels, demonstrated by CT. These have not been reported commonly from tropical patients, but this may represent infrequency of scanning rather than absence of the pathology.

It is important to note that the incidence of intracranial tuberculomas complicating tuberculous meningitis is 4%–28%: at least one such case has developed during adequate antituberculous treatment. The presence of an intracranial tuberculoma may be indicated by clinical deterioration and may be thought to be due to a resistant organism. When a clinical change occurs, MRI or CT scanning is indicated.

Tuberculomas

Any granuloma may develop into a tuberculoma, forming a mass anywhere on or within the meninges. A tuberculoma can be single and nodular, or there may be more than one (Fig. 100). Provided they are treated, almost all will resolve without residual imaging abnormality; untreated many will continue to become calcified granulomas (Figs. 96, 100 A, B). A few, during development, may become tuberculous cerebral abscesses. These have thin walls and are smooth but can be multiloculated. They are indistinguishable from pyogenic abscesses except histologically. Many have surrounding edema which can be demonstrated by CT (hypodense) and T2-weighted MRI (hyperintense) (Fig. 101).

The tuberculoma without central necrosis is hypodense on CT and hyperintense on T1-weighted MR scans. T2-weighted scans show hypointensity. On both CT and MRI there is contrast enhancement. If there is central necrosis, this does not enhance and there will be a ring mass (Fig. 101 C, D). The central necrosis of tuberculoma may have two origins. The granulation tissue may undergo necrosis or it may result from cellular components which can be caseous or quite liquefied, more like pus. Both types of necrosis may be present in the same abscess and will react differently on CT and MRI. When due to a granuloma, the necrotic area does not enhance, is hyperdense on CT, and is hypointense on T2-weighted MRI. When liquid, it is hypodense on CT and hyperintense on T2-weighted MRI (Fig. 102).

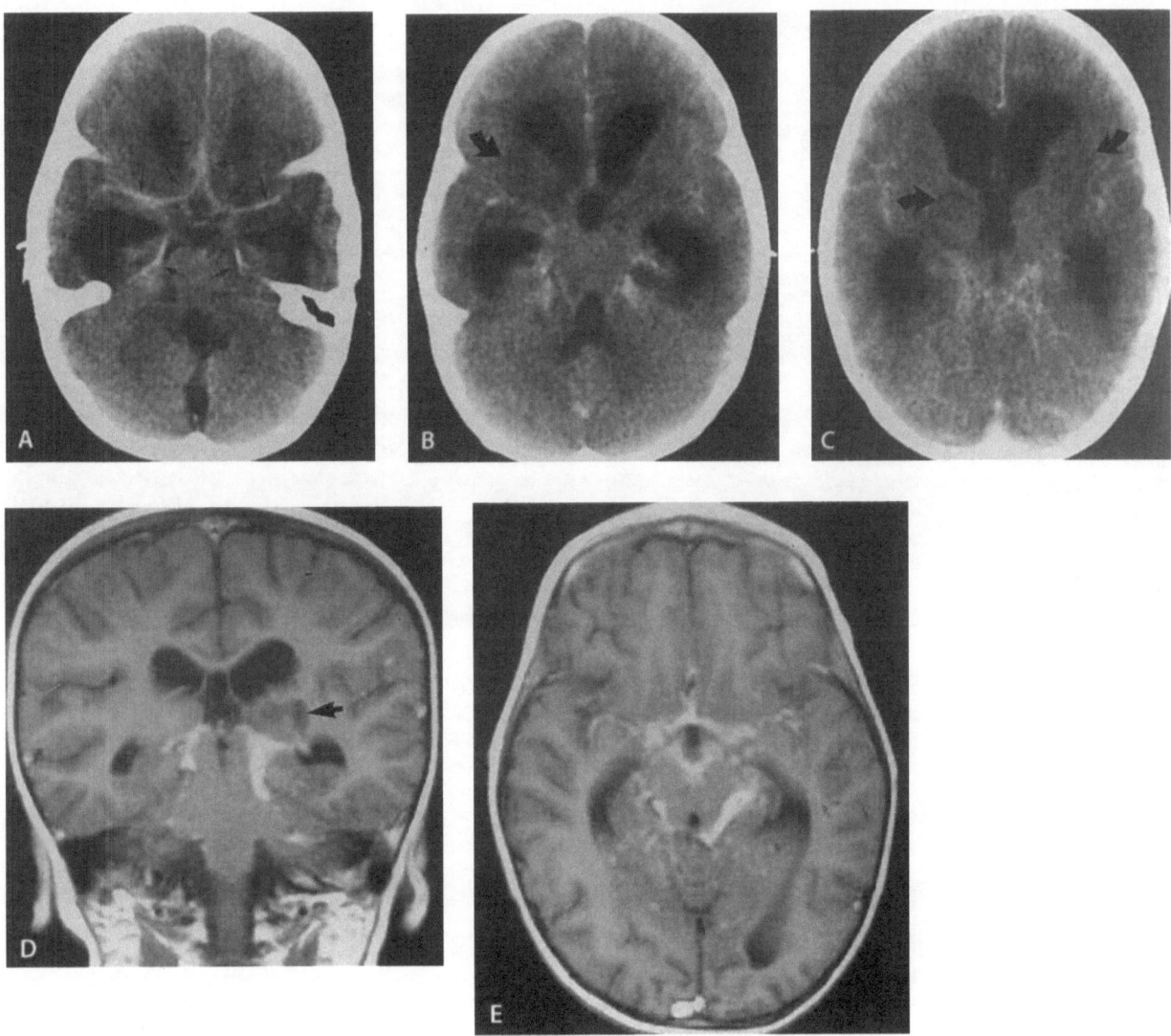

Fig. 97. A-C. Three important findings on CT scanning of tuberculous meningitis: **A** basal leptomeningeal enhancement (*small arrows*); **B, C** infarction in the basal ganglia (*curved arrow*) and hydrocephalus. **D, E** MRI with gadolinium enhancement in another child showing basal leptomeningeal enhancement, as well as a cavitated infarct deep in the left gray matter (*arrow*) and hydrocephalus. (From Cremin and Jamieson 1995)

Some tuberculomas under treatment may resolve in 3 or 4 months, but others may take over a year. Of those which are treated, about 5% probably calcify; this is a surprisingly small figure, because CT and MRI scans undertaken for other reasons have shown granulomas in patients with no previous history or knowledge of a CNS tuberculous infection: natural resistance must result in calcification more often than occurs with therapy.

Cerebral arteriography can demonstrate the mass lesion of a tuberculoma, depending on its size and situation. They are almost all avascular, but a small minority have a ring blush.

The differential diagnosis for tuberculous meningitis in the early stages includes any pyogenic infec-

tion. Tuberculomas can be mistaken for an intracranial neoplasm, or almost any other disease process causing mass lesions in the brain (such as subacute trauma, hemorrhage, pyogenic abscesses, or hydatid disease).

Careful clinical history, the clinical examination, and a high index of suspicion are very important. Mycotic infections can give rise to identical clinical and imaging findings.

Tuberculosis of the skull vault results in multifocal, lytic, and well-circumscribed fluctuant swellings (Fig. 95). The majority are symptom free, and they almost always occur under the age of 20 years. A few may discharge to the surface and then become secondarily infected. Internal extension is very com-

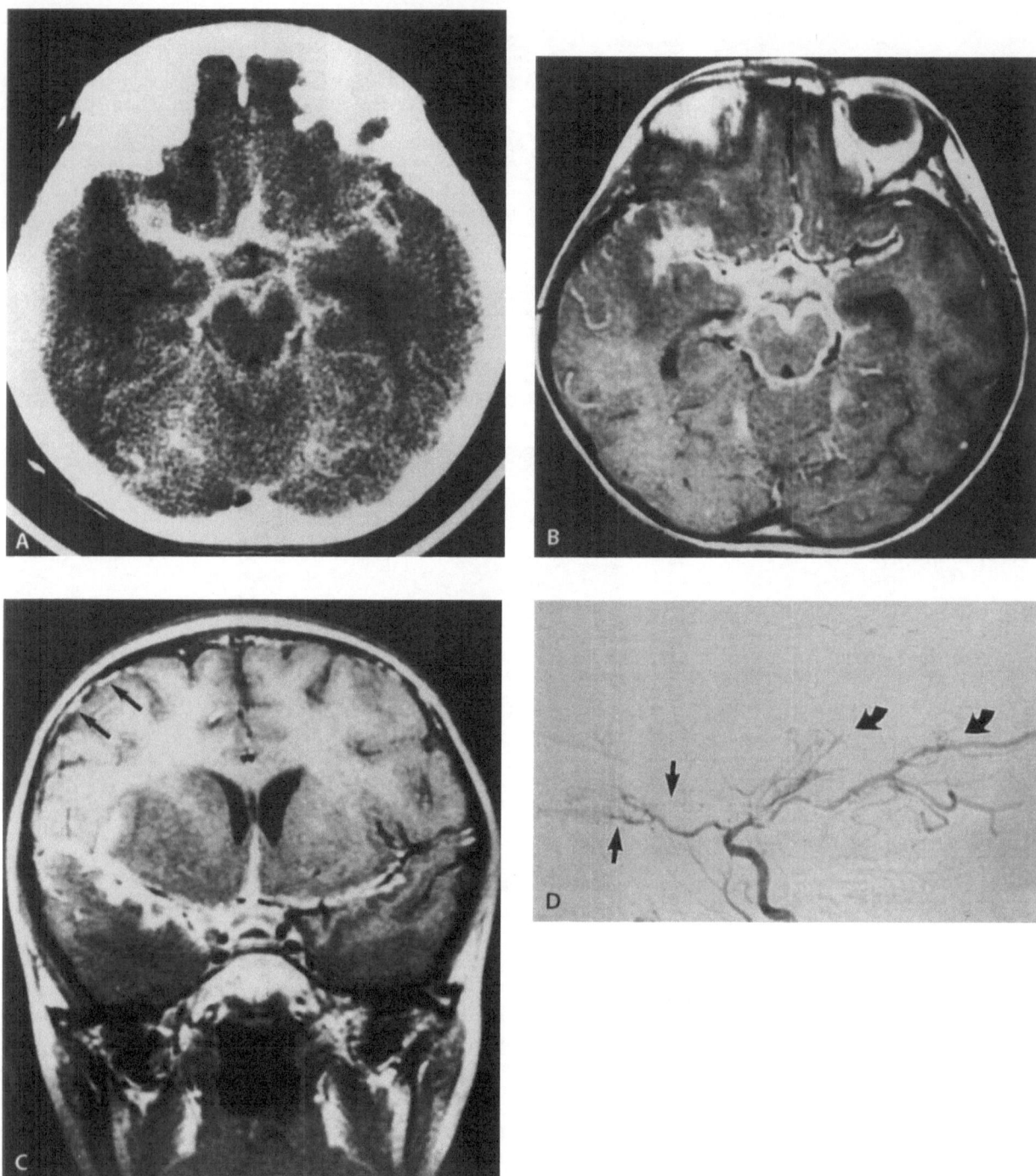

Fig. 98 A–D. A 9-year-old girl from the Pacific Islands with tuberculous meningitis. **A** Contrast-enhanced CT scan shows edema in the right temporal lobe and inflammatory reaction in the basal cisterns and sylvian fissures, with has enhanced. **B, C** T1-weighted MR images with contrast show similar enhancement of the basilar cisterns and the sylvian fissures, encasing the middle cerebral arteries. **D** Another child; a right carotid angiogram shows nearly complete occlusion of the artery as it traverses the meninges. An alternative collateral supply has been derived from the middle meningeal artery (*long arrows*), the ophthalmic artery (*short arrows*), and the posterior system (*curved arrows*). (**A–C** courtesy of Dr. Cheryl Sisler, Hawaii; **D** from Cremin and Jamieson 1995)

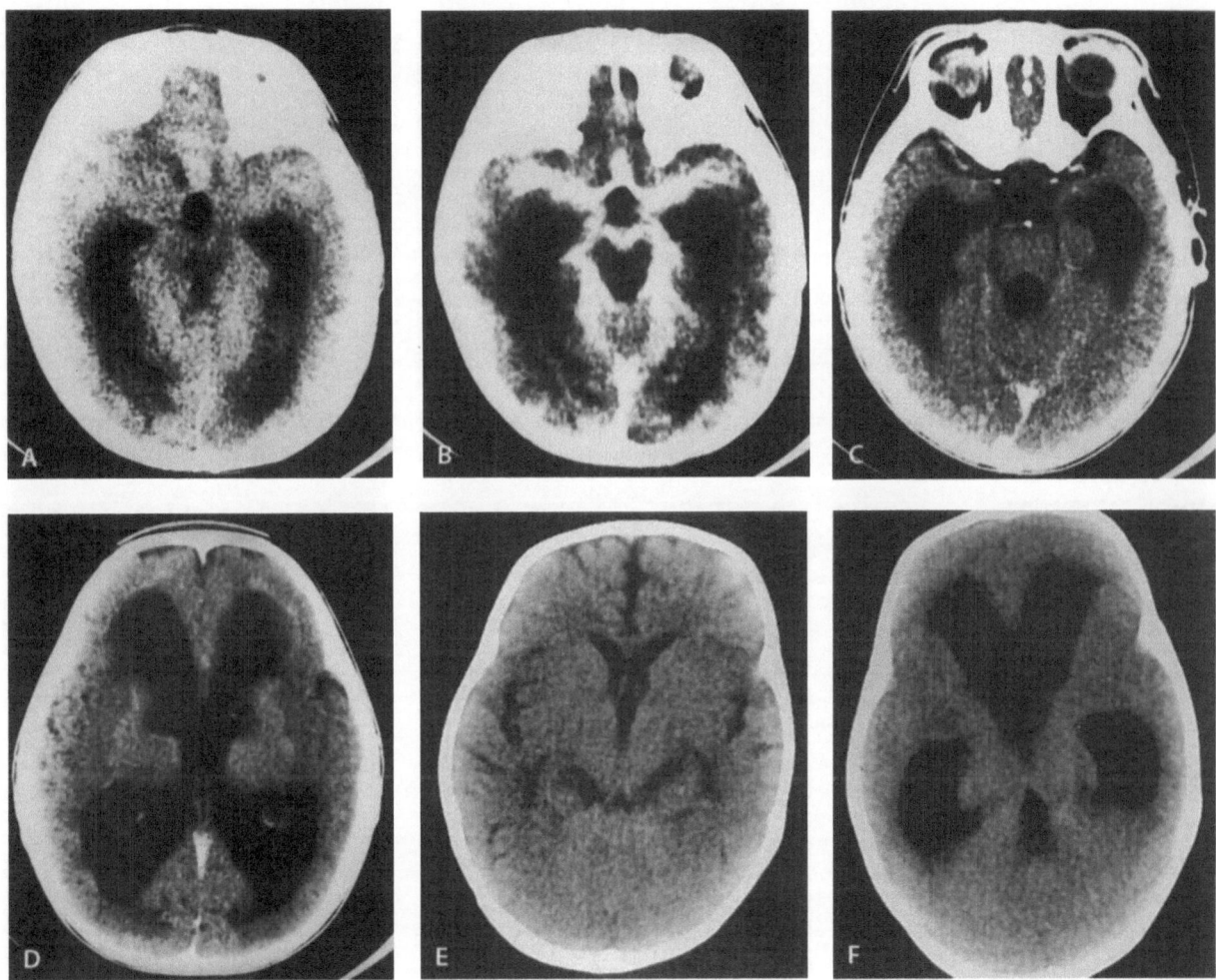

Fig. 99 A-F. Ventricular dilatation is common with tuberculous meningitis, but is not always permanent. A A 14-month-old boy from the Pacific Islands with tuberculous meningitis and positive CSF culture for *M. tuberculosis*; a nonenhanced CT shows ventricular enlargement. B With contrast there is enhancement of the basilar cisterns and sylvian fissures. C A 23-month-old girl, also from the Pacific Islands, was treated successfully for tuberculous meningitis when she was 6 months old. The enhanced CT scan shows persistent ventricular enlargement, but there is no enhancement of the basal cisterns. D The lateral, third and fourth ventricles, also enlarged, showing that there is communicating hydrocephalus. E This is the admission CT scan of another child with active tuberculous meningitis, already on treatment. F A repeat scan of the same child 1 month later. Another scan after 3 months did not show any significant change. (E, F from Cremin and Jamieson 1995)

mon. These are the "puffy tumours" described by Sir Percival Pott. The soft tissue swelling can be quite significant. They can be imaged with ultrasonography, CT, or MRI. It is important to distinguish them from histiocytosis or a low-grade pyogenic infection. These and other tuberculous lesions within the bones of the skull have been described on pp. 115, 116.

(Text continues on p. 134)

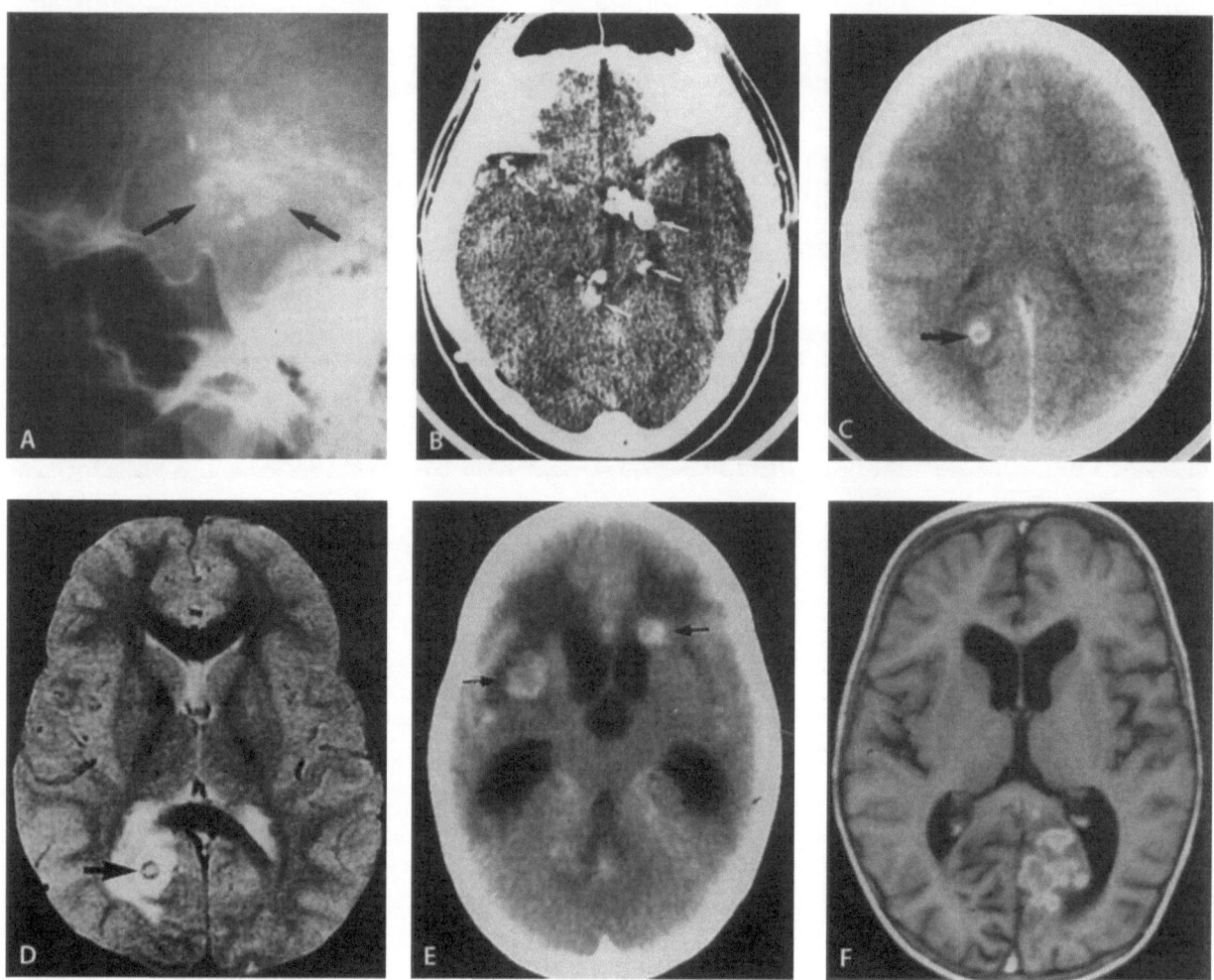

Fig. 100 A–F. Tuberculous granulomas. **A** The lateral radiograph of the skull of a 35-year-old man from the Pacific Islands who had tuberculous meningitis when he was 19 year old. There is a group of calcified granulomas (*arrows*) which are better seen on the nonenhanced CT (**B**: *arrows*). There are more than would be expected from the plain radiograph. **C** The contrast-enhanced CT scan of another child from South Africa showing that the granuloma has a hypodense center and surrounding edema (*arrow*). **D** T2-weighted MR scan of the same patient shows that the caseous necrotic center is T2 hyperintense, the rim of the granuloma is hypointense (*arrow*), and the edema is T2 hyperintense. **E** Several granulomas (*arrows*) are shown on a contrast-enhanced CT of another patient. There is also basal enhancement and hydrocephalus. **F** A gadolinium-enhanced T1-weighted MR scan of another patient shows a cluster of granulomas in the left occipital lobe. (**A, B** courtesy of Dr. Cheryl Sisler, Hawaii; **C–F** from Cremin and Jamieson 1995)

Fig. 102 A–C. Tuberculous abscesses do not image in the same way as tuberculomas. **A** A contrast-enhanced CT scan shows rim enhancement with a hypodense center. There is also a group of enhancing granulomas posterior to the abscess. **B** A T2-weighted MR scan without enhancement shows that the abscess has a hypointense rim and a hyperintense center. **C** A T1-weighted MR scan of the same patient, with gadolinium enhancement, shows that the center of the abscess is hypointense and the wall of the abscess enhances. (From Cremin and Jamieson) ▶

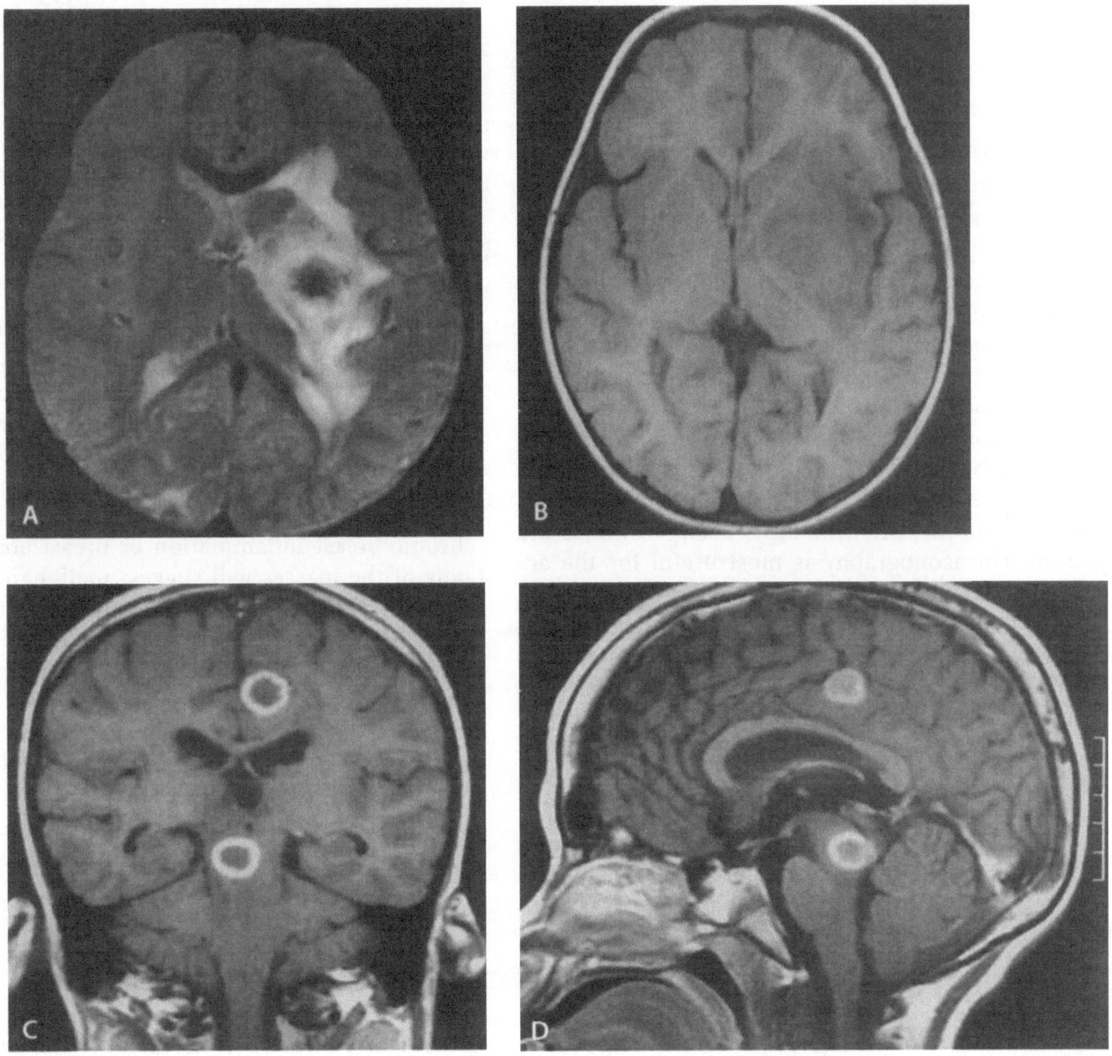

Fig. 101 A–D. Some tuberculomas have been described as "gummatous" and become quite large. **A** T2-weighted MR scan of a large tuberculoma. There is hyperintense edema surrounding a predominantly hypointense granuloma. **B** On a T1-weighted MR scan, the tuberculoma is isointense. **C, D** Gadolinium-enhanced MR scans of another patient show that the central necrosis is hypointense but there is strong rim enhancement. Both tuberculomas imaged in the same way. (From Cremin and Jamieson 1995)

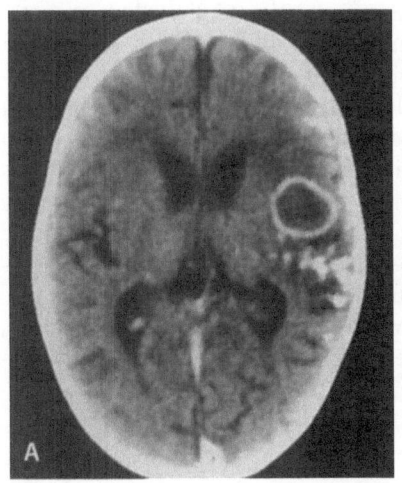

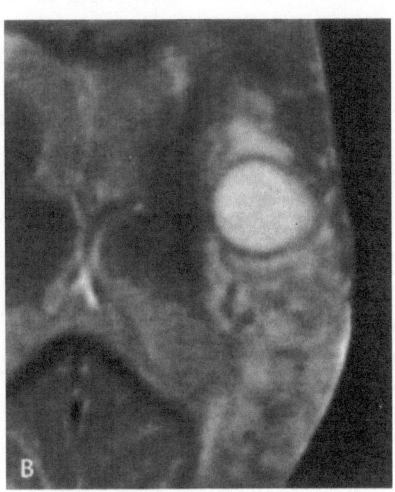

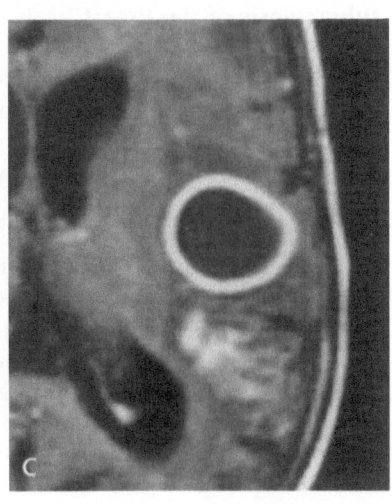

Tuberculosis Involving Other Sites

Tuberculous Lymphadenopathy

Abdominal tuberculosis (see p. 69)

Tuberculous lymphadenitis (scrofula) is by no means rare in many parts of the tropics; one study in Uganda showed that 41% of all enlarged cervical nodes were tuberculous. While tuberculosis is the most common cause of lymphadenopathy under the age of 30 years (unless the patient has Burkitt's lymphoma) there will also be a significant number of elderly patients in whom cervical lymphadenopathy is tuberculous in origin. Ultrasonography can accurately demonstrate the lymphadenopathy (as can CT or MRI), but cannot establish the etiology. Most lymph nodes will appear as hypoechoic masses with a regular outline; often multiple enlarged nodes are present. Ultrasonography is most useful for the accurate follow-up of resolution during treatment. During the acute stage there are no changes of radiological significance; when the infection has healed, calcification frequently follows. Calcified lymph nodes may be seen not only in the neck, but in the axilla, around the shoulder, in the inguinal region, in the popliteal fossa, and elsewhere (Fig. 103). All must be differentiated from other causes of soft tissue calcification, e.g., parasites, or, in the limbs, from tumoral calcinosis. This is usually possible because calcified tuberculous granulomas are irregular in size and shape but, more importantly, their anatomical situation may suggest the diagnosis.

Calcified tuberculomas may be seen in the liver and spleen, but have little clinical and no radiological significance (see p. 71). A tuberculous granuloma can calcify anywhere and only histological examination will differentiate its nature. Even this may be difficult because of the fibrosis.

Tuberculosis of the Breast

Although not common anywhere, tuberculosis of the breast is not rare and significant series have been reported. The most common presentation is a mass in the breast of a 30- to 40-year-old woman. It is usually painful, but if there is a sinus tract leading from the "mass" (which is the tuberculous breast abscess) to the skin (Fig. 104), the surface ulceration is often painless unless secondarily infected. Many tuberculous breast lesions are quite chronic and have been felt by the patient for some months; a few may be more acute, with a history of only a few days, and such rapid onset occurs particularly in the lactating breast. The clinical diagnosis is usually of a carcinoma or, in the more acute cases, a pyogenic breast abscess.

Mammography shows the palpable mass as a diffuse density, which, when there is fluid, may change in shape and density on the two standard mammographic projections. There may be one or more sinus tracts connecting the mass to the thickened, overlying skin. In some patients the skin bulges (Fig. 104 C), probably at the stage before the sinus tract has formed. The underlying breast stroma is coarse, and may be reticulated. There is almost always nipple retraction. The breast size is often reduced. Ultrasonography can be used to confirm the fluid, but does not give much further information.

In some patients there will be minimal regional lymphadenopathy, but in others the lymph nodes are normal. The usual mammographic diagnosis is "chronic breast inflammation or breast abscess," but many of the masses will suggest malignancy. Recognition of the sinus tract and the skin thickening should suggest tuberculosis, particularly when there is little surface pain clinically. Unfortunately, many sinus tracts become secondarily infected and there is then a pyogenic breast abscess: only histology will show the underlying tuberculosis. (The differential diagnosis, apart from malignancy and pyogenic abscess, will include, in the painless lesion, syphilis, mycotic infections, and parasites such as guinea worm.)

Tuberculosis of the Parotid Gland

Tuberculosis of the parotid gland is uncommon, but can be the presenting symptom of tuberculosis. There may be associated tuberculous cervical lymph nodes, but in the majority of patients there is no systemic symptom of tuberculosis and the chest radiographs are normal. Clinically, parotid tuberculosis resembles a tumor, and biopsy to establish the diagnosis is usually necessary. However, CT scanning may show characteristic thick-walled, smooth, round, rim-enhancing lesions, usually with central lucency. The clinical mass will be shown on scanning to be multiloculated; the thick enhancing rim surrounding the low-density central tissue is often corrugated and irregular. If the enhanced rim is thin, biopsy becomes essential. Contrast sialography does not add significant information and may be contraindicated because in some cases it has exacerbated the inflammatory reaction.

There may be similar lesions in the lymph nodes, and there may be lymph node tissue within the parotid glands.

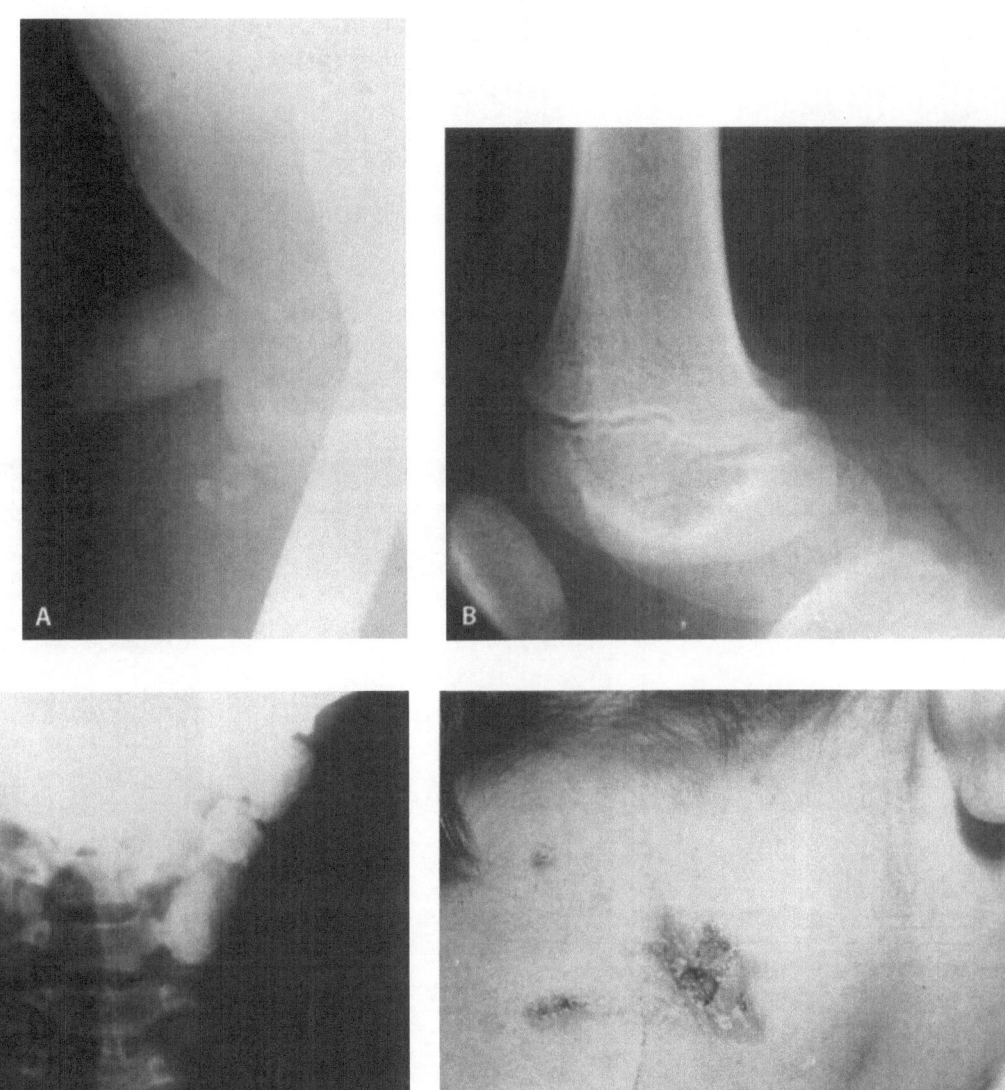

Fig. 103 A–D. Calcification in tuberculous lymphadenitis is not difficult to recognize in the chest or abdomen, but may occur unexpectedly wherever there are lymph nodes. A Calcified lymph nodes in the groin of a child and B in another patient, behind the knee. These small calcifications must be differentiated from parasites and early tumoral calcinosis. There is seldom a clinical history of significance. C Heavily calcified lymph nodes in the neck of a child from the Pacific Islands with a discharging sinus (D). This is the scrofula well known for centuries and in this child was, in fact, caused by *M. scrofulaceum*

Ocular Tuberculosis

Ocular tuberculosis is rare and usually associated with pulmonary or skeletal infection. Scleral tuberculosis has been reported but has no imaging significance.

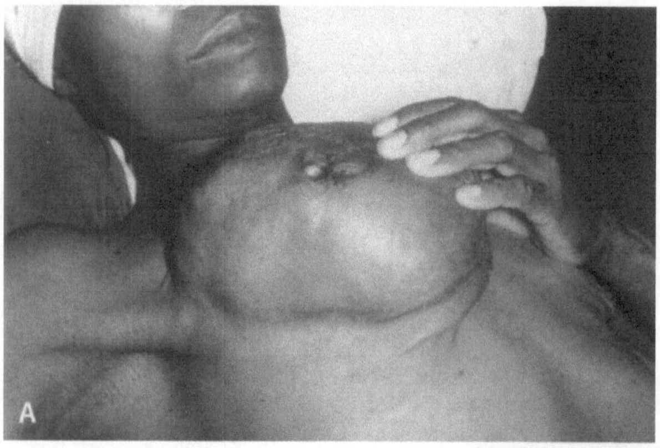

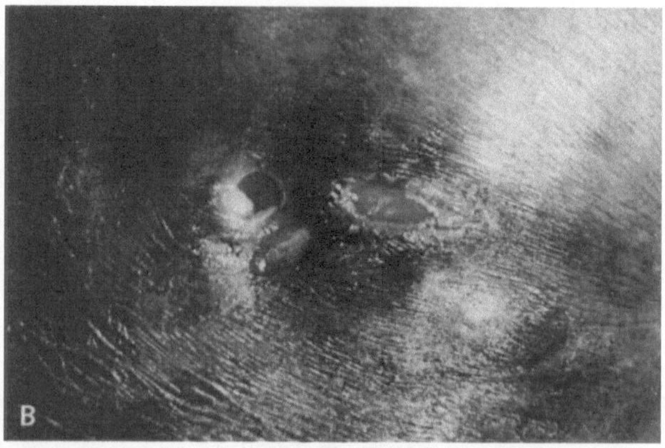

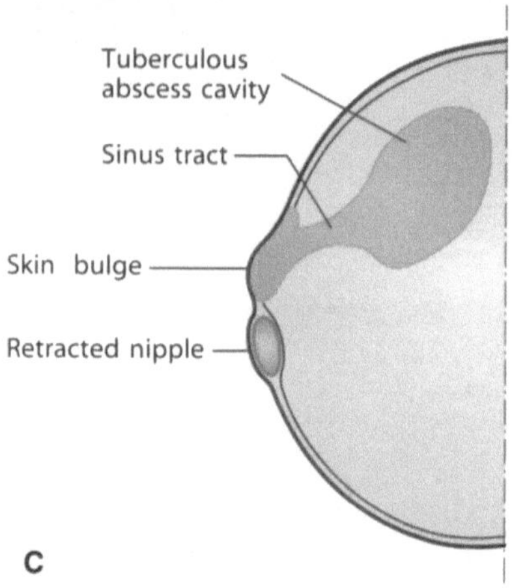

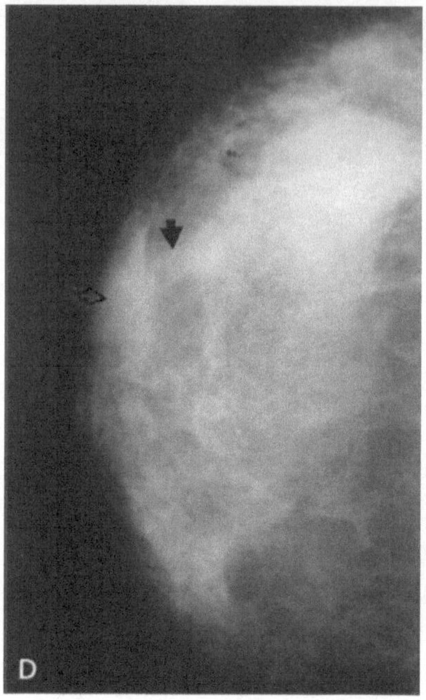

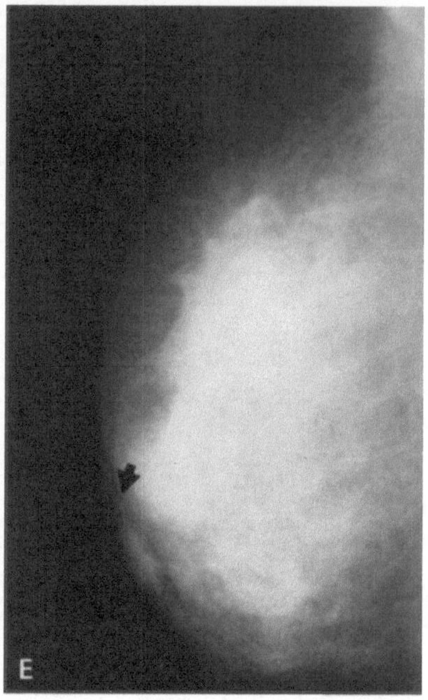

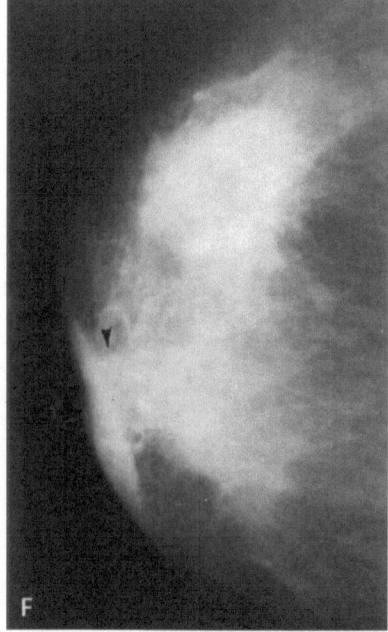

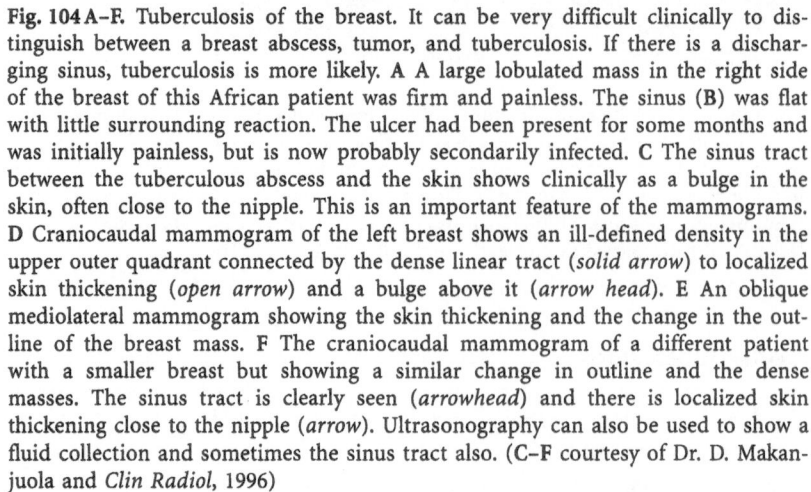

Fig. 104 A–F. Tuberculosis of the breast. It can be very difficult clinically to distinguish between a breast abscess, tumor, and tuberculosis. If there is a discharging sinus, tuberculosis is more likely. **A** A large lobulated mass in the right side of the breast of this African patient was firm and painless. The sinus (**B**) was flat with little surrounding reaction. The ulcer had been present for some months and was initially painless, but is now probably secondarily infected. **C** The sinus tract between the tuberculous abscess and the skin shows clinically as a bulge in the skin, often close to the nipple. This is an important feature of the mammograms. **D** Craniocaudal mammogram of the left breast shows an ill-defined density in the upper outer quadrant connected by the dense linear tract (*solid arrow*) to localized skin thickening (*open arrow*) and a bulge above it (*arrow head*). **E** An oblique mediolateral mammogram showing the skin thickening and the change in the outline of the breast mass. **F** The craniocaudal mammogram of a different patient with a smaller breast but showing a similar change in outline and the dense masses. The sinus tract is clearly seen (*arrowhead*) and there is localized skin thickening close to the nipple (*arrow*). Ultrasonography can also be used to show a fluid collection and sometimes the sinus tract also. (**C–F** courtesy of Dr. D. Makanjuola and *Clin Radiol*, 1996)

Bibliography

Abdelwhab IF, et al: Atypical skeletal tuberculosis mimicking neoplasm. *Br J Radiol* 64:551–555, 1991

Abrams IS, Holden WD: Tuberculosis of the gastrointestinal tract. *Arch Surg* 89:282–288, 1964

Adams WM, Fitzgerald R: Abdominal tuberculosis: unusual findings on CT. *Clin Radiol* 50:734–735, 1995

Aderle WI: Radiological patterns of pulmonary tuberculosis in Nigerian children. *Tuberculosis* 61:157–163, 1980

Ahmed FB, et al: Tuberculous enteritis. *BMJ* 313:215–217, 1996

Aldossary J, Grant CS: Abdominal tuberculosis: a frequent diagnostic challenge. *Ann Trop Med Parasitol* 79:163–167, 1985

Allen CM, Craze J, Grundy A: case report: tuberculous broncho-oesophageal fistula in the acquired immuno-deficiency syndrome. *Clin Radiol* 43:60–62, 1991

Anderson DH, et al: Pulmonary lesions due to opportunistic myobacteria. *Clin Radiol* 26:461–471, 1975

Anscombe AR, Keddie NC, Schofield PF: Caecal tuberculosis. *Gut* 8:337–341, 1967

Apps MCP, Harrison NK, Blanth EIA: Tuberculosis of the breast. *BMJ* 288:1874–1875, 1984

Aston NO, DeCosta AM: Tuberculous perforation of the small bowel. *Postgrad Med J* 61:251–252, 1985

Avery GR, Chippendale AJ: Case of the month: a single diagnosis. that is the question. *Br J Radiol* 68:665–666, 1995

Bahdur P, et al: Tuberculosis of the mammary gland. *J Indian Med Assoc* 80:8–12, 1983

Bailey IC: The Pattern and presentation of cerebral tumours in Uganda. *East Afr Med J* 48:565–575, 1971

Baker RD: Endocardial tuberculosis. *Arch Pathol* 19:611–635, 1935

Ball PAJ: Abdominal tuberculosis. In: Davey WW (ed) *Companion to surgery in Africa*. E & S Livingstone, Edinburgh, 1968, pp 267–270

Balthazar EJ, Gordon R, Mulnick D: Ileocecal tuberculosis: CT and radiologic evaluation. *Am J Roentgenol* 154:499–503, 1990

Bankier AA, et al: Update: abdominal tuberculosis. Unusual findings on CT. *Clin Radiol* 50:223–228, 1995

Barnes P, et al: An unusual case of tuberculous peritonitis in a man with AIDS. *West J Med* 144:467–469, 1986

Barson AJ, Kirk RS: Colonic tuberculosis with carcinoma. *J Pathol Bacteriol* 101:289–292, 1970

Bastani B, et al: Tuberculous peritonitis. Report of 30 cases and review of the literature. *Q J Med* 56:549–557, 1985

Bateson EM, Woo-Ming M: Destroyed lung: a report of cases in West Indians and Australian aborigines. *Clin Radiol* 27:223–226, 1976

Bell GM, de Klerk JN: Tuberculose van die penis (corpora cavernosa). *S Afr Med J* 50:1489–1490, 1976

Belli L: Studio ecografico dei versamenti tuberculori. *Radiol Med (Torino)* 83:659–661, 1992

Betoulieres P: The lymphangiographic appearances of lymph node tuberculosis. *J Radiol Elec* 49:1–5, 1968

Bevans M, Wilkins SA: Tuberculous endocarditis. *Am Med J* 24:843–849, 1942

Bhansali SK: Abdominal tuberculosis. Experiences with 300 cases. *Am J Gastroenterol* 67:324–337, 1977

Bhargave S, et al: Case report: tuberculosis of the parotid gland – diagnosis by CT. *Br J Radiol* 69:1181–1183, 1996

Black GA, Carsky EW: Duodenal tuberculosis: *AJR* 131:329–330, 1978

Bohrer SP: Tuberculous synovitis with widening of the intercondylar notch of the distal femur. *Br J Radiol* 42:703–704, 1969

Bolton JM, Snelling MR: Review of tuberculosis among the Orang Asli (Aborigines) in West Malaysia. *Med J Malaysia* 30:10–29, 1975

Brauner M, et al: Sonography and computed tomography of macroscopic tuberculosis of the liver. *J Clin Ultrasound* 17:563–568, 1989

Brenner SM, Annes G, Parker JG: Tuberculous colitis simulating nonspecific granulomatous disease of the colon. *Am J Dig Dis* 15:85, 1970

Brody JM, et al: Gastric tuberculosis: a manifestation of the acquired immuno-deficiency syndrome. *Radiology* 159:347–348, 1986

Brombart M, Massion J: Radiological differences between ileocecal tuberculosis and Crohn's disease. *Am J Dig Dis* 6:589–603, 1961

Burack WR, Hollister RM: Tuberculous peritonitis: a study of forty-seven proved cases encountered by a general medical unit in twenty-five years. *Am J Med* 28:510–521, 1960

Burke GJ, Zafar SA: Problems in distinguishing tuberculosis of bowel from Crohn's disease in Asians. *BMJ* 4:395–397, 1975

Calder JF, Norredam K: Chylous ascites due to tuberculosis. *East Afr Med J* 49:684–686, 1972

Camiel MR: Ileocecal tuberculosis. *Radiology* 44:344–351, 1945

Carazo-Martinez O, et al: Realtime ultrasound evaluation of tubercular pleural effusions. *J Clin Ultrasound* 17:407–410, 1989

Cardocio SA, Locker GY: Elevated serum CA 125 secondary to tuberculous peritonitis. *Cancer* 72:2016–2018, 1993

Cardozo LJ, et al: Tuberculous meningitis in adult Africans; problems of diagnosis and management. *East Afr Med J* 53:136–142, 1976

Carroll SR, Newsom SWB, Jenner JR: Treatment of septic arthritis due to *Mycobacterium kansasii*. *BMJ* 289:591–592, 1984

Casselman ES, et al: CT of meningitis in infants and children. *J Comput Assist Tomogr* 4:211–216, 1980

Chan HS, Pang J: Isolated giant tuberculomata of the liver detected by computed tomography. *GI Radiol* 14:305–307, 1989

Chandracharoensin C, Viranuvalti V: Tuberculous abscess of the retrosternal thyroid gland. *Diagn Imaging (Netherlands)* 50:29–31, 1981

Charatan FB: Quoted in the Proceedings of the National Academy of Science (March, 1994). "Peruvian mummy shows that TB preceded Columbus." *BMJ* 308:808, 1994

Chatfield WR, et al: The investigation and management of infertility in East Africa (a study of 200 cases). *East Afr Med J* 47:212–216, 1970

Chatterji DN: Tuberculosis in children (pattern of intrathoracic lesions in hospitalised patient). *J Indian Med Assoc* 67:217–223, 1976

Chanuhry UR, Ghani MU: Diagnosis of tuberculous basal meningitis by CT scan and transcranial Doppler (TCD) studies. Proc. 19th Int. Cong. Radiol., PP 1.2 Beijing, 1996

Chazan BI, Aichison JD: Gastric tuberculosis. BMJ 2:1288–1290, 1960

Choe KO, Jeon P: Reexpansibility of the lung after decortication in tuberculous empyema. Proc. 19th Int. Cong. Radiol, SS 3.16. Beijing, 1996

Citron KM: BCQ – vaccination against tuberculosis: international perspectives. *BMJ* 306:222–223, 1993

Clark RA, et al: Hematogenous dissemination of *Mycobacterium tuberculosis* in patients with AIDS. *Rev Infect Dis* 13:1089–1092, 1991

Cohen C: Tuberculous mastitis: *S Afr Med J* 52:12–16, 1977

Collins CM, Yates MD, Grange JM: Names for mycobacteria. *BMJ* 288:463–464, 1984

Compere EL, Garrison M: The correlation of pathologic and roentgenologic findings in tuberculous and pyogenic infection of the vertebrae. The fate of the intervertebral disc. *Ann Surg* 104:1038–1067, 1936

Cook SM, Frank TS: Detection and characterization of atypical mycobacteria by the polymerase chain reaction. *Mod Pathol* 6:104 A, 1993

Cope AP, Heber M, Wilkins EGL: Valvular tuberculous endocarditis: a case report and review of the literature. *J Infect* 21:293–296, 1990

Corr P: MRI of tuberculous spondylitis. *Imaging S Afr Mag* 17–20, 1990

Corr P, Handler L, Davey H: Pott's paraplegia and tuberculous spondylitis. Evaluation by magnetic resonance. *Neuroradiology* 33 (Suppl):109–110, 1991

Cremin BJ: Radiological imaging of urogenital tuberculosis in children, with emphasis on ultrasound. *Pediatr Radiol* 17:34–38, 1987

Cremin BJ: CT in tuberculous spondylitis. *Clin Radiol* 49:433–435, 1994

Cremin BJ, Jamieson DM, Hoffman EB: CT and MRI in the management of advanced spinal tuberculosis. *Pediatr Radiol* 23:298–300, 1993

Cremin BJ, Jamieson DM: *Childhood tuberculosis*. Springer, Berlin Heidelberg New York, 1995

Cremin BJ, Levinsohn MW: Multiple bone tuberculosis in the young. *Br J Radiol* 43:638–645, 1970

Dahlene DH, et al: Abdominal tuberculosis: CT findings. *J Comput Assist Tomogr* 8:443–445, 1984

Damrongsak C, et al: Functional hepatogram and coeliac arteriogram in tuberculosis of the liver. *J Med Assoc Thai* 60:285–288, 1977

Danziger J, et al: Cranial and intracranial tuberculosis. *S Afr Med J* 50:1403–1405, 1976

DaRoche-Afodu JT: Tuberculous thyroiditis. *West Afr Med J* 23:159–161, 1975

Das MK, Indudhara R, Vaidyanalthan S: Sonographic features of genitourinary tuberculosis. *Am J Roentgenol* 158:327–329, 1991

Dastur HM, Shah MD: Intramedullary tuberculoma of the spinal cord. *Indian Pediatr* 5:468–471, 1968

Davey WW: Tuberculous pericarditis. In: Davey WW (ed) Companion to surgery in Africa. E & S Livingstone, Edinburgh, 1968, pp 310–312

De Labarthe B, et al: Acute, lethal miliary tuberculosis due to *Mycobacterium bovis*. *Poumon Coeur* 34:139–142, 1978

Denath FM: Abdominal tuberculosis in children. CT findings. *Gastrointest Radiol* 15:303–306, 1990

Dent DM, Webber BL: Tuberculosis of the breast. *S Afr Med J* 51:611–614, 1977

Denton T, Hossain J: A radiological study of abdominal tuberculosis in a Saudi population. *Clin Radiol* 47:409–414, 1993

Dineen P, Homan WP, Grafe WR: Tuberculous peritonitis: 43 years of experience in diagnosis and treatment. *Ann Surg* 184:717–722, 1976

Dixon JM: Breast diseases. *Br Med J* 309:947–949, 1994

Dong-Ho-Lee, et al: Sonographic findings of intestinal tuberculosis. *J Ultrasound Med* 12:537–540, 1993

Dorff GJ, et al: Musculoskeletal infections due to *Mycobacterium kansasii*. *Clin Orthop* 136:244–246, 1978

Downey DB, Nakielny RA: Aphthoid ulcers in colonic tuberculosis. *Br J Radiol* 58:561–562, 1985

Doy RW: Tuberculosis presenting as an acute surgical emergency. *Cent Afr J Med* 18:91–93, 1972

Ehlers MRW: The biology of *mycobacterium tuberculosis* and the host-pathogen relationship. In: Cremin BJ, Jamieson DM (eds) Childhood tuberculosis. Springer, London Berlin Heidelberg New York, 1995, pp 7–17

El Masri SH, et al: Abdominal tuberculosis in Sudanese patients. *East Afr Med J* 54:319–326, 1977

Er-hu J, et al: CT diagnosis of tracheobronchial tuberculosis. Proc. 19th Int. Cong. Radiol. 33. 3.15. Beijing, 1996

Fisher D, Hiller N: Case report: giant tuberculous cystic lymphangioma of posterior mediastinum, retroperitoneum and groin. *Clin Radiol* 49:215–216, 1994

Foster DR: Miliary tuberculosis following intravesical BCG treatment (letter). *Br J Radiol* 70:429–431, 1997

Fowler WC: A preliminary report on the treatment of tuberculosis with turtle vaccine. *Tubercle* 12:12–17, 1930

Fox E: Tuberculosis. In: Strickland GT (ed) Hunter's tropical medicine, 7th edn. Saunders, Philadelphia, 1991, pp 458–483

Gaines W, Steinbach HL, Lowenhaupt E: Tuberculosis of the stomach. Radiology 58:808, 1952

Ganguli PK: *The radiology of bone and joint tuberculosis with special reference to tropical countries*. Asia Publishing House, London, 1963

Garber EK, Bluestone R: Tuberculous arthritis of the knee with a cold abscess communicating with the knee joint. *Skeletal Radiol* 6:76, 1981

Gawne-Cain ML, Hansell DUR: The pattern and distribution of calcified mediastinal lymph nodes in sarcoidosis and tuberculosis: a CT study. *Clin Radiol* 51:263–267, 1996

Gerritsen J, Knol K: Hypercalcaemia in a child with miliary tuberculosis. *Eur J Pediatr* 148:650–651, 1989

Gershon-Cohen J, Kremens V: X-ray studies of the ileocecal valve in ileocecal tuberculosis. *Radiology* 62:251–254, 1954

Gilinsky NH, et al: Abdominal tuberculosis: a 10 year review. *S Afr Med J* 64:849–857, 1983

Godfrey-Fausset P, et al: Evidence of transmission of tuberculosis by SNA finger printing. *Br Med J* 305:221–223, 1992

Goldblatt M, Cremin BJ: Osteo-articular tuberculosis: its prevention in coloured races. *Clin Radiol* 29:669–677, 1978

Granet E: Intestinal tuberculosis: a clinical, roentgenological and pathological study of 2086 patients affected with pulmonary tuberculosis. *Am J Dig Dis* 2:209, 1935

Grassi R, et al: Case report: massive enterolithiasis associated with ileal dysgenesis. *Br J Radiol* 70:207–209, 1997

Graybill JR, Geiger J: Naturally acquired tuberculin hypersensitivity in Ethiopia. *East Afr Med J* 49:597–603, 1972

Greenberg SD, et al: Active pulmonary tuberculosis in patients with AIDS: spectrum of radiographic findings (including a normal appearance). *Radiology* 193:115–119, 1994

Gulati P, et al: Impact of MRI in the management of spinal tuberculosis. Proc. 18th Int. Congr. Radiol. Singapore, S410.186, 1994

Gupta RK, et al: MR Imaging of intracranial tuberculomas. *J Comput Assist Tomogr* 12:280–285, 1988

Halleran WJ, Martin NL: Non-tuberculous mycobacterial arthritis. *Kans Med Soc J* 83:284–286, 1982

Hancock SM: Hyperplastic tuberculosis of the distal colon. *Br J Surg* 46:63–68, 1958

Hanson J: CT of tuberculous peritonitis. *Am J Roentgenol* 144:931–933, 1985

Hayashi H, et al: Case report: tuberculous pericarditis. *Br J Radiol* 71:680–682, 1998

Herlinger H: Angiography in the diagnosis of ileocecal tuberculosis. *Gastrointest Radiol* 2:371–376, 1978

Hermon-Taylor J, et al: *Mycobacterium paratuberculosis* cervical lymphadenitis, followed five years later by terminal ileitis similar to Crohn's disease. *Br Med J* 316:449–453, 1998

Hersch C: Tuberculosis of the liver: a study of 200 cases. *S Afr Med J* 38:857–863, 1964

Hodgson G, et al: the x-ray appearances of tuberculosis of the spine. The intervertebral disc. Charles C. Thomas, Springfield, Ill. 1969, pp 34–40

Hoffman EG, Crosier JH, Cremin BJ: Imaging in children with spinal tuberculosis. *J Bone Jt Surg [Br]* 75:233–239, 1993

Hollins FR: Tuberculosis in the African community in the city of Salisbury (Rhodesia). *Cent Afr J Med* 22:25–28, 1976

Hoon JR, Dockerty MG, Pemberton JdeJ: Collective review: ileocecal tuberculosis including comparison of this disease with nonspecific regional enterocolitis and noncaseous tuberculous enterocolitis. *Int Abstr Surg* 91:417, 1950

Hossain J, Kiniroms M, Lewis RR: Laparoscopy does have a role in tuberculous peritonitis (letter). *Br Med J* 313:1145–1146, 1996

Hsu LCS, Leong JCY: Tuberculosis of the lower cervical spine (C2 to C7) - a report of 40 cases. *J Bone Jt Surg [Br]* 66:1–5, 1984

Hughes HJ, Carr DT, Geraci JE: Tuberculous peritonitis: a review of 34 cases with emphasis on the diagnostic aspects. *Dis Chest* 38:42–50, 1960

Hulnick DH, et al: Abdominal tuberculosis: CT evaluation. *Radiology* 157:199–204, 1985

Jain R, et al: Sonographic diagnosis of early abdominal tuberculosis. Proc. 18th Int. Congr. Radiology. Singapore, S 213.184. 1994

Jain SK, et al: A clinico-radiological study of secondary mycoses in pulmonary tuberculosis. *Indian J Med Sci* 45:81–84, 1991

Jain VK, et al: The far cry of a TB brain. Report of a case of tuberculous meningitis, multiple tuberculomas and tubercular abscess. *Clin Neurol Neurosurg* 91:171–176, 1989

Jamieson DH, Cremin BJ: High resolution CT of the lungs in acute disseminated tuberculosis, and a pediatric radiology perspective of the term miliary. *Pediatr Radiol* 23:380–383, 1993

Jena A, et al: Demonstration of intramedullary tuberculomas by magnetic resonance imaging: a report of 2 cases. *Br J Radiol* 64:555–557, 1991

Jinkins JR: Computed tomography of intracranial tuberculosis. *Neuroradiology* 33:126–153, 1991

Kapoor R, et al: Ultrasound detection of tuberculomas of the spleen. *Clin Radiol* 43:128–129, 1991

Kawamori Y, et al: Macronodular tuberculoma of the liver: CT and MR findings. *Am J Rad* 158:311–313, 1992

Kedar RP, et al: Sonographic findings in gastrointestinal and peritoneal tuberculosis. *Clin Radiol* 49:24–29, 1994

Kent PW: The pattern of tuberculosis in East Africa. *East Afr Med J* 48:450–455, 1971

Kipps A, et al: Negative tuberculin tests. *S Afr Med J* 40:108–114, 1966

Klimach OE, Ormerod LP: Gastrointestinal tuberculosis: a retrospective review of 109 cases in a district general hospital. *Q J Med* 56:569–578, 1988

Kolawale TM, Lewis EA: Radiologic study of tuberculosis of the abdomen. *Am J Roentgenol* 123:348–358, 1975

Konstam PG, Bleskovsky A: The ambulant treatment of spinal tuberculosis. *Br J Surg* 50:26–38, 1962

Kool HES: HIV seropositivity and tuberculosis in a large general hospital in Malawi. *Trop Geogr Med* 42:128–132, 1990

Kulholma P, et al: Tuberculous peritonitis simulating peritoneal carcinomatosis. *Acta Obstet Gynaecol Scand* 61:491–494, 1982

Kurpiewska-Radzimska D, Makuchowa K: Multifocal pseudocystic bone tuberculosis in children. *Chir Narzadow Ruchu Ortop Pol* 42:293–298, 1977

Lachman E, Moodley J, Pitsoes B: Peritoneal tuberculosis imitating ovarian carcinoma. "Special category". *Acta Obstet Gynecol Scand* 64:677–679, 1985

Lagundoye SB, Singh SP: Tuberculosis of the mastoid masquerading as a glomus jugulare tumour: a case report. *Niger Med J* 8:161–163, 1978

Lawn SD, et al: Pulmonary tuberculosis: radiological features in West Africans co-infected with HIV. *Br J Radiol* 72:339–344, 1999

Le CT, Yoiung LW: Radiological case of the month. Non-tuberculous mycobacterial infection. *Am J Dis Child* 139:608–608, 1985

Leader M, Revell P, Clark G: Synovial infection with *M. kansasii*. *Am Rheum Dis* 43:80–82, 1984

Lee KS, et al: The evaluation of tracheo-bronchial disease with helical CT, with multiplanar and three dimensional reconstruction: correlation with bronchoscopy. *Radiographics* 17:555–567, 1997

Lennard-Jones JE: Clues to the elucidation of Crohn's disease. *Trans Coll Med S Afr* 40:34–41, 1996

Lesar MS, et al: Pericardial tuberculoma. *Radiology* 136:309–310, 1981

Lester FT, Tsega E: Tuberculous peritonitis in Ethiopian patients. *Trop Geogr Med* 28:169–174, 1976

Levine C: Primary macro-nodular hepatic tuberculosis. US and CT appearances. *Gastrointest Radiol* 15:307–309, 1990

Levine R, et al: Tuberculous abscess of the pancreas. Case report and review of the literature. *Am J Dig Dis* 377:1141–1144, 1992

Lewis EA, Kolawole TM: Tuberculous ileocolitis in Ibadan: a clinico-radiological review. *Gut* 13:644–653, 1972

Lobo PI: The first reported case of sarcoidosis in an East African. *East Afr Med J* 49:45–48, 1972

Locke EA: Secondary hypertrophic osteoarthropathy and it's relation to simple club fingers. *Arch Intern Med* 15:659–713, 1915

Ly HM, et al: Skin test responsiveness to a series of new tuberculins of children living in three Vietnamese cities. *Tubercle* 70:27–36, 1989

Mahendra KK, et al: Lymphangiography in the tubercular abdomen. *Indian J Radiol* 26:287–289, 1972

Makanjuola D, et al: Mammographic features of breast tuberculosis: the skin bulge and sinus tract sign. *Clin Radiol* 51:354–358, 1996

Makanjuola D, et al: Radiological evaluation of complications of intestinal tuberculosis. *Eur J Radiol* 26:261–268, 1998

Malone JL, et al: Intracranial tuberculoma developing during therapy for tuberculous meningitis. *West J Med* 152:188–190, 1990

Manni JJ: Tuberculous laryngitis. *Trop Geogr Med* 34:109–112, 1982.l

Marsot-Dupuch K, et al: Tuberculosis of the ear; contribution of radiology. *Ann Radiol (Paris)* 26:473–476, 1983

McDonal Jb, Middleton PJ: Tuberculosis of the colon simulating carcinoma. *Radiology* 118:293, 1976

Medical Research Council (UK): Controlled trial of short course regimens of chemotherapy in the circulatory treatment of spinal tuberculosis. *J Bone Jt Surg [Br]* 75:240–248, 1993

Mehta JB: Abdominal tuberculosis mimicking neoplasia on computerized tomography. *South Med J* 78:1385–1386, 1985

Menzies IM, Fitzgerald JM, Mulpeter K: Laparoscopic diagnosis of ascites in Lesotho. *Br Med J* 291:473–475, 1985

Metcalf C: A history of tuberculosis. In: Cooradia HM, Benetar SR (eds) *A century of tuberculosis*. Oxford University Press, Cape Town, 1991, pp 1–31

Miles MM, Shaw RJ: The effect of inadvertent intradermal administration of high dose percutaneous BCG vaccine. *Br Med J* 312:1205, 1996

Milika-Cabanne N, et al: Radiographic abnormalities in tuberculosis and risk of co-existing human immuno-deficiency virus infection. Methods and preliminary results from Bujumbura, Burundi. *Am J Resp Crit Care Med* 152:794–799, 1995

Milika-Cabanne N, et al: Radiographic abnormalities in tuberculosis and risk of co-existing human immuno-deficiency virus infection. Results from Dar es Salamm. Tanzania. *Am J Resp Crit Care Med* 152:786–793, 1995

Miller WT. Spectrum of pulmonary non-tuberculous mycobacterial infection. *Radiology* 191:343–350, 1994

Mitchell RS, Bristol LJ: Intestinal tuberculosis: an analysis of 346 cases diagnosed by routine intestinal radiography on 5529 admissions for pulmonary tuberculosis, 1924–1949. *Am J Med Sci* 227:241–249, 1954

Moore EH: Atypical mycobacterial infections in the lung: CT appearances. *Radiology* 187:777–782, 1993

Morgan M: Breast tuberculosis. *Surg Gynecol Obstet* 52:593–603, 1931

Morse D, Brothwell DR, Ucko PJ: Tuberculosis in Ancient Egypt. *Am Rev Respir Dis* 90:525–535, 1964

Mortenson W, et al: Radiologic aspects of BCG-osteomyelitis in infants and children. *Radiology* 124:858, 1977

Moskovic E: Macronodular hepatic tuberculosis in a child: computed tomographic appearances. *Br J Radiol* 63:656–658, 1990

Naim-Ur-Rahman J: Atypical forms of spinal tuberculosis. *J Bone Joint Surg [Br]* 62:162–165, 1980

N'Dhatz M, et al: Les aspects de la radiographie thoracique chez les tubercleux infectes par le VIH en Cote d'Ivoire. *Rev Pneumol Clin* 50:317–322, 1994

Obisesan AA, et al: Radiological features of tuberculosis of the spine in Ibadan, Nigeria. *Afr J Med Sci* 6:55–67, 1977

Okoro EO, Komolafe OF: Gastric tuberculosis: unusual presentations in two patients. *Clin Radiol* 54:257–259, 1999

Onuora VA: Nonthyroid neck masses in tropical Africans. *Trop Geogr Med* 39:256–259, 1987

Palmer PES: Pulmonary tuberculosis. In: Middlemiss H (ed) *Tropical Radiology*. Heinemann, London, 1961, pp 191–200

Palmer PES: Pulmonary tuberculosis-usual and unusual radiographic presentations. *Semin Roentgenol* 14:204–243, 1979

Palmer PES, Daynes G: Transkei silicosis. *S Afr Med J* 41:1182–1188, 1967

Palmer PES, Rothman WT: The incidence and pattern of chest disease in an African hospital. *Clin Radiol* 15:211–218, 1963

Paterson DE: Tuberculosis of the upper alimentary tract. In: Middlemiss H (ed) *Tropical radiology*. William Heinemann Medical Books, London, 1961, pp 223–224

Pathria MN: Septic arthritis. Proc. 18th. Int. Congr. Radiol. Singapore, 203–204, 1994

Perneger TV, et al: Does the onset of tuberculosis in AIDS predict shorter survival? Results of a cohort study in 17 European countries over 13 years. *Br Med J* 311:1468–1471, 1995

Phemister DB, Hatcher CH: Correlation of the pathological and roentgenological findings in the diagnosis of tuberculous arthritis. *Am J Roentgenol Radium Ther Nucl Med* 29:736–740, 1993

Pinto RS, Zausner J, Bernbaum ER: Gastric tuberculosis. Report of a case with discussion of angiographic findings. *Am J Roentgenol* 110:808, 1970

Pomerance A: Tuberculoma of the intraventricular septum. *Br Heart J* 25:412–414, 1963

Premkumar A, Lattimer J, Newhouse JH: CT and sonography of advanced renal tract tuberculosis. *Am J Roentgenol* 148:65–69, 1987

Primack L, et al: Pulmonary tuberculosis and *mycobacterium avium-intra cellulare*: a comparison of CT findings. *Radiology* 194:413–417, 1995

Puliyel JM, et al: Adverse local reaction from accidental BCG overdose in infants. *Br Med J* 313:528–529, 1996

Raffa H, Mosieri J. Constrictive pericarditis in Saudi Arabia. *East Afric Med J* 67:609–613, 1990

Rajshekhar V, et al: Differentiating solitary small cysticercus granulomas and tuberculomas in patients with epilepsy. Clinical and computerized tomography criteria. *J Neurosurg* 78:402–407, 1993

Ramages LJ, Gertler R: Aural tuberculosis. A series of 25 patients. *J Laryngol* 99:1073–1076, 1985

Rathakrishan V: Osteoarticular tuberculosis. A radiological study in a Malaysian hospital. *Skeletal Radiol* 18:267–272, 1989

Reed DH, Nash AF, Valabhji R: Radiological diagnosis and management of a solitary tuberculous hepatic abscess. *Br J Radiol* 63:902–904, 1990

Reede DL, Bergeron RT: Cervical tuberculosis adenitis. CT manifestations. *Radiology* 154:701–704, 1985

Reeve PA et al: Clubbing in African patients with pulmonary tuberculosis. *Thorax* 42:986–987, 1987

Richter C, et al: Clinical features of HIV seropositive and HIV seronegative patients with tuberculous pleural effusion in Dar-es-Salaam, Tanzania. *Chest* 106:1471–1475, 1994

Richter C, et al: Chest radiography and beta-2-microglobulin levels in HIV-seronegative and HIV-seropositive African patients with pulmonary tuberculosis. *Trop Geogr Med* 46:283–287, 1994

Rochat T: Tuberculose 1992: Nouveaux aspects clinique, epidemiologiques et diagnostiques. *Schweiz Med Wochenschr* 123:140–147, 1993

Rosenzweig DT, Stead WW: The role of tuberculosis in the pathogenesis of bronchiectasis. *Am Rev Respir Dis* 93:769–785, 1966

Saini JS et al: Scleral tuberculosis. *Trop Geogr Med* 40:350–352, 1988

Saunders NA: State of the art: typing *Mycobacterium tuberculosis*. *J Hosp Infect* 29:169–176, 1995

Schaffer R, Becker JA, Goodman J: Sonography of the tuberculous kidney. *Urology* 22:209–211, 1983

Schatz A, Bugie E, Waxman SA: Streptomycin, a substance exhibiting antibiotic activity against gram positive and gram negative bacteria. *Proc Soc Exp Biol Med* 55:66–69, 1944

Schofield PF: Abdominal tuberculosis. *Gut* 26:1275–1278, 1985

Schulze K, Warner HA, Murray O: Intestinal tuberculosis. *Am J Med* 63:735–745, 1977

Scott JW et al: Tuberculous false aneurysm of the abdominal aorta with rupture into the stomach: a case report with review of the literature. *Am Heart J* 37:820–827, 1949

Segal T: Intestinal tuberculosis; Crohn's disease and ulcerative colitis in an urban black population. *S Afr Med J* 65:37–44, 1984

Serrano-Heranz A, et al: Tuberculous cardiac tamponade and AIDS. *Eur Heart J* 16:430–432, 1995

Shanan DH, Kibel MA: Tuberculosis. In: Stanfied P, et al. (eds) *Diseases of children in the subtropics and tropics,* 4th edn. Edward Arnold, London, 1991, pp 519–552

Sharif HS et al: Role of CT and MR imaging in the management of tuberculous spondylitis. *Radiol Clin North Am* 33:787–804, 1995

Sharma N, Prescott S: BCG vaccine in superficial bladder cancer. *Br Med J* 308:801–802, 1984

Smith RL, et al: Factors affecting the yield of acid-fast sputum smears in patients with HIV and tuberculosis. *Chest* 106:684–686, 1994

Staples WG et al: Disseminated tuberculosis with bone marrow necrosis and lymphoma. *S Afr Med J* 52:680–683, 1977

Stassa G: Tuberculous Peritonitis. *Am J Roentgenol* 101:409–500, 1967

Stewart CA et al: Unusual pattern of lung uptake of technetium 99m sulfur colloid seen on the liver scan of a patient with pulmonary tuberculosis. *Clin Nucl Med* 14:271–274, 1989

Stoker DJ: Spinal infection. Proc. 18th Int. Congr. Radiol. Singapore, 1994, pp 199–202

Suri S, et al: Computed tomography in abdominal tuberculosis: a pictorial revue. *Br J Radiol* 72:92–98, 1999.

Sutinen S: Evaluation of activity in tuberculous cavities of the lung. *Scand J Resp Dis Suppl* 67:5–78, 1968

Swensen SJ, Hartmann TE, Williams DE: Computerized tomography diagnosis of *Mycobacterium avium-intracellulare* complex in patients of *Mycobacterium avium-intracellulare* complex in patients with bronchiectasis. *Chest* 105:49–52, 1994

Tabar L, et al: Tuberculosis of the breast. *Radiology* 118:587–589, 1976

Takhtani D, et al: Radiology of pancreatic tuberculosis; a report of three cases. *Am J Gastroenterol* 91:1832–1834, 1996.

Tam PKH, Saing H, Lee JMH: Colonoscopy in the diagnosis of abdominal tuberculosis in children. *Aust Paediatr J* 22:143–144, 1986

Tan TCF, et al: Tuberculoma of the liver presenting as a hyperechoic mass on ultrasound. *Br J Radiol* 10:1293–1295, 1997

Tandon RK, et al: A clinico-radiological reappraisal of intestinal tuberculosis – changing profile. *Gastroenterol Jpn* 21:17–22, 1986

Templeton AC, Schmanz R: Cervical lymphadenopathy in Uganda. *East Afr Med J* 47:582–587, 1970

Thijn CJP, Steensma JT: *Tuberculosis of the skeleton: focus on radiology,* 1st edn. Springer, Berlin Heidelberg New York, 1990, p 12.1

Ting YM, et al: Lung carcinoma superimposed on pulmonary tuberculosis. *Radiology* 119:307, 1976

Umerah BC: Radiological patterns of spinal tuberculosis in the African. *East Afr Med J* 54:598–605, 1977

Vaidya MG, Sodhi JS: Gastrointestinal tract tuberculosis: a study of 102 cases including 55 hemicolectomies. *Clin Radiol* 29:189–195, 1970

Villoria MF, et al: MR imaging and CT of central nervous system tuberculosis in the patient with AIDS (review). *Radiol Clin North Am* 33:805–820, 1995

Vogel H, Rodrigues GHC: Peritonitis tuberculosa des Kleinkindes. *Roentgenol* 36:118–119, 1983

Voigt M, et al: Diagnostic value of ascites adenosine deaminase in tuberculous peritonitis. *Lancet* I:751–754, 1989

Wainwright J: Tuberculous endocarditis. A report of 2 cases. *S Afr Med J* 56:731–733, 1979

Walley J, Porter J: Chemoprophylaxis in tuberculosis and HIV infection. *Br Med J* 310:1621–1622, 1995

Weaver P, Lifeso RM: The radiological diagnosis of tuberculosis of the adult spine. *Skeletal Radiol* 12:178–186, 1984

Werbeloff L, et al: The radiology of tuberculosis of the gastrointestinal tract. Br J Radiol 46:329–336, 1973

Westall J: Tuberculosis leveling off worldwide. *Br Med J* 314:921, 1997

Wilms GE, et al: Computed tomographic findings in bilateral adrenal tuberculosis. *Radiology* 146:729–730, 1983

Witcombe JB, Cremin BJ: Tuberculosis erosion of the sphenoid bone. *Br J Radiol* 51:347–350, 1978

World Health Organisation: Tuberculosis, a global emergency. *World Health Forum* 14:438, 1993

World Health Organisation: Need for action against tuberculosis. *World Health Forum* 16:218, 1995

Yang PJ, et al: Brain abscess: an atypical CT appearance of CNS tuberculosis. *Am J Neuroradiol* 8:919–920, 1987

Yang SO: MR imaging diagnosis of musculoskeletal tuberculosis. Proc. 18th Int Congr Radiol Singapore C 19:222, 1994

Yang ZH, et al: DNA fingerprinting and phenotyping of *Mycobacterium tuberculosis* isolates from human immunodeficiency virus (HIV) – sero-positive and HIV-sero-negative patients in Tanzania. *J Clin Microbiol* 33:1064–1069, 1995

Yaniv E, Avedillo H: Parotid tumour as a presenting symptom of tuberculosis – a report of two cases. *S Afr Med J* 68:613–615, 1990

Young D: Gene deletion behind drug resistance: quoted in editorial. *Br Med J* 305:441, 1992

Zheng JW, Zhang QH: Tuberculosis on the parotid gland. A report of 12 cases. *J Oral Maxillofac Surg* 53:849–851, 1995

Zvetina JR, et al: *Mycobacterium intracellulare* infection of the shoulder and spine. *Skeletal Radiol* 8:111–113, 1982

Subject Index

Abdominal lymph nodes 67
Abscess
    adrenal 81
    brain 116, 128
    lung 13
    paravertebral 85, 94
    tubo-ovarian 81
Apple-core lesion of colon 57, 65
Arthritis, myocobacterial 104
Asthenia (tuberculous) 2
Auto-amputation of kidney 75

Bacille Calmette-Guerin (BCG) 45
Battey bacillus *(Mycobacterium intracellulare)* 48
BCG *(bacille Calmette-Guerin)*
    osteomyelitis 46
    vaccination 44, 45
Bladder 81
Brain
    abscess 125
    calcification 125, 128
    cyst 128
Breast 134
Bronchial blockage 19
Bronchiectasis 20
Bronchography 20
Bronchopneumania 13
"Button sequestrum" of skull 115

Calcification
    bladder 81
    bowel 51, 57, 67
    brain 125, 128
    lungs 11, 27
    lymph nodes 27, 71, 134
    soft tissues, joints 109
Caecum 58
Caries (spinal tuberculosis) 85
Cavitation (pulmonary) 35
Chest wall 45
Cecum (also colon) 58
Central nervous system 125
Clubbing (hypertrophic pulmonary osteoarthropathy) 47
Colitis 61
Colon 61
    apple-core or napkin ring tumors
        tuberculoma 61
    pseudotumor
        tuberculoma 61
    "tumors" (pseudo)
        tuberculoma 61
Constrictive pericarditis 24, 28

Consumption (pulmonary tuberculosis) 1–49
Crohn's disease
    of colon
        versus tuberculosis 61
    of small intestine
        versus tuberculosis 51
Cysts
    in bone 110
    in lung 44

Definition 2
Destroyed lung 19
Digits 110, 118
Duodenum 51

Enterolith, calcified in bowel 51
Epidemiology 3
Esophagus 49

Fistula
    broncho-pleural 45
Fungus in tuberculous cavities 20, 35, 41

Genital 81
Geographical distribution 2
Granulomas, pulmonary 3, 10

Heart (cardiovascular system) 24
    pericardial calcification 24, 28
Hectic fever 2
Histopathology 3
Hydrosalpinges 81
Hyperergic (immune) tuberculosis 34
Hyperplastic (GI) 58
Hypertrophic pulmonary osteoarthropathy 47

Ileocecal lesion 58, 61
Ileum 51
Ileus, small bowel due to peritonitis 51
Isoniazid cyst 44

Joints (non-spinal) 104
Johne's bacillus (M. paratuberculosis) 48
Kidneys 75
Koch's disease 2

Large intestine 61
Larynx 5
Liver
    abscess 71
    calcifications 71, 73

Lung
  abscess and other cavities 3, 35
  bronchiectasis 20
  bronchopneumonia 13
  calcifications 11, 29, 45
  consolidation, lobar 8
  miliary pattern 30
    silicosis 33
Lymphadenopathy (Lymphnodes)
  mediastinal and/or hilar 8
  mesenteric 52, 67
  peripheral (scrofula) 134

Mesentery 67, 81
Miliary 30
Mycobacterium bovis 2
Mycobacteria, other than tuberculosis (MOTT) 48
Mycobacterium tuberculosis 2
Myelography 101

Non-immune infections 6

Occular 135
Osteomyelitis 84, 104, 110
Osteoporosis 110

Pancreatic ducts and pancreas 71
Paraplegia 94
Paraspinal (paravertebral) abscess 85–94
Parotid gland 134
Pathology 3
Pericardial calcification 24, 28
Pericardial effusion 24
Periosteal reaction 110
Peritonitis 67, 81
Phalanges (including dactylitis) 110, 118
Phthisis (pulmonary tuberculosis) 2–49
Pleura and chest wall 24, 45
Pneumoconiosis, associated with
  grind-stones 33
Pneumonia
  lobar 8
  broncho 13
Pott's disease (spinal tuberculosis) 85
Pott's "puffy tumor" of skull 131
PPD conversion (chest x-ray) 47
Primary infections 6
Pseudotumor of bowel 61
Pseudotumor of cecum due to extrinsic pressure from
  lymphadenopathy 58
Psoas (paraspinal) abscess 94
Reactivation 34
Rectum proctitis 65
Renal 75
Ribs 45, 116

Sacroiliac joints 105
Scalp 115
Sclerosis 89
Scrofula (tuberculosis) 2, 48, 134
Scrotum 81
Secondary infections 34
Seminal vesicles 81
Sensitivity (resistance) 3

Sequestra
  button of skull 115, 125
  kissing sequestra Knee 110
Silicosis (pneumoconiosis)
  in grindstone manufacturers 33
  in women (transkei silicosis) 33
Skeleton (appendicular or peripheral) 84, 104
Skull 115, 120
  button sequestra 115, 124
Small intestine 51, 67
  ileus 51
  peritonitis 67
Soft tissues
  calcification (scrofula) 134
  nodules, masses or ulcerating lesions 134
Spine 85
Spleen
  calcification 71
Spondylitis 85, 93
Stomach 50
Stricture or stenosis
    colon, due to adhesions 58, 61
Stierlin's sign (cecum) 58
Struma (tuberculosis) 2
Synonyms 2

Tabes
  mesenterica 2
  pulmonalis 2
Testes 81
Trachea 5
Transkei silicosis 33
Treatment
  tuberculous spine 94
tuberculoma
  cerebral 125
  colon 61
  liver 71
  pulmonary 43
  spleen 71
Tuberculosis
  adrenal 81
  alimentary tract 49
    cecum 58
    colon 61
    duodenum and small intestine 51
      calcified enteroliths 51, 57, 67
    esophagus 49
    ileocecal region 61
    hepatosplenornegaly 32
Tuberculosis, alimentary tract (continued)
    large intestine 61
    liver 71
    pancreas 71
    peritoneum 49, 67
    small intestine 51
    spleen 71
    stomach 50
    rectum 65
  BCG
    osteomyelitis 46
    vaccination 4, 45
  Bones and joints, non-spinal 84, 104, 110, 124
    arthritis 104

chest wall 45
cystic and periosteal tuberculosis 110
dactylitis 110, 118
differential diagnosis 120
healing 116
hypertrophic pulmonary osteoarthropathy 47
knees
long bones 110, 112
osteoarticular
osteomyelitis
  acute 84, 104
  BCG 46
  multifocal 115, 121
osteoporosis 110
periosteal reaction 110
peripheral 104
ribs 45, 116
scintigraphy 107
sequestra 110, 114
  button sequestrum 115
skull 115, 121, 123, 125
  button sequestrum 115
  Pott's puffy tumor 121, 131
spine 85–104
synovial 107
vertebrae 85–104
breast 134
cardiovascular system (pericarditis) 24
central nervous system 125
brain abscess 125
calcification 126, 128
hydrocephalus 128
meningitis 128
tuberculomas of brain 125, 128
congenital tuberculosis 45
genitourinary tract
bladder 81
fallopian tubes 81
kidneys 75
scrotum 81
testes 81
ureters 75
  calcified 80
uterus 81
hyperergic 4, 33
joints 84, 107
ankylosis 110
arthritis 84
"kissing sequestra" 110
osteoarticular 107
osteoporosis 107
synovial 107
larynx 5
lymphadenopathy
abdomen 51–67
calcification 51, 51, 134
chest 8, 15
neck 15, 134
peripheral 134

Mycobacterium bovis 2
non-tuberculous (MOTT) 48
M. tuberculosis: Kochs 2
ocular 135
pericanditis 24
peritoneum and abdominal lymph nodes 67
pulmonary (respiratory tract) 5–49
abscesses (thin-walled) 13
bronchiectasis 20
bronchopneumonia 13
calcification 11, 24, 27
cavitation 35
chest wall involvement 45
chronic pleural disease 27, 45
destroyed lung 19
fibrosis 39
hilar and mediastinal adenopathy 15
immunization with BCG 4, 45
isoniazid cysts 44
larynx and trachea 5
lobar pneumonia 8
miliary 30
pericardial effusion 24
primary (non-immune) tuberculosis 6
progressive primary infection 8, 13
PPD conversion 47
resistance to infection 3
secondary (immune or adult) tuberculosis 34
silicosis and tuberculosis 33
sputum 35
susceptibility 3
transplacental infection 3
tuberculoma of lung 42
upper respiratory tract 5
resistance to tuberculosis 3
sclerosis 89
spine 85–104
clinico-pathological-radiological correlation 85
healing 89
intervertebral discs 85
myelography 101
paraplegia 101
paravertebral abscess 85, 94
psoas abscess 97
sclerosis 89
treatment, alternatives 94
vertebral body 86
scrofula 134
thyroid 49
trachea 5
Tuberculous joint debris 107, 109
Tuberculous peritonitis causing small bowel ileus 67
Tuberculous spondylitis 85, 104
Tubo-ovarian tuberculosis 81, 83

Ureters 75
calcified 80
urinary tract 75
uterus 80